Plasma Source Mass Spectrometry
The New Millennium

Plasma Source Mass Spectrometry

The New Millennium

Edited by

Grenville Holland
Department of Geological Sciences, University of Durham, UK

Scott D. Tanner
PE-Sciex, Concord, Ontario, Canada

ROYAL SOCIETY OF CHEMISTRY

Sep/Ae
chem

The proceedings of the 7th International Conference on Plasma Source Mass Spectrometry: The Millennium Conference held at the University of Durham on 10–15 September 2000.

Special Publication No. 267

ISBN 0-85404-895-2

A catalogue record for this book is available from the British Library

Published by The Royal Society of Chemistry,
Thomas Graham House, Science Park, Milton Road,
Cambridge CB4 0WF, UK
Registered Charity No. 207890

For further information see our web site at www.rsc.org

Printed by Athenaeum Press Ltd, Gateshead, Tyne & Wear, UK

Preface

The new Millennium, which the title of this book celebrates, opens a further chapter in the record of Plasma Source Mass Spectrometry. One hundred years ago, at the turn of the last century, analytical science was in its infancy. Within the electromagnetic spectrum the wavelength properties of light were beginning to be appreciated; knowledge of the X-rays and their associated fluorescence had been uncovered; but an understanding of the isotopes and their properties still lay over the horizon.

How will scientists one hundred years from now look back at us? Will they too see us as beginners in the world of analytical chemistry, as novices still coming to terms with easily available technology? Or will they see the last two decades as a period of great advance in the fields of analytical science, rather as we now recognise the 1960s for its conquest of space, the one small step that was one giant leap for mankind?

Long journeys always begin with one small step. The history of Plasma Source Mass Spectrometry, when it is finally told, will be no exception. Durham's biannual International Conference provides a forum for discussion, a chance to speculate about the future amongst a gathering of friends. The Conference also offers delegates that vital opportunity to publish the results of their recent research and to look to the future. For this opportunity we thank the Royal Society of Chemistry whose quiet patience we both appreciate.

Like all its predecessors the 7th International Conference on Plasma Source Mass Spectrometry not only contained good science but it also maintained its traditions of warm hospitality and sociability. PerkinElmerSCIEX once more underpinned the Conference and we also received invaluable support from CPI international, Micromass, Leco Instruments, ThermoFinnigan and Varian. We would like to express our sincere thanks to each and everyone of these companies for their vital contribution. We also both owe a particular debt of gratitude to Trevor Morse who ran the day-to-day administration of the Conference with such efficiency; and we are also grateful to the many delegates who combined to make this particular Conference so successful.

In a Conference that never once flagged there were highlights. On Tuesday, September 11th we took our traditional dinner aboard the North Yorkshire Moors Railway Pullman steam-hauled by the *Sir Nigel Gresley*. This particular day also coincided with Naoki Furuta's 50th birthday, so it was he who took to the footplate on our return journey to Grosmont. With Japanese technology at the helm we not only arrived safe and sound but also on time. Another tradition was also fulfilled on the Thursday evening: the Conference banquet in the Great Hall of Durham Castle followed by a concert. An unexpected highlight of this concert was the stunning performance from a local physics graduate and mezzo-soprano, Dawn Furness, who, midway through, stepped from the audience and quite simply took our breath away with Handel's 'Where 'ere you walk'.

We have, as usual, made few alterations to the papers submitted. They have been edited and formatted but the authors' conclusions have been left intact and as they intended. However, this should not be construed as an editorial endorsement of these conclusions.

Grenville Holland   Scott D. Tanner
University of Durham PerkinElmerSCIEX, Toronto
England   Canada

January 2001

Contents

4. Applications

5. Isotope Ratio Measurement

6. Speciation

Section 1

Sample Preparation and Introduction

SAMPLE INTRODUCTION, PLASMA-SAMPLE INTERACTIONS, ION TRANSPORT AND ION-MOLECULE REACTIONS: FUNDAMENTAL UNDERSTANDING AND PRACTICAL IMPROVEMENTS IN ICP-MS

John W. Olesik, Carl Hensman, Savelas Rabb and Deanna Rago

Laboratory for Plasma Spectrochemistry, Laser Spectroscopy and Mass Spectrometry, Department of Geological Sciences, Ohio State University, Columbus, OH 43210, USA

1 INTRODUCTION

Improved fundamental understanding of the processes that control ICP-MS signals can lead to practical improvements in analytical capabilities. Here we discuss high efficiency analyte transport at sample uptake rates as high as 0.8 mL/min and the unique behavior of selenium which when present at high concentrations is vaporized before other elements. We also explore the use of ion-molecule reactions in a dynamic reaction cell to overcome spectral overlaps and examine the ability of the quadrupole-based cell to prevent formation of undesired product ions.

Droplet-droplet collisions and coagulation in the spray chamber appear to be the main processes that limit analyte transport efficiency[1,2]. When the sample uptake rate is increased, transport efficiency naturally decreases.[1,3,4] By promoting evaporation within the spray chamber, it is possible to attain virtually 100% sample transport efficiency[1,2,5]. The deleterious effects of solvent vapor loading prevent use of high analyte transport rates unless most of the sample solvent vapor is removed prior to entering the plasma. Even water vapor loads greater than about 60 mg/min can cause deleterious effects in the plasma.[1,5] However, if solvent vapor is efficiently removed, analyte from up to 0.8 mL/min of sample can be introduced into the ICP and ionized.

Most of the mass dependent matrix effects that occur when the sample contains high concentrations of dissolved solids to be due to space charge induced loss of ions.[6-15] However, Farnsworth has reported measurements of ion concentrations between the sampler and skimmer orifices that suggest that significant matrix effects are occurring during sampling of ions in contrast to previous assumptions.[16]

Within the ICP itself, ions are generated earlier (closer to the load coil) when the sample contains high concentrations of dissolved solids[17-19]. This is consistent with shifts in emission[20,21] and laser induced fluorescence profiles[20,22] Lazar and Farnsworth[18] recently proposed that the earlier appearance of ions is due to an increase in the size of the desolvated particle, so that the time required for desolvation of a fixed size aerosol droplet is smaller when the dissolved solids concentration is high. Here we confirm that the initial appearance of ions in the plasma is consistent with the proposed explanation for a variety of different dissolved solids. Previously, we have seen no evidence of element dependent initial appearance of sample ions in the plasma; all analyte ions appeared at the same time. However, the behavior due to the presence of high concentrations of selenium oxide is very different.   We show that this is due to vaporization of the selenium prior to vaporization of analyte species in the particle.

Spectral overlaps have been a major problem in ICP-MS, particularly for quadrupole mass spectrometers, since inception of the technique[23-27]. Several approaches have been used to overcome potential errors due to spectral overlaps. Sector based mass spectrometers can provide resolution of 10,000 or more to separate many, but not all, spectral overlaps.[28] Mathematical corrections can be used if the ratio of the spectral overlap ion to the analyte ion signals is not too large.[29] Removal of the solvent before the sample is introduced into the plasma and mixed gas plasmas have also been used to reduce spectral overlaps.[30] "Cold" or "cool" plasmas have also been used[31-33] although this approach can suffer from much more severe chemical matrix effects than normal or "hot" plasmas.[34]

Attempts to use collisionally induced dissociation (CID) to break molecular ions after sampling from the ICP in reaction or collision cells were not successful because such high energies are required to break most molecular ion bonds that scattering losses are too large.[35,36]   However, Rowan and Houk described charge transfer reactions that likely occurred due to low-energy collisions while Douglas reacted O_2 with Ce^+ to form CeO^+.

Ion-molecule reactions, promoted in a pressurized cell after sample ions pass through the skimmer and ion optics, can very efficiently remove many ions that would result in spectral overlaps in ICP-MS . Barinaga, Eiden and Koppenaal [37-41] described the use of ion-molecule reactions (using H_2 or H_2O) in an ion trap or reaction cell to overcome argide ions. Turner et al [42] described an rf-only hexapole "collision cell" originally proposed to remove argide ions due to collisions with He but later attributed to reactions with H_2.

Tanner and co-workers[43-50] have described the design and use of a quadrupole reaction cell, as was used in this work. The most effective reaction gas and pressure will depend on the chemistry of the element of interest and the spectral overlap ion. We will show some examples in this chapter. Signals due to spectral overlap ions can be reduced by over eight orders of magnitude in some cases. However, it is important to prevent the formation of product ions that will produce new spectral overlaps. We will discuss how the bandpass of a quadrupole reaction cell can be used to prevent the production of new spectral overlap ions by making one of the reactants unstable in the cell.

2 RESULTS AND DISCUSSION

2.1 High Efficiency Sample Introduction

If a pneumatic, concentric nebulizer is used in a conventional spray chamber, analyte transport efficiency will naturally decrease from better than 50% at a sample uptake rate of 20 :L/min to less than 2% at a sample uptake rate of 1 mL/min[1,2]. As the sample uptake rate is increased the droplet number density in the spray chamber increases, droplet-droplet collisions become more likely and drop-drop coagulation leads to loss of small droplets into large ones.[2,3] The large droplets have a low probably of survival through the spray chamber and into the ICP.

If droplet evaporation can be made rapid enough in the spray chamber, desolvation of the droplets will be virtually complete before droplets impact the walls of the spray chamber[1, 5]. The nebulizer gas jet entrains a huge amount of gas into it. Therefore, if heated gas can be fed into the nebulizer gas and aerosol jet, it can be efficiently mixed with the aerosol to promote rapid evaporation of the droplets in the spray chamber. This is the basis of the spray chamber design described by Debrah and Legere[51,52], and shown in Figure 1. A Meinhard HEN nebulizer was used with a nebulizer gas flow rate of approximately 0.3 L/min. A make-up gas (approximately 0.6 L/min) was added to the spray chamber and heated as it flowed along the spray chamber wall, which was heated to a temperature of approximately 180E C. Previously, we showed that the ICP-MS signal increases linearly and proportionally as the sample uptake rate is increased from 25 to 250 :L/min, using this system[1]. This will produce an improvement in ICP-MS sensitivity by a factor of 5 to 10 compared to a conventional concentric nebulizer/spray chamber.

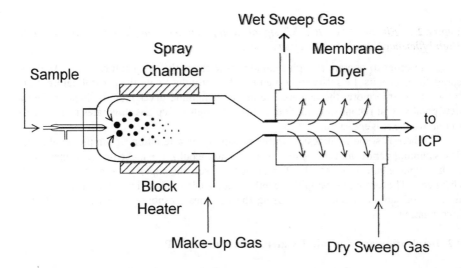

Figure 1 *High efficiency sample introduction system consisting of a Meinhard® HEN nebulizer, heated spray chamber with heated make up gas entrained into the aerosol and nebulizer gas stream and a Nafion® membrane desolvator*

In order to obtain virtually 100% analyte transport rate with sample uptake rates greater than about 50 :L/min, desolvation of the sample aerosol and removal of the solvent vapor produced is essential[5]. Solvent vapor loading has severe effects on the plasma (even extinguishing it) at water vapor loads of 50 to 60 mg/min.[5]

If desolvation and solvent vapor removal are efficient enough, sample uptake rates up to 0.8 mL/min can be used without distinguishing the plasma (Figure 2). Sensitivities greater than 1.6×10^9 cps/ppm can be obtained for U^+ at a sample uptake rate of 0.6 mL/min. With a conventional nebulizer/spray chamber this instrument typically has a sensitivity of approximately 0.04×10^9 cps/ppm. Therefore, a factor of 40 increase in The Rh^+ sensitivity was 0.6×10^9 cps/ppm at a sample uptake rate of 0.6 mL/min, compared to a typical sensitivity of 0.03×10^9 cps/ppm for a conventional nebulizer/spray chamber with an uptake rate of 1 mL/min, a factor of 20 increase in sensitivity can be obtained using the High Efficiency Sample Introduction System (HESIS).

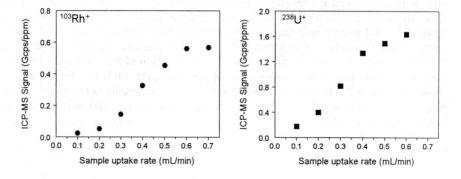

Figure 2 $^{103}Rh^+$ *and* $^{238}U^+$ *ICP-MS signals as a function of sample uptake rate using the High Efficiency Sample Introduction System*

In order to operate at high sample uptake rates, the temperature of the Ar in the spray chamber must be sufficiently high to efficiently desolvate the aerosol before it hits the wall. Furthermore, the temperature in the membrane dryer must be kept above the dew point so that recondensation does not occur. High sweep gas flow rates must be used to remove the solvent vapor.

There is some nonlinearity in the sensitivity as a function of sample uptake rate. The sampling depth and total center gas flow rate (nebulizer + make up) was fixed as the sample uptake rate was varied. The amount of solvent vapor entering the ICP likely changed. The optimum sampling depth or total center gas flow rate may vary as a function of sample uptake rate, causing the nonlinear response. This requires further investigation.

2.2 Desolvation and Chemical Matrix Effects in the ICP

Desolvation of aerosol droplets, vaporization of desolvated particles, atomization and ionization in the ICP as well as ion transport from the ICP to the MS detector can be studied with previously unattainable clarity by introducing the sample as isolated, monodisperse droplets.[17] The monodisperse dried microparticulate injector (MDMI) device provides such a sample introduction system.

When the sample contains a high concentration of dissolved salts, several of the processes that occur in the plasma are affected. Atoms and ions appear earlier (closer to the load coil) when the sample contains a high concentration of dissolved solids. Lazar and Farnsworth[18] proposed that for two identically sized aerosol droplets, the atoms and ions appear earlier because less solvent must be evaporated from the droplet to obtain a desolvated particle. Laser induced fluorescence and emission profiles also shift closer to the load coil when a conventional nebulizer/spray chamber sample introduction system is used[17,19].

The square of the aerosol droplet diameter should decrease linearly with time as droplets desolvate in the plasma[53]. The size of the desolvated particle that results after the solvent completely evaporates from the droplet in the plasma can be predicted from the density of the original solid sample, assuming a spherical particle.

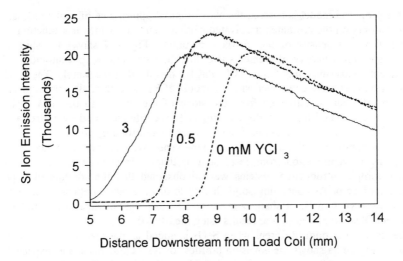

Figure 3 *Sr⁺ signal as a function of distance downstream from the load coil for solutionscontaining 0, 0.5 and 3 mM YCl₃*

If the model of Lazar and Farnsworth[18] is correct, the shift to earlier appearance times should therefore be linearly dependent on the difference in the square of estimated desolvated particle diameters with an without the addition of matrix. Figure 3 shows the earlier appearance of Sr⁺emission as the concentrations of YCl₃ added to the sample is increased from 0 to 3 mM.

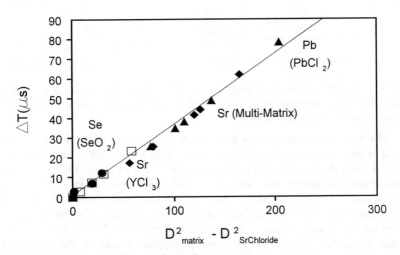

Figure 4 *Shift to earlier appearance times for analyte emission due to high concentrations of dissolved solids as a function of the difference in the square of the desolvated particle diameter with matrix minus the square of the desolvated particle diameter without matrix*

If the proposed model is correct, the shift in initial appearance of analyte signals should depend only on the estimated desolvated particle diameter, which is a function of the concentration and density of the matrix compound. Figure 4 shows the shift in appearance times for a variety of elements (Se, Sr, Pb) with various concentrations of several different solids (including SeO_2, YCl_3, $PbCl_2$,) dissolved in the sample. The good linear correlation is entirely consistent with the model of Farnsworth et al. Furthermore, a desolvation rate can be estimated from the slope of the line fit to the data. The desolvation rate calculated from the slope is similar to the rate measured by Kinzer and Olesik [54] using resonance Mie scattering from desolvating aerosol droplets.

In experiments carried out over the last nine years, we had seen no evidence of element dependent vaporization from desolvated particles. However, long ago using conventional sample introduction systems we had observed that the addition of large concentrations of Se or As containing solids had little or no effect on emission or laser induced fluorescence signals, in sharp contrast to all other matrices that were studied. McGowan and Olesik reported similar results using the MDMI.[55]

The situation is quite different when SeO_2 is added as the matrix to a solution containing Sr. Figure 5a shows the emission profiles for Se. These behave as expected; the initial appearance of Se signal moves closer to the load coil as the Se concentration is increased. In contrast, the location (time) of the initial appearance of the Sr^+ emission does not change as the Se concentration in the solution is increased (Figure 5b).

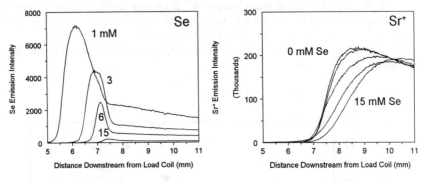

Figure 5 a. *Se emission intensity as a function of distance downstream of the end of the load coil as the Se concentration was increased.* **b.** *Sr emission intensity spatial profile as the Se concentration was increased*

The data are consistent with desolvation of the water from the aerosol droplet, followed by Se vaporization from the desolvated particle and finally, vaporization of the Sr. This is likely due to sublimation of the SeO_2 at a temperature of 340^0C. In order to test this hypothesis further, we measured droplet and particle sizes at the exit of the MDMI furnace. The carrier gas flow rate through the furnace was varied in order to change the time for desolvation and sublimation in the furnace. The droplet or particle size was measured using a phase Doppler particle analyzer (PDPA, Aerometrics) in a reflective mode. Figure 6 shows the measured droplet or particle diameters for three different solutions, each with 10 mM $SrCl_2$. One of the solutions also contained 75 mM selenium oxide. One solution also contained 30 mM YCl_3. The selenium oxide and yttrium chloride concentrations were chosen to produce similar sized desolvated particles.

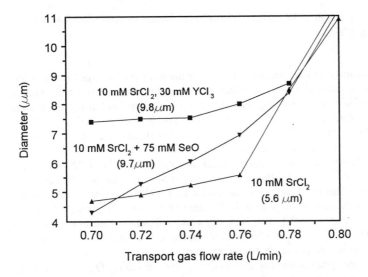

Figure 6 *Droplet/particle size measured as a function of transport gas flow rate for three solutions (labeled in figure)*

The particle or droplet diameter for the solution that contains only $SrCl_2$ decreases continuously as the carrier gas flow rate is decreased to 0.76 L/min. Then it levels off at a diameter of about 5 :m. This is consistent with completion of desolvation, leaving a particle that does not vaporize at the temperatures attained in the furnace.

The particle diameter behavior for the solution that also contains YCl_3 is similar, except that the diameter levels off earlier (at a higher carrier gas flow rate) at a larger size (about 7.5 :m). This is as one would expect as the desolvated particle size should be larger when the YCl_3 is added to the sample. So, desolvation should be completed earlier and the desolvated particle diameter should be larger than for the solution without YCl_3.

The particle diameter for the solution that contains selenium oxide continuously decreases as the carrier gas flow rate is decreased. At a carrier gas flow rate of 0.70 L/min, the measured particle size is similar for the solutions with and without selenium oxide. The observed behavior is consistent with desolvation of the aerosol droplet followed by sublimation of the selenium oxide.

2.3 Ion-molecule Reactions to Overcome Spectral Overlaps in ICP-MS

As noted previously, several different approaches have been used in attempts to overcome spectral overlaps including high resolution mass spectrometers, solvent removal, "cold" plasmas and mathematical corrections or spectral fitting. An alternative is to use ion-molecule reactions to either remove the spectral overlap ion or to form a new species that contains the analyte, so that it can be measured at a different mass. Ion-molecule reactions can be highly efficient and selective, if a sufficient number of reactive collisions occurs in the reaction cell and the ion energies are low enough (close to thermal). For example, ammonia can be used to reduce the Ar^+ signal by more than eight orders of magnitude through a charge transfer reaction. Detection limits for Ca^+ at mass 40 are then on the ppt level.

2.3.1. Use of ion molecule reactions to remove the ion causing a spectral overlap.

In order for a reaction gas to be effective it must be selective and react rapidly enough. It must either react efficiently with the spectral overlap ion but not the analyte ion or vice versa. Ammonia is highly reactive with argide species (including $^{40}Ar^+$, $^{38}Ar^1H^+$, $^{40}Ar^{16}O^1H^+$, $^{40}Ar^{12}C^+$, $^{40}Ar^{16}O^+$, $^{40}Ar^{14}N^+$) but only slightly reactive with analyte ions at the corresponding nominal masses ($^{40}Ca^+$, $^{39}K^+$, $^{55}Mn^+$, $^{52}Cr^+$, $^{56}Fe^+$, $^{54}Fe^+$). Ammonia also reacts rapidly with Cl containing species including ClN^+, ClO^+, ClO_2^+ and Cl_2^+. While ammonia is a highly effective reaction gas to overcome a wide variety of spectral overlaps, it is not universal. Analyte ions with a higher ionization energies that are within 0.4 eV of the ionization potential of ammonia (10.14 eV), such as Se (I.P. = 10.36 eV) and As (I.P. = 9.8 eV), also typically react efficiently with ammonia via a charge transfer reaction. Group 5a and 5b elements (including V^+ and As^+) can react with ammonia through condensation and clustering reactions.

We have investigated a number of reaction gases (including O_2, N_2, CO, CH_4, C_2H_2 and NH_3) to determine how to best overcome the Ar_2^+ spectral overlap with the major isotope of Se^+ at mass 80. Relative reaction rates between ions and reaction gases can be estimated from measurements of ICP-MS signal versus reaction gas flow rate, as shown for O_2, NH_3 and CO, in Figure 7. For a bimolecular reaction, the logarithm of the ion concentration (and therefore, the ICP-MS signal) should decrease linearly with the reaction gas flow rate (assuming that the pressure in the cell increases linearly with reaction gas flow rate).

All of the gases investigated reacted with the Ar_2^+ ion. Ethane had the fastest reaction rate with Ar_2^+, followed closely by CO. However, both also react significantly with Se^+. N_2^+, O_2^+ and NH_3 all have similar reaction rates with Ar_2^+. However, while N_2 and O_2 appear to react slowly with Se^+, NH_3 reacts rapidly with Se^+, so that it is not an effective reaction gas to measure Se^+.

2.3.2. Measurement of reaction product ion for analysis.

In some cases, it may be possible to efficiently form a reaction product ion at a mass where there is no other ion that causes a spectral overlap. While the reaction between Se^+ and O_2 is relatively slow, it is fast enough to produce significant quantities of SeO^+, as shown in Figure 8. Therefore, when O_2 is used as the reaction gas Se^+ can be measured at mass 80 with the Ar_2^+ dramatically reduced by reaction with oxygen or as SeO^+ at mass 96, where the background is low if the sample does not contain Mo, Ru or Zr, each of which has a minor isotope of mass 96. Se detection limits at mass 80 and mass 96 were both estimated to be 5 to 7 ppt.

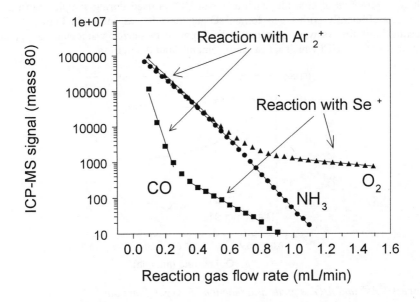

Figure 7 *Reaction profiles for Ar_2^+ and Se^+ with three different gases (CO, NH₃, O₂)*

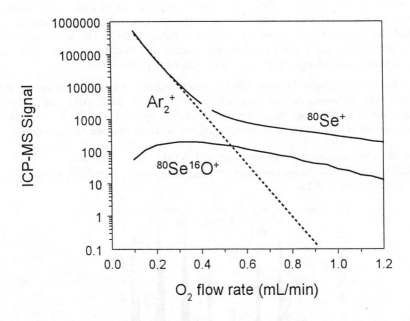

Figure 8 *ICP-MS signal at m/z 80 and 96 as a function of oxygen flow rate. The sample contained 1 ppb Se*

The overlap of ArCl$^+$ with As$^+$ at mass 75 is a problem when the sample contains high concentrations of chloride, such as when HCl is used during sample preparation. As$^+$ reacts efficiently with O_2 to form AsO$^+$ at mass 91, as shown in Figure 9. The detection limit for As was estimated to be 1 ppt in deionized water and 13 ppt in a solution of 1% v/v HCl which appeared to contain about 0.3 ppb As.

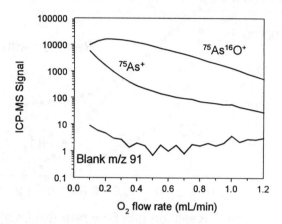

Figure 9 $^{75}As^+$ *and AsO$^+$ signals as a function of oxygen flow rate*

2.3.3. Prevention of product ion formation using the quadrupole cell bandpass.

A potential problem when a reaction gas is used is the formation of product ions that could result in a new spectral overlap. The PerkinElmer Sciex ELAN 6100 DRC instrument is unique in that it uses a quadrupole reaction cell in front of the analyzer quadrupole mass spectrometer. The center mass/charge value for the quadrupole reaction cell and the analytzer quadrupole as scanned synchronously. The bandpass of the reaction cell can be controlled to prevent the reactant ion from being stable in the cell when the center mass/charge value equal to the mass of a product ion.

Ammonia is one of the most effective reaction gases to remove ions such as Ar$^+$, molecular ions containing Ar including ArO$^+$, ArOH$^+$, ArH$^+$, ArC$^+$, ArN$^+$ and ArH$_2$$^+$, as well as other molecular ions including Cl$_2$$^+$, ClO$_2$$^+$, CO$_2$$^+$, CN$^+$ and HCN$^+$. However, ammonia forms cluster ions with many elemental ions. If the reaction gas flow rate is

Figure 10 *Mass spectra showing formation of $V(NH_3)_n{}^+$, $V(NH_2)(NH_3)_n{}^+$, $V(NH)(NH_3)_n{}^+$under conditions favorable for reaction between V^+ and NH_3 with a reaction cell q value of 0.25*

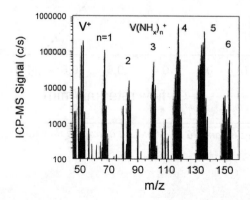

sufficiently high and the a and q parameters are set for a wide bandpass, cluster ions may be formed at many different masses. An example is shown in Figure 10.

If the value of q is increased from 0.25 to 0.75, the low mass bankpass of the quadrupole reaction cell is increased so that the V^+ ions are not stable in the cell. Then the spectrum shown in Figure 11 is obtained. Signals due to the $V(NH_x)_n^+$ ions have all be reduced from as high as 600,000 counts/s to less than 100 c/s. The user specifies the value of q for the analyte mass. The value of q is proportional to the radio frequency voltage applied to the quadrupole rods and inversely proportional to the frequency and ion mass. Therefore, the value of q for a reactant ion of lower mass than the product ion is larger than the value of q for the analyte ion by a factor of $m_{analyte}^+/m_{reactant}^+$, where $m_{analyte}^+$ is the mass of the analyte ion and $m_{reactant}^+$ is the mass of the reactant ion. Table 1 shows the relative signals for $Ni(NH_3)^+$ at m/z 77, $Ni(NH_3)_2^+$ at m/z 92 and $Ni(NH_3)_3^+$ at m/z 111 as a function of the $q_{analyte}^+$. The estimated low mass cutoff (ions below this m/z ratio do not pass through the quadrupole cell) and the calculated q for the Ni^+ reactant ion are also listed.

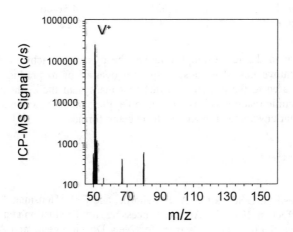

Figure 11 *Mass spectrum from V solution with same experimental conditions as the spectrum in Figure 10 except that the reaction cell q value is 0.75*

As seen in Table 1, the $Ni(NH_3)_2^+$ cluster ion signal was reduced by up to a factor of about 500,000 when the $q_{analyte}^+$ was 0.75, which corresponds to a value of q for Ni^+ of 1.2. The $Ni(NH_3)_n^+$ signal decreases significantly as the low mass bandpass is increased to 60 and above.

Table 1 *Analyte and Reactant Ion Q Values and Normalized Ni(NH₃)ₙ⁺ Signals*

Analyte m/z = 77					
$q_{analyte}^+$	0.5	0.6	0.65	0.7	0.75
q_{Ni}^+	0.64	0.77	0.83	0.90	0.96
m_{low}	42	51	55	59	63
Normalized signal	1	0.3	0.06	0.005	0.0003
1/Normalized signal	1	3	16	192	3297

Analyte m/z=92						
$q_{analyte}^+$	0.25	0.5	0.6	0.65	0.7	0.75
q_{Ni}^+	0.40	0.79	0.95	1.03	1.11	1.19
m_{low}	25	50	61	66	71	78
Normalized signal	1	0.1	0.0005	0.0001	7.3e-06	1.9e-06
1/Normalized signal	1	7	2090	8817	136662	516277

Analyte m/z=111			
$q_{analyte}^+$	0.25	0.5	0.65
q_{Ni}^+	0.46	0.93	1.20
m_{low}	30	61	79
Normalized signal	1	0.001	4.5e-06
1/Normalized signal	1	704	2.2e+05

As shown in the few examples above, the dynamic reaction cell can be used effectively to remove ions that cause a spectral overlap or to produce an ion that is quantitatively related to the analyte at a different mass than the elemental analyte ion. Because the dynamic reaction cell is a quadrupole, the bandpass can be adjusted to very effectively prevent unwanted product ions from being formed.

3 REFERENCES

1.   J. W. Olesik, I. I. Stewart, J. A. Hartshorne, and C. E. Hensman, "Sensitivity and matrix effects in ICP-MS: Aerosol Processing, Ion Production and Ion Transport" in Plasma Source Mass Spectrometry: New Developments and Applications, S. D. Tanner and G. Holland, p. 1-19, Royal Society of Chemistry, Cambridge, 1999.

2.   J. W. Olesik, J. A. Hartshorne, and B. Etkin, *Analytical Chemistry*, in preparation, 2000.

3.   J. A. Hartshorne, "Studies of Charactistics and Aerosol Formation in Inductively Coupled Plasma Emission Spectrometry: Sample Introduction for Speciation by Carbon Phase Liquid Chromatography", M. S. Thesis, The Ohio State University, Columbus, 1995.

4.   J. W. Olesik and L. C. Bates, *Spectrochim. Acta B*, 1995, **50B**, 285.

5.   C. E. Hensman and J. W. Olesik, *J. Anal. Atom. Spectrom.*, in preparation, 2000.

6.   H. Niu and R. S. Houk, *Spectrochim. Acta B*, 1996, **51B**, 779.

7.   G. R. Gillson, D. J. Douglas, J. E. Fulford, K. W. Halligan, and S. D. Tanner, *Anal. Chem.*, 1988, **60**, 1472.

8.   D. J. Douglas and S. D. Tanner, "Fundamental Considerations in ICPMS", in Inductively Coupled Plasma Mass Spectrometry, A. Montaser, ed., Wiley-VCH, New York, 1997.
9.   J. W. Olesik and M. P. Dziewatkoski, *J. Am. Soc. Mass Spectrom.*, 1996, **7**, 362.
10.   I. I. Stewart and J. W. Olesik, *J. Am. Soc. Mass Spectrom.*, 1999, **10**, 159.
11.   S. D. Tanner, *Spectrochim. Acta B*, 1992, **47B**, 809.
12.   S. D. Tanner, L. M. Cousins, and D. J. Douglas, *Appl. Spectrom.*, 1994, **48**, 1367.
13.   S. D. Tanner, "Ion Optics for ICP-MS: Modeling, Intuition or Blind Luck" in Plasma Source Mass Spectrometry: Developments and Applications, J. G. Holland and S. D. Tanner, ed., Royal Society of Chemistry, Cambridge, 1997.
14.   L. A. Allen, J. J. Leach, and R. S. Houk, *Anal. Chem.*, 1997, **69**, 2384.
15.   N. Praphairaskit and R. S. Houk, *Anal. Chem.*, 2000, **72**, 2356.
16.   P. B. Farnsworth, Federation of Analytical Chemistry and Spectroscopy Societies Meeting, paper 221, Nashville, TN, USA, 2000.
17.   J. W. Olesik, *Appl. Spectrosc.*, 1997, **51**, 158A.
18.   A. C. Lazar and P. B. Farnsworth, *Appl. Spectrosc.*, 1999, **53**, 457.
19.   J. W. Olesik, J. A. Kinzer, and M. P. Dziewatkoski, "Generation and transport of ions: ICP-MS from single droplets and microsecond time scales to practical measurements", in Plasma Source Mass Spectrometry: Developments and Applications, J. G. Holland and S. D. Tanner, ed., Royal Society of Chemistry, Cambridge, 1997.
20.   J. W. Olesik and E. J. Williamsen, *Appl. Spectrosc.*, 1989, **43**, 1223.
21.   M. W. Blades and G. Horlick, *Spectrochim. Acta, Part B*, 1981, **36B**, 881.
22.   G. Gillson and G. Horlick, *Spectrochim. Acta, Part B*, 1986, **41B**, 619.
23.   S. H. Tan and G. Horlick, *Appl. Spectrosc.*, 1986, **40**, 445.
24.   M. A. Vaughan and G. Horlick, *Applied Spectroscopy*, 1986, **40**, 434.
25.   K. E. Jarvis, A. L. Gray, and R. S. Houk, Handbook of inductively coupled plasma mass spectrometry, Blackie ; Chapman and Hall, 1992.
26.   A. Montaser and G. W. Golightly, Inductively Coupled Plasmas in Analytical Atomic Spectrometry, VCH Publishers, 1992.
27.   A. Montaser, Inductively Coupled Plasma Mass Spectrometry, Wiley-VCH, 1997.
28.   N. M. Reed, R. O. Carins, and R. C. Hutton, *J. Anal. At. Spectrom.*, 1994, **9**, 881.
29.   J. L. M. de Boer, *Spectrochim. Acta, Part B*, 1997, **52B**, 389.
30.   L. C. Alves, D. R. Wiederin, and R. S. Houk, *Anal. Chem.*, 1992, **64**, 1164.
31.   S.-J. Jiang, R. S. Houk, and M. A. Stevens, *Anal. Chem.*, 1988, **60**, 1217.
32.   K. Sakata and K. Kawabata, *Spectrochim. Acta Part B*, 1994, **49B**, 1027.
33.   S. D. Tanner, M. Paul, S. A. Beres, and E. R. Denoyer, *At. Spectrosc.*, 1995, **16**, 16.
34.   S. D. Tanner, *J. Anal. At. Spectrom.*, 1995, **10**, 905.
35.   J. T. Rowan and R. S. Houk, *Appl. Spectrosc.*, 1989, **43**, 976.
36.   D. J. Douglas, *Can. J. Spectrosc.*, 1989, **34**, 38.
37.   C. J. Barinaga and D. W. Koppenaal, *Rapid Commun. Mass Spectrom.*, 1994, **8**, 71.
38.   D. W. Koppenaal, C. J. Barinaga, and M. R. Smith, *J. Anal. At. Spectrom.*, 1994, **9**, 1053.
39.   G. C. Eiden, C. J. Barinaga, and D. W. Koppenaal, *J. Anal. At. Spectrom.*, 1996, **11**, 317.
40.   G. C. Eiden, C. J. Barinaga, and D. W. Koppenaal, *Rapid Commun. Mass Spectrom.*, 1997, **11**, 37.

41.   G. C. Eiden; C. J. Barinaga, and D. W. Koppenaal, *J. Anal. At. Spectrom.*, 1999, **14**, 1129.

42.   P. Turner, T. Merren, J. Speakman, and C. Haines, "Interface Studies in the ICP-Mass Spectrometer", in Plasma Source Mass Spectrometry: Developments and Applications, J. G. Holland and S. D. Tanner, ed., Royal Society of Chemistry, Cambridge, 1997.

43.   S. D. Tanner, V. I. Baranov, and U. Vollkopf, *J. Anal. At. Spectrom.*, 2000, **15**, 1261.

44.   U. Vollkopf, V. Baranov, and S. D. Tanner, "ICP-MS multielement analysis at sub-ppt levels applying new instrumental design concepts" in Plasma Source Mass Spectrometry, J. G. Holland and S. D. Tanner, eds., Royal Society of Chemistry, p. 63, Cambridge, 1999.

45.   S. D. Tanner and V. I. Baranov, "Fundamental Processes Impacting Performance of an ICPMS Dynamic Reaction Cell" in Plasma Source Mass Spectrometry, J. G. Holland and S. D. Tanner, eds., p. 46, Royal Society of Chemistry, Cambridge, 1999.

46.   S. D. Tanner and V. I. Baranov, *J. Am. Soc. Mass Spectrom.*, 1999, **10**, 1083.

47.   S. D. Tanner and V. I. Baranov, *At. Spectrosc.*, 1999, **20**, 45.

48.   V. I. Baranov and S. D. Tanner, *J. Anal. At. Spectrom.*, 1999, **14**, 1133.

49.   V. I. Baranov and S. D. Tanner, "Selective thermochemistry in a dynamic reaction cell" in Plasma Source Mass Spectrometry, J. G. Holland and S. D. Tanner, eds., p. 34-45, Royal Society of Chemistry, Cambridge, 1999.

50.   S. D. Tanner and V. I. Baranov, "Bandpass reactive collision cell", PCT Int. Appl., 9856030, 1998.

51.   E. Debrah and G. Legere, "Design and performance of a novel high efficiency sample introduction system for plasma source spectrometry" in Plasma Source Mass Spectrometry, J. G. Holland and S. D. Tanner, eds., p. 20-26, Royal Society of Chemistry, Cambridge, 1999.

52.   E. Debrah and G. Legere, *At. Spectrosc.*, 1999, **20**, 73.

53.   N. C. Clampitt and G. M. Hieftje, *Analytical Chemistry*, 1972, **44**, 1211.

54.   J. A. Kinzer and J. W. Olesik, *Appl. Spectrosc.*, 2000, submitted.

55.   G. J. McGowan and J. W. Olesik, Federation of Analytical Chemistry and Spectroscopy Societies Meeting, 1994.

MICROWAVE DIGESTION OF OILS FOR ANALYSIS OF PLATINUM GROUP AND RARE EARTH ELEMENTS BY ICP-MS

S. J. Woodland, C. J. Ottley, D. G. Pearson and R. E. Swarbrick

Department of Geological Sciences, University of Durham, South Road, Durham City, DH1 3LE, UK.

1 INTRODUCTION

The platinum group elements (PGEs) consist of Os, Ir, Ru, Pt and Pd. Re is also considered along with this group due to similarities in its chemical behaviour. The PGEs are capable of exhibiting both siderophile and organophile behaviour and thus offer great potential for tracing the sources of oils in sedimentary basins. This organophile behaviour may cause significant PGE enrichment in oils over typical crustal values (Table 1), rendering the PGE signature of an oil robust to chemical effects during its migration from source to trap. Hence the PGE signatures of the oil-source would be preserved despite migration of the oil through overlying crustal rocks. This in itself could prove a useful exploration tool. Development of such fingerprinting however relies on the availability of a routine, precise and sensitive analytical technique.

1.1 Previous work

The potential for enrichment of PGEs within organic material has been well documented, for example organic-rich fractions from the Witswatersrand basin have been measured to contain greater than 300ppb of Os and Pd [1]. Likewise, organic-rich black shales deposited in anoxic marine environments are also commonly enriched in PGEs when compared to lithic-type sediments [2, 3] (Table 1). To date however there are very little data on precious metal abundances within oils, this is predominantly due to analytical difficulties and a lack of suitably sensitive instrumentation prior to the advent of ICP-MS [4].

Knowledge of the metal content within oils is extremely useful not just from a tracer point of view but also because such impurities have to be removed during refining. Extremely high metal contents may make it less profitable to drill certain reservoirs. Crude oils contain many metal ion-organic complexes particularly of nickel and vanadium and hence may well contain PGEs which often exhibit similar behaviour. One recent study indicated that Os may be slightly enriched in certain oils (0.6-1.5ppb) [5]. Other workers however have found much lower Os levels (<<1ppb) [6]. If enrichments of other PGEs are also present within oils then they could potentially be used to characterise specific oil reserves and to trace the source of oil spills.

Table 1 *Typical concentrations (ppb) of PGEs within geological materials where andesites and lithic sediments [3] are representative of crustal-type material and Mn-rich [3] and anoxic black shales [2] are typical of oil source rocks*

	Os	Ir	Ru	Pt	Pd	Re
Andesite	0.02	0.004	0.09	1.90	0.66	0.36
Lithic Sediment		0.002	0.017	0.211	0.029	0.137
Mn-rich shale	1.073	0.294	0.302	3.037	6.019	0.041
Anoxic black shale		0.108		1.17	4.28	

2 METHODOLOGY

Metal and trace element concentrations have previously been obtained on oils using ICP-MS[4]. The method most commonly used is to dissolve the oil sample using an organic solvent (such as xylene or kerosene) and then to aspirate this mixture directly into the plasma[4]. This technique is not readily applicable to PGE analyses however, as the low concentrations typical of geological materials require that a preconcentration technique must be employed. Dilution of the oil sample at the outset with large volumes of organic solvent would reduce the PGE signal to an unquantifiable level.

2.1 Sample Digestion

Silicate rock samples to be preconcentrated for PGEs are either fused with nickel-sulphide[7] or digested using inverse aqua regia within a sealed, high-pressure glass Carius tube[8]. The Ni-S fusion is considered unsuitable for oil analysis due to the high blanks associated with this method[9]. Carius tube digestions were attempted, however, the high volatile content of oils makes this a highly dangerous procedure and extreme care has to be taken to prevent explosion of the Carius tubes. The authors found that at temperatures considered "safe" for attempted digestions (~100°C) the oil was not oxidised and hence was not in a suitable form to continue with further preconcentration.

Digestion was thus undertaken using a Prolabo[R] focussed microwave digestion system which has a total output power of 200watts. Isotope dilution was implemented as part of the procedure. As such a known volume of isotopically enriched PGE spike was allowed to equilibrate with the oil in a sealed vessel at low temperature on a hotplate prior to microwave digestion, then any later sample losses can be corrected for. The digestion is a simple procedure and involves 3 principle steps.

Firstly, the oil is decarbonised by reaction with HNO_3. This involves the break-up of the C-H bonds in the oil and the solution changes from an immiscible layer of oil over acid into a thick black solution. Nitrous gases are constantly evolved during this step and removed from the microwave condenser via a scrubber, this procedure should be carried out within a fume hood. H_2SO_4 can also be utilised in conjunction with HNO_3, however we chose to use only HNO_3 in order to minimise the blank. Next the solution is oxidised by reaction with HNO_3 and $HClO_4$ acid and the C is removed from the solution as CO_2. When this step is completed the solution should be clear (although it often still retains an orange or yellow colouration). The solution can now be dried down and the residue reacted with HCl in order to convert the PGEs to chloride form, this is essential for the next step in the preconcentration procedure. Full details of the microwave procedure are illustrated schematically below (Fig 1).

The procedure can be halted at Step 4 and materials diluted in 0.5M HNO_3 for trace elements analysis by ICP-MS or it can be continued though to Step 5 if PGEs are required.

2.2 Anion Exchange Preconcentration

The procedure for PGE pre-concentration following digestion is adapted from that used by Pearson and Woodland[10]. Following digestion solutions are diluted to 10ml with 0.5M HCl and chlorinated 12 hours prior to loading on anion exchange columns. Chlorination oxidises the samples and is essential to ensure the efficient binding of the PGEs, particularly Ir, with the Biorad AG1-X8 anion-exchange resin (#100-200). The bulk of the oil matrix elements wash straight through the column, however, an initial elution step using dilute HF / HNO₃ is undertaken to remove elements such as Zr and Hf which may generate later oxide interference problems during mass spectrometry[10].

Sulphurous acid is now used to reduce the columns and allow removal of the PGEs in successive acid fractions. Ir and Pt are stripped from the column using 6N HCl, and Re and Pd are collected using warm 12N HNO₃. 100% yields are not obtained but this is compensated for by the use of isotope dilution.

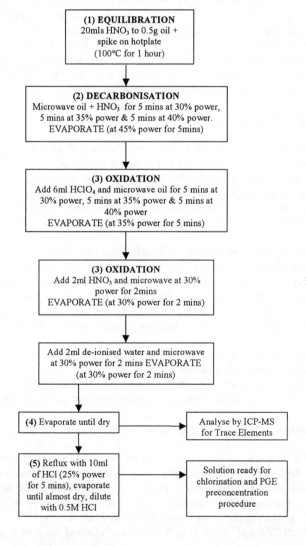

(1) EQUILIBRATION
20mls HNO₃ to 0.5g oil +
spike on hotplate
(100°C for 1 hour)

(2) DECARBONISATION
Microwave oil + HNO₃ for 5 mins at 30% power,
5 mins at 35% power & 5 mins at 40% power.
EVAPORATE (at 45% power for 5mins)

(3) OXIDATION
Add 6ml HClO₄ and microwave oil for 5 mins at
30% power, 5 mins at 35% power & 5 mins at
40% power
EVAPORATE (at 35% power for 5 mins)

(3) OXIDATION
Add 2ml HNO₃ and microwave at 30%
power for 2mins
EVAPORATE (at 30% power for 2 mins)

Add 2ml de-ionised water and microwave
at 30% power for 2 mins EVAPORATE
(at 30% power for 2 mins)

(4) Evaporate until dry

Analyse by ICP-MS
for Trace Elements

(5) Reflux with 10ml
of HCl (25% power
for 5 mins), evaporate
until almost dry, dilute
with 0.5M HCl

Solution ready for
chlorination and PGE
preconcentration
procedure

Figure 1 *Microwave digestion procedure for oil and other organic materials*

2.3 Instrumentation

All analyses in this study were carried out using a Perkin Elmer-Sciex Elan 6000 quadrupole ICP-MS. Isotope ratios were measured and overall concentrations calculated following corrections for mass bias and isobaric oxide interferences. Isotope pairs which suffer least from isobaric interferences are used[10]. The collected PGE fractions are diluted as little as possible for ICP-MS analysis; to 1ml for cross-flow nebulisation (using a Scott-type spray-chamber) and 0.5ml for desolvating micro-concentric nebulisation (MCN). This ensures that optimal signal and hence high sensitivity is achieved during the PGE isotope ratio measurements.

The MCN is advantageous in that it reduces oxide interferences (particularly of Zr and Hf, on Pd and Ir) and enhances sensitivity (samples need be less dilute for analysis due to the low sample uptake rate). However, severe memory effects have been observed particularly for Pd and Re when using the MCN. For inter-sample washout using dilute HNO_3, the Re background gradually diminishes over a period of between 5 and 10 minutes to an acceptable level (to <10ppt). However, if the washout solution is then changed for example to IPA, recommended for cleaning the membrane within the MCN, the background dramatically increases once again. This jump-up in background is recurrent when the washout solvent is varied and suggests that Re (possibly complexed with organic phases) is adhering to the membrane within the MCN. Hence Re and Pd analyses were conducted using a cross-flow nebuliser.

2.4 Advantages of microwave digestion

The technique outlined above has a very low analytical blank since the only source of contamination is from the acids used, the anion exchange resin and from the glass digestion vessels. To minimise contribution from these sources, only high purity acids are used, the resin is thoroughly cleaned with $12N\ HNO_3$ and the glassware is cleaned by refluxing with aqua regia between each use. If all precautions are followed strictly, blank should be <10ppt for all PGEs[10]. Organic solvents used in other dilution methods may adversely affect the blank.

Traditional analyses of oil by ICP-MS, where oil-solvent mixtures are aspirated directly into the plasma, require the addition of oxygen to the nebuliser gas. If insufficient oxygen is added however, C will build up on the sampler cones, so reducing signal and sensitivity and eventually leading to blockages. With our method, the sample is de-carbonised (hence excess oxygen is not required) and is introduced within a dilute-acid matrix which is much less viscous than an organic solvent matrix. This has several advantages:
a)   ionisation efficiency and hence detection limits are improved.
b)   memory effects are less severe.
c)   blocking of the sampler and skimmer cones is much less likely to occur.
d)   we do not have problems finding oil-organic solvent matrix matched standards.

2.5 Disadvantages of microwave digestion

One major drawback of our technique is that the range of PGEs that can be quantified is very limited. Rh cannot be measured using isotope dilution as it is mono-isotopic, and Os and Ru are both lost as volatile oxide species during digestion of the oil in the unsealed microwave environment. The borosilicate glass used for these vessels also apparently contributes Zr to the sample. This is undesirable as ZrO^+ species ($^{90}Zr + ^{16}O$ and $^{92}Zr + ^{16}O$) interfere on both ^{106}Pd and ^{108}Pd during ICP-MS analysis and must be corrected for. The anion exchange procedure has been designed to try and minimise the Zr carry over by eluting Zr and Hf with a mixed acid solution (see above). This problem could be overcome by using more expensive quartz digestion tubes. Additionally, since the desolvating MCN is not suitable for Pd and Re measurement, the use of a dilute acid matrix for sample introduction into the ICP-MS may exacerbate the oxide interference problem not experienced when using a purely organic solvent.

One issue which must be seriously addressed when using an isotope dilution technique is whether spike-sample equilibration has been properly achieved. In the case of the oils studied we tried to address this issue by means of running replicate samples of one particular oil (Ecofisk, see Table 2), both with spike-sample equilibration (sealed vessel on a hotplate) and without. This oil sample was not in hindsight ideal, as the PGE concentrations were extremely low and there was considerable variability in the PGE data obtained. However, it is unclear whether this is an equilibration problem or a sampling heterogeneity problem; different fractions of the oil with different viscosities may have separated with time and may not have been completely homogenised prior to sampling for analysis. This is an issue which will be addressed in future work.

3 RESULTS

Samples were selected from a variety of oil fields around the UK, predominantly from the North Sea, but also to the N.W. of Britain, e.g. the West of Shetlands Field. All oils were probably sourced from the Kimmeridge Clay, but were retrieved from reservoirs of varying lithology. Using the techniques outlined above a suite of 10 oils were analysed for rare earth elements (REEs), base metals and PGEs.

3.1 REE Patterns

All oils were found to have extremely low concentrations of REE. Concentrations for 9 of the samples were so low (<1ppb), that the signatures were not distinguishable above background. The one exception to this is for an oil sample taken from the West of Shetland Field (Fig 2). The REE pattern for this oil sample has a steep slope resulting from relative LREE enrichment compared to the HREE. This sample also has a positive Eu anomaly. This is thought to be a genuine effect as the sample has been corrected for oxide interferences.

Positive Eu anomalies are common within many geological samples and often characterise the presence of feldspars. In the case of an oil reservoir, it may indicate presence of clay minerals derived from weathering of feldspars in the source rock, or the extensive interaction of the oil with such a unit during migration from source to trap.

The low REE concentrations within these oils makes them highly susceptible to contamination as the oil migrates through crustal rocks which have much higher REE concentrations. This highlights the fact that lithophile elements are not suitable elements with which to fingerprint oil source rocks.

3.2 PGE Patterns

Most of the 10 oils analysed had very low concentrations of PGEs, especially for Ir and Pd (<10ppt) and are therefore below the detection limit for this technique (Table 2). Ir, Pt, Pd and Re data were obtained for several samples however and the PGE patterns for these samples are illustrated in Fig 3. Pt and Re concentrations tend to be higher, Re concentrations are the most constant and generally around the 1-2ppb range. This is consistent with the organophilic character of Re and also with the fact that Re is less siderophile than the other PGEs and therefore more enriched in crustal rocks. Ir and Os are often considered to behave identically during geochemical processes, from which we can presume that Os content in these oils would have been extremely low. As such the · Re/Os ratios of the oils would be uniformly high and therefore the oils would lend themselves well to provenance studies using the Re-Os isotope system[12].

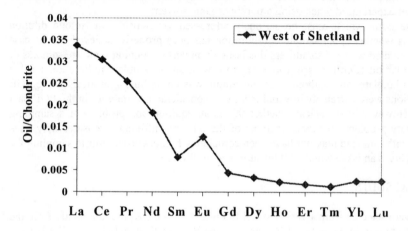

Figure 2 *REE pattern for oil sample from the West of Shetland (chondrite normalisation values* [11]*)*

Table 2 *PGE concentrations within oil samples analysed during this study. <dl is below detection limits, typically <10ppt (Errors on Ekofisk are 1 x Stdev of the mean and are based on 6 replicate analyses)*

Sample Location	Ir (ppb)	Pt (ppb)	Pd (ppb)	Re (ppb)
Dauntless	0.31	319.78	0.14	1.52
West of Shetland	0.01	1.38	0.03	2.07
Abbot	0.01	15.72	<dl	1.38
Scott	<dl	1.45	<dl	1.06
Marmion	<dl	0.59	<dl	0.94
Hannay	<dl	1.05	<dl	1.08
Telford	0.31	8.35	0.28	1.06
Scheilhalion	0.02	0.97	0.13	0.91
Hudson	<dl	1.38	0.07	2.4
Mobil	<dl	1.1	0.06	<dl
Ekofisk	0.02 ± 0.02	1.32 ± 1.05	0.13 ± 0.09	0.66 ± 0.69

The Pt data is the most heterogeneous between oils from different locations and in certain samples is anomalously high, with concentrations peaking at 320ppb Pt within the oil from the Dauntless Field. This data is very preliminary and the reason for the substantial Pt enrichment compared to other oils is not yet fully understood. Further investigations are necessary to determine whether the Pt-enrichment can be explained by a geological phenomenon and to eliminate the possibility that the Pt may have been picked up as a contaminant during the drilling process.

3.3 Correlation between PGE and Base Metal Enrichment

The oils which are moderately enriched in PGEs also have the highest Fe, Mn and Cu contents of the 10 oils analysed in this study. Interestingly though, the Dauntless oil sample with the anomalously high Pt contains low concentrations of Fe and Mn. This negative correlation is important, as Fe and Mn are 2 of the elements most susceptible to contamination by drilling fluids[4]. This could indicate that the Pt peak is a primary feature of the crude oil rather than being caused by late-stage contamination.

Although deficient in Fe and Mn, the Dauntless oil sample is enriched in Ni, V and Mo, as well as Pt, compared to the other oils analysed. Recent research[4] has revealed that high concentrations of Ni and V frequently occur in oils as these elements form stable porphyrins and that high levels of V and Mo are characteristic of sediments deposited in highly anoxic marine environments. Hence, with further work we might be able to infer that Pt is also capable of forming such stable organo-metallic compounds in reducing environments.

It is likely that most of the metal content of oils is concentrated in the asphaltene fraction[5]. Future work will therefore investigate if there is any correlation between asphaltene content and PGE systematics in the samples we have studied.

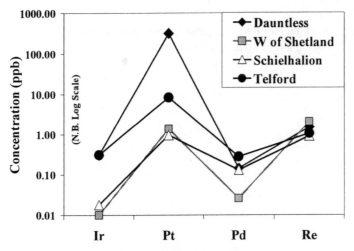

Figure 3 *PGE patterns for oils analysed during this study*

4 CONCLUSIONS

During this study a method which allows the low-blank digestion of organic materials, such as oil, plastic and synthetic fibres using a focussed microwave digestion system has been developed. Following this digestion procedure, samples can be analysed directly by ICP-MS (or ICP-AES) for major and trace elements, or further pre-concentration procedures can be adopted for precious metal analysis.

Using this method we have undertaken a preliminary analysis of REE and PGE concentrations in oil samples from 10 different locations. Both REE and PGE concentrations are generally very low, however, certain samples do contain quantifiable PGE concentrations which have occasional high Pt anomalies. The low REE concentrations in oils effectively preclude them from use as tracer elements since they are prone to contamination as oils migrate through the crust. Thus, if the Pt enrichment is a genuine oil-source rock signature rather than an artefact of drilling contamination, Pt may prove to be an extremely useful tracer element for fingerprinting specific oil fields.

5 REFERENCES

1.   B. Schaefer et al., 2000, *Goldschmidt Journal of Conference Abstracts*, 5 (2), 881.
2.   G. Ravizza and D. Pyle, 1997, *Chem. Geol.*, **141**, 251-268.
3.   S. J. Woodland, D. G. Pearson and M. Thirlwall, 2000, *J. Pet.*, in press.
4.   S. Olsen, 1998, Ph.D. thesis, Newcastle University, 284pp.
5.   A. Barré, A. Prinzhofer, C. J. Allègre, 1995, *Terra Nova*, 7 (Suppl. No.1), 199.
6.   R. Creaser Pers. Comm. 2000.
7.   S. E. Jackson et al., 1980, *Chem. Geol.*, **83**, 119-132.
8.   S. B. Shirey and R.J.Walker, *Anal. Chem.*, 1995, **34**, 2136-2141.
9.   S. J. Woodland and D. G. Pearson, 1999, *In* J.G. Holland, and S.C. Tanner, Eds., Developments in ICP-MS. Royal Society of Chemistry.
10.   D. G. Pearson and S. J. Woodland, 2000, *Chem. Geol.*, **165,** No.1-2, 87-107.
11.   W. F. McDonough and S. S. Sun, 1995, *Chem. Geol.*, **120,** 223-253.
12.   S. B. Shirey and R. J. Walker, 1998, *Annu. Rev. E.P.S.L.*, **26,** 423-500.

6 ACKNOWLEDGEMENTS

We would like to thank Mobil North Sea Ltd, Phillips Petroleum Company (UK) Ltd and Amerada Hess for providing the oil samples for this work.

USN-ICPMS: A POOR MAN'S HIGH RESOLUTION

David J Gray, Charles LeBlanc, Wilson Chan, Bernie Denis

Elemental Research Inc., Vancouver, Canada.

ABSTRACT

Solving the problem of molecular ion interference in ICPMS analysis has been the goal of much instrument development over the past decade. Solutions have varied from front-end sample introduction techniques, such as hydride generation, cold plasma, and membrane desolvation, to analyzer modifications such as collision/reaction cells, and high-resolution magnetic sector devices. In this work, we have evaluated a USN-membrane desolvation device for a variety of analytes (including Si, P, S, V, Ni) and matrices (human plasma, mussel shells) and compared its performance between two different ICPMS instruments. Data will be presented for standard figures of merit (LOD, LOQ, sensitivity etc..). Instrumentation optimization procedures for routine operation will also be discussed.

1 INTRODUCTION

The interferences present in ICPMS spectra using conventional pneumatic nebulisation for liquid sample introduction have been well documented in the literature[1,2]. They manifest themselves principally from three sources:

(i)   isobars - where isotopes of different elements have coincident unit masses (e.g. ^{116}Cd and ^{116}Sn);

(ii)   argon plasma gas/aqueous phase – molecular ion combinations of O, N, H and Ar species (e.g. $^{40}Ar^{16}O^+$ interference on $^{56}Fe^+$);

(iii)   matrix interferences – from doubly charged species of the major matrix components, oxides, nitrides, hydrides and argides (e.g. In the analysis of a copper-nitrate solution, $^{63}Cu^{40}Ar^+$ interferes with $^{103}Rh^+$).

The fundamental limitation of conventional ICPMS instruments with respect to addressing the problems of spectral interference derives from the single mass resolution capabilities of most commercially available systems. There are thus two possible solutions:

(i)   remove or substantially reduce the interfering species prior to detection, or:

(ii)   replace the single mass resolution device with an analyzer of suitable resolving power to separate the interfering signal from the analyte signal[3,]. While the latter may be a technical preference, it is not necessarily an economic option for the many commercial laboratories with existing conventional quadrupole ICPMS systems installed.

There are many methods of removing spectral interferences. Examples include:

(i)   "cold plasma"[4]– where the forward rf power to the plasma is reduced to between 500-600W. When instruments are operated in this mode, the backgrounds on the alkali metals and Fe are substantially reduced. This technique has found many applications in the semiconductor industry where ultra-pure reagents and trace metal control in clean room fabrication facilities are critical to the production of defect free silicon chips. The drawbacks to the technique are that it is extremely intolerant of complex samples with moderate total dissolved solids and many elements are not efficiently ionized under the low power conditions;

(ii)   Hydride generation[5] – in this method, vapor phase hydride products, formed by reacting an acidified sample with for example, sodium boro-hydride, are passed through a suitable gas-liquid separator before introduction into the plasma. The technique is limited to the hydride forming elements (Ge, As, Se, Sb, Sn, Pb, Bi) and in particular eliminates the spectral interference $^{40}Ar^{35}Cl$ from ^{75}As.

(iii)   Reaction Cells[6]– these devices are based on the principle of collisional dissociation and neutralisation of unwanted molecular ion species with a suitable reaction gas in a multipole cell just prior to the ion beam entering the quadrupole. This device is not, to the authors' knowledge, retro-fitable to older instruments.

      The technique we chose to evaluate as a general means of reducing molecular interference in liquid sample analysis was ultra-sonic nebulisation coupled to a membrane desolvator. A schematic of this technique is illustrated in figure 1. The liquid sample is pumped onto the vibrating surface of a piezoelectric transducer, which transforms the liquid into a very fine aerosol.

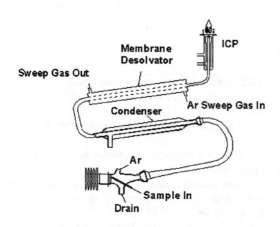

Figure 1 *Schematic of a USN-membrane desolvation system (Courtesy of Cetac)*

      After nebulization, the aerosol is carried by the argon gas flow through a heated tube where the solvent is vaporized. The vaporized solvent and dry aerosol particles are then carried through an air condenser and then a Pelltier cooled condenser where most of the solvent vapor is condensed and removed from the argon stream. The argon gas carrying the aerosol is then passed through a porous tube, the heated PTFE membrane of the membrane desolvation unit. In this porous tube, most of the remaining solvent vapor passes through the membrane and is carried off by a sweep gas. This dry argon gas containing the analyte aerosol is then directed to the injector of the ICP.

2 INSTRUMENTATION

Ultra-Sonic Nebulizer/Desolvating Membrane: The instrument used in this work was a U-6000AT$^+$ (Cetac, NE, USA). The nominal settings used for the various controllable parameters are detailed in Table 1.

Table 1 *Instrument settings for the U-6000AT$^+$ instrument.*

Parameter	Setting
Condenser	-2°C
Heated U Tube	140°C
Sweep Gas Flow	2 L/min
Sample Argon Flow (nebulizer flow)	0.6 L/min

2.1 ICP-MS

Two ICPMS instruments were utilized during this work: an Elan 6000 (Sciex, ONT, Canada), and a PQ3 (TJA Solutions, Cheshire, UK). The settings for these instruments in both conventional and USN mode are detailed in tables 2 and 3.

Table 2 *Instrument settings for the ICPMS in conventional mode*

Parameter	Elan Setting	PQ3 Setting
RF Power	1100 Watts	1350 Watts
Nebuliser Ar	0.8-0.9 L/min	0.8-0.9 L/min
Scan Mode	Scanning	Scanning
Dwell Time	20 ms/channel	20 ms/channel
#repeats	3/scan	3/scan

Table 3 *Instrument settings for the ICPMS in USN mode*

Parameter	Elan Setting	PQ3 Setting
RF Power	1000 Watts	1000 Watts
Nebuliser Ar	0.6 L/min	0.6 L/min
Scan Mode	Scanning	Scanning
Dwell Time	20 ms/channel	20 ms/channel
#repeats	3/scan	3/scan

2.2 Sample delivery

In both conventional and USN modes, the liquid samples were delivered to the nebuliser via a standard peristaltic pump set up to deliver a flow rate of 1 mL/min. A glass concentric "high TDS tolerant" Meinhard nebuliser was used for conventional analysis on both instruments. All instrument rinse and sample dilutions discussed were prepared from high purity nitric and hydrochloric acids (Seastar, BC, Canada) using 15 and 50 mL metal free polypropylene tubes (Elkay), and all dilutions and sample aliquots were performed using air displacement pipettes (Eppendorf) and an autodilutor (Gilson 401). Deionised water with a resistivity >18MΩcm was supplied sequentially through a still (Corning-MP-3A) and deionisation unit (Barnstead – Nanopure UV).

3 RESULTS

3.1 Basic Performance

Initially, the Elan 6000 instruments torch position is adjusted (in the horizontal and vertical plane) to maximize the count rate obtained for [103]Rh. This is followed by an automatic optimization of the nebuliser gas flow and lens voltage to yield maximum sensitivity at [103]Rh whilst minimizing both doubly charged ion formation (69/138 ratio) and oxide formation (156/140 ratio) using the auto-tune routine that comes with the instrument software. The voltage range for the lens is –24 to +20 V. The standard setting is typically between 6 and 7.5 V, however the voltage rises to 8.5 to10 V through use. As this voltage rises it is more difficult to meet factory specifications for sensitivity in the low mass range, for oxide formation and for doubly charged ion formation. At this point, the lens is replaced with a cleaned lens assembly.

   Initially, the PQ3 torch position is automatically optimized (in the horizontal and vertical plane) to maximize sensitivity at [103]Rh. While the Elan 6000 ion optics comprise of a single "ion lens", the PQ3 system is a multiple lens stack assembly. After torch position optimization, the individual lens settings are adjusted manually to optimize sensitivity across the mass range while monitoring [45]Sc, [103]Rh and [208]Pb. It is found that when the auto-tune routine within the software is used for lens assembly optimization, it tends to overly bias the instruments sensitivity in the particular mass range chosen for optimisation. Finally, the nebuliser flow rate and sampling depth of the torch are adjusted to minimize oxides and doubly charged ion formation.

   Table 4 illustrates ten randomly selected daily performance checks from both instruments in conventional mode over a six-month period. These checks are used as system suitability verification in our laboratory before client samples are analysed. The solutions used to assess the daily performance were all prepared weekly by serial dilution of a custom made NIST traceable multi-element stock. The 10 µg/mL stock solution was prepared in 5%HNO3

Table 4   *A comparison of two ICPMS systems' performance in conventional mode over a 6-month period. Isotope responses are in counts per second*

Analyte (1 ng/mL)	PQIII response Average N=10	%RSD	ELAN 6000 response Average N=10	%RSD
[9]Be	2884	32%	663	25%
[45]Sc	32318	21%	13856	17%
[103]Rh	45245	15%	20108	13%
[159]Tb	54456	28%	28506	25%
[208]Pb	13889	19%	15299	19%
[69/138]Ba[++]	2.68%	38%	1.92%	20%
[(156/140)]CeO	1.51%	49%	2.68%	12%

The overall response of the two instruments is comparable in the medium to high mass range of the spectrum, although there are significant differences below mass 40. This is in part due to the difference in lens stack designs, as the multiple lens assembly offers greater flexibility to control the response across the mass range. This greater flexibility also manifests itself in the slightly greater relative standard deviations, listed in table 4, for the PQ3 machine. The doubly charged ion formation on the PQ3 is higher on average and we believe this is attributable to the higher plasma operating power (hence plasma temperature) used. By the same token, the average CeO formation rate is correspondingly lower with these nominal settings. In our measurements, oxide formation rates of 2-3% (as CeO) are therefore to be expected for conventional ICPMS analysis, consistent with manufacturers specifications.

3.2 Optimization procedure

The USN-ICPMS optimization procedure for both instruments was similar. Initially, each system was optimized by adjusting the torch X-Y positions and lens' voltage(s), to maximize the signal obtained for ^{103}Rh. The final step is to adjust the sweep gas from its nominal setting of 2 L/min to between 1.8-2.3 L/min. This latter parameter affects both signal intensity and oxide formation. For the USN-PQ3 system, there is an extra step performed before sweep gas adjustment; optimisation of the plasma sampling depth (or Z-position) relative to the sampler cone. It has been observed for the PQ3 that the torch is in optimum position 0.5 cm further back from the sampler cone in USN-mode as compared to conventional mode.

The promises of USN sample introduction with membrane desolvation are higher sensitivity and reduced molecular interference. Table 5 illustrates the data obtained from the same "daily performance" solutions run on both instruments while coupled to the U-6000AT$^+$.

Table 5   *A comparison of the performance of two ICPMS systems in USN- mode*

Analyter (1ng/mL)	Elan 6000 Average	Enhancement Factor	PQ3 Average	Enhancement Factor
9Be	2057	3	89875	31
45Sc	85932	6	738192	23
103Rh	155320	8	1130943	25
159Tb	242520	8	2424042	45
208Pb	149052	10	663769	48
Ba++	2.60%	1.29	1.26%	0.47
CeO	0.10%	0.04	0.07%	0.05

It can be seen that there is a 20-25-fold reduction in oxide formation on both systems. There is little change in the doubly charged ion formation for the Elan since the RF powers used are similar in both conventional and USN modes. By the same token, since the forward power is less in the USN-PQ3 system as compared to conventional mode, the doubly charged ion formation is lower.

An improvement in sensitivity of 3-10 fold is observed for the USN-Elan system, which is within the expected range claimed by the USN vendor. By contrast, the observed enhancement factors found for the USN-PQ3 system are in the 25-50 range. The bulk of this extra sensitivity is achieved when the plasma sampling depth is increased during optimization. Without the latter step, only the moderate enhancements of 3-10 found on the USN-Elan system are observed. Once in operation the Elan 6000 ICPMS has no control over the plasma sampling depth other than via changing the nebulizer flow rate. Unfortunately this also affects the ultra-sonic nebulisation and desolvation efficiency. Based on the higher sensitivity obtained, the PQ3/U-6000AT$^+$ system was used for the remainder of the investigations reported here

3.3 Results on Silicon, Phosphorus and Sulphur

The spectra obtained from 28-34 amu probably constitute one of the most interfered mass ranges in conventional ICPMS analysis. In nominally blank nitric acid solutions, all three isotopes of Si and S, as well as mono-isotopic P, are background limited to a great extent. In the case of Si, the interference is so significant that proper isotope ratio measurements are practically impossible by this technique, while for Sulphur the situation is compounded by a low ion yield due to its high ionization potential (10.36 eV)[9] placing detection limits at the 1000-10000 ng/mL range. Table 6 lists the common interferences on these elements.

Table 6   *Some of the common spectral interferences observed on Si, P and S isotopes during conventional ICPMS analysis.*

Isotope	Interference
^{28}Si	^{14}N$_2$, ^{12}C^{16}O
^{29}Si	^{14}N^{15}N, ^{13}C^{16}O
^{30}Si	^{14}N^{16}O, ^{15}N$_2$
^{31}P	^{14}N^{16}O^{1}H, ^{15}N^{16}O
^{32}S	^{16}O$_2$
^{33}S	^{15}N^{18}O, ^{16}O^{17}O
^{34}S	^{16}O^{18}O, ^{17}O$_2$

Figure 2 is a bar graph showing the response of the PQ3 unit for a nominally blank 2%HNO$_3$ solution in both USN and conventional "Meinhard" modes.

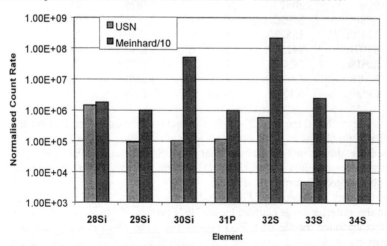

Figure 2 *2%HNO3 blank spectra by both USN and conventional ICPMS. It should be noted that for clarity, the count rate for conventional mode has been divided by a factor of ten.*

The relative abundance of the three Si isotopes are 92.23% (mass 28), 4.67% (mass 29) and 3.10% (mass 30) [7]. It can be seen in figure 2 that the Si background count rates are drastically skewed from natural abundance in conventional mode and this prohibits any accurate assessment of Si isotope ratios in a real sample, especially given that the response at mass 30 is already close to the saturation point of the detector. On the other hand, the background noise in USN mode is reduced overall. As an experiment, we decided to evaluate the accuracy of Si isotope ratio measurements by USN-ICPMS given this improved background. A 10µg/mL Si and Mg solution was prepared in de-ionised water by dilution from a NIST traceable 1000 µg/mL stock. The ^{25}Mg and ^{26}Mg isotopes were used as a measurement of mass bias and the associated "ΔM" correction applied to the Si isotopes. As can be seen in Table 7, the accuracy and precision of the measured abundance is very good.

Table 7   *Measured abundance of Silicon Isotopes in a 10 µg/mL solution By USN-ICPMS*

Isotope	^{28}Si	^{29}Si	^{30}Si
	92.13%	4.50%	3.07%
	92.33%	4.60%	3.07%
	92.41%	4.62%	3.13%
	92.10%	4.63%	3.07%
	92.31%	4.64%	3.13%
	92.41%	4.65%	3.17%
	92.10%	4.60%	3.17%
	92.16%	4.67%	3.13%
	92.30%	4.69%	3.11%
	92.29%	4.66%	3.02%
Average	92.25%	4.66%	3.09%
Std. Dev.	0.11%	0.03%	0.05%
%RSD	0.1%	0.6%	1.5%
Theory	92.23%	4.67%	3.10%

The average count rate obtained for this 10 µg/mL solution at mass 28 was approximately 1.4×10^8 cps, with a blank count rate of 1×10^6 cps. While P is mono-isotopic and isotope ratios are not a factor, this element is nevertheless background limited by interference from the $^{14}N^{16}OH^+$ ion. Background count rates of 1×10^7 are typical in conventional mode, with real detection limits in the 10 ng/mL range or even higher for samples of unknown acidity and nitrogen content. As figure 2 shows, the background for P is reduced significantly in USN mode. Since sensitivities are higher in this mode, detection limits are correspondingly lower (by approximately 2 orders of magnitude). A calibration curve for P by USN-ICPMS in the range 1-100 ng/mL is shown in figure 3 demonstrating good linearity (i.e. $R^2>0.99$) over this range.

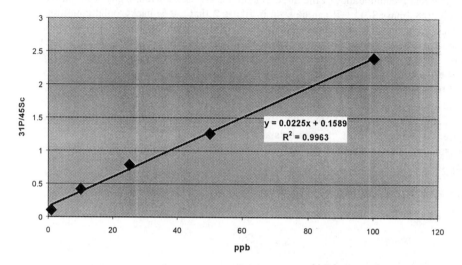

Figure 3 *Calibration curve for P in 2%HNO₃ in the range 1 – 100 ng/mL by USN-ICPMS*

All of the isotopes of sulphur are severely interfered during conventional ICPMS analysis. The background count rates were lowered by 3-4 orders of magnitude in USN mode as shown in figure 2. This resulted in an absolute detection limit based on 3x the standard deviation of 10 blank replicates of 0.5 ng/mL, a 2000-20000 fold improvement! A calibration curve for S by USN-ICPMS from 1-100 ng/mL is shown in figure 4 demonstrating that the technique is also linear over this range.

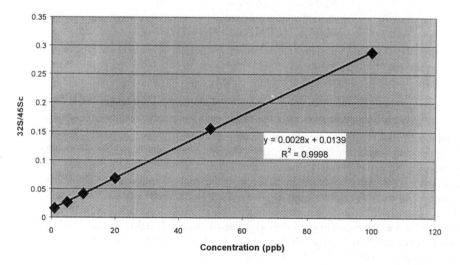

Figure 4 *Calibration curve of S in 2%HNO₃ by USN-ICPMS from 1-100 ng/mL*

3.4 Interference in a high Calcium matrix

There are numerous problems associated with the ICPMS assay of "hard tissue" samples such as bone tissue, fish scales, shells and coral. The high calcium (carbonate) content of these samples can lead to cone blockage and signal suppression. Other difficulties with the analysis of these sample types are related to the interferences present (see table 8) on some key analytes such as Ni, Co and Sr, used for discriminating between fish populations in population dynamic studies and monitoring environmental temperature effects over time[7]. In some cases, it is possible to correct for interferences in ICPMS spectra by measuring the level of interference present in a standard solution, say 10 µg/mL Ca, then arithmetically correcting the interfered peaks by measuring the Ca present in a sample. Unfortunately this is rendered impossible in biota due to the potentially high levels of Sr present in natural Ca-rich samples. Sr forms doubly charged species that interfere directly with the measurable isotopes of Ca. (i.e. $^{88}Sr^{++}$, $^{86}Sr^{++}$ and $^{84}Sr^{++}$ interfere with $^{44}Ca^{+}$, $^{43}Ca^{+}$ and $^{42}Ca^{+}$ respectively.). One is therefore left with the task of eliminating the interferences by other means.

Table 8   *Some common interferences in biota sample analysis by ICPMS*

Interference	Analyte
$^{40}Ar^{12}C$	^{52}Cr
$^{40}Ar^{13}C$	^{53}Cr
$^{42}Ca^{16}O$	^{58}Ni
$^{43}Ca^{16}O$	^{59}Co
$^{44}Ca^{16}O$	^{60}Ni
$^{44}Ca^{18}O$, $^{48}Ca^{14}N$	^{62}Ni
$^{46}Ca^{40}Ar$	^{86}Sr
$^{48}Ca^{40}Ar$	^{88}Sr

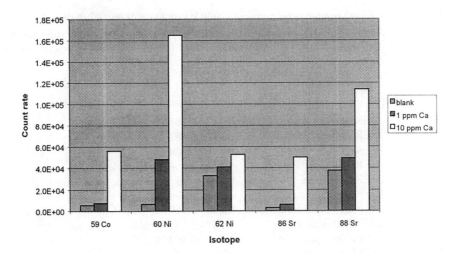

Figure 5 *Background effects of Ca levels by conventional ICPMS*

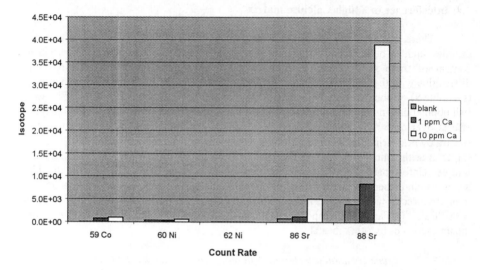

Figure 6 *Effects on Background of varying Ca levels during USN-ICPMS analysis*

The background count rates of three solutions of 2%HNO₃ (a blank, 1 μg/mL Ca and 10 μg/mL Ca) were analysed by both conventional and USN ICPMS. These results are detailed in figures 5 and 6. As would be expected, in USN mode the Ca levels have little effect on the observed backgrounds whereas the nitrides and oxides of Ca are evident in conventional mode. By the same token, it is evident that the argides of Ca are present to a significant extent in both modes of operation. This is expected since the desolvation unit will not remove the Ca analyte from the sample aerosol, and the argides of Ca form in the plasma ion source.

Table 9   *QC and sample results for USN-ICPMS analysis of mussel shells. The prefixes A and B refer to two separate sampling sites. All results are in μg/g relative to the solid sample except the TM-24 samples, which are in ng/mL*

Sample ID	Cr	Co	Ni
A-ERI500 replicate 1	0.412	0.408	0.565
A-ERI500 replicate 2	0.441	0.382	0.360
A-ERI500 replicate 3	0.471	0.384	0.268
A-ERI500 replicate 4	0.347	0.364	0.231
A-ERI500 Spike	9.65	11.3	10.9
B-ERI500 replicate 1	0.550	0.294	0.218
B-ERI500 replicate 2	0.648	0.359	0.191
B-ERI500 replicate 3	0.649	0.338	0.330
B-ERI500 replicate 4	0.650	0.392	0.279
B-ERI500 Spike	10.8	10.4	10.1
TM-24 replicate 1	5.32	19.0	3.54
TM-24 replicate 2	5.15	18.5	3.49
TM-24 certified values	5.60	19.4	3.50

It was then decided to assess the ability of the USN-PQ3 system to determine the concentration of selected elements in samples of mussel shells collected from two separate sites off the coast of Vancouver Island, BC, Canada (Courtesy of 2WE Associates). Four separate 0.05g fragments of each mussel shell were digested by reacting with 1mL HNO_3 in a metal free polypropylene tube at 95°C. The samples were made up to 10mL final volume with deionised water then analysed at a X5 dilution by USN-ICPMS. Two fragments were spiked with an equivalent concentration of 10 µg/g of each analyte relative to the solid sample to assess recovery. Indium was added as an internal standard at 1ng/ml in the final test solution. Finally, two certified reference water solutions (TM-24 supplied by NWRI) were included in the analyses to verify calibration. All of these results are tabulated in table 9 below where it can be seen that spike recoveries and CRM values are within 10% of theoretical.

3.5 Vanadium In Human Plasma

Human plasma contains a relatively high chloride content (~103.7mmol/L [8]). As a result, analysis of this matrix by conventional ICPMS is prone to molecular interferences from combinations of Cl and O, as well as Cl and Ar. An example of this is the $^{35}Cl^{16}O$ molecular ion, which interferes with ^{51}V, limits the quantitation and detection limits attainable by conventional ICPMS. Also, due to the large total dissolved solid content, this

Table 10 *Comparison of V determination parameters in human plasma by conventional and USN ICPMS*

Parameter	Conventional	USN Mode
Blank count rate (average)	5350	289
25 ng/mL V count rate	4501	13055
LOD (3σ) ng/mL	7.5	0.25
LOQ (10σ) ng/mL	25	0.75

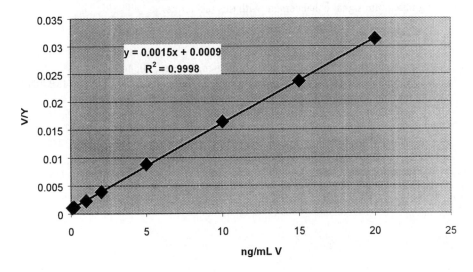

Figure 7 *Calibration curve of V in human plasma by USN-ICPMS*

matrix suppresses the ICPMS signal in both conventional and USN modes. The effect is more evident in the latter mode since it is inherently more efficient at transporting sample to the instrument.

We performed the following study to determine whether USN could perform better on this relatively difficult matrix. The plasma was first thawed to room temperature from its storage temperature of -70°C, then 100 μL aliquots were transferred to 15 mL metal free polypropylene tubes. These aliquots were then digested with 200 μL of HNO_3 at 95°C in a water bath for 2 hours. After cooling to room temperature, 100 μL of Y internal standard solution (100 ng/mL) and appropriate amounts of a NIST traceable Vanadium calibration solution (Spex Certiprep) were then added. The samples were then brought to a final volume of 2.5 mL (giving a dilution factor of 25 relative to the initial plasma sample) with de-ionised water using a Gilson autodilutor.

Ten blanks and two 8-point calibration curves (in the range 0.1-20 ng/mL vanadium in plasma) were prepared and subsequently analysed by both conventional and USN-ICPMS. A summary of the results, and an example calibration curve for USN mode is shown below in table 10 and figure 7. The net result is that USN produces lower quantitation and detection limits for V in human plasma than conventional ICPMS, as a result of reduction in the ^{51}V background by removal of the $^{35}Cl^{16}O$ interference.

4 CONCLUSIONS

Ultrasonic nebulisation with a desolvating membrane is a cost effective way of removing many molecular ion interferences arising from the sample solvent during ICPMS analysis. While we have successfully applied it to a variety of analytes and sample types, it is clearly not a panacea for all situations. Some obvious drawbacks are its inability to affect the formation of molecular ions from analytes and the argon plasma gas (e.g. $^{65}Cu^{40}Ar$ on ^{103}Rh), loss of volatile elements through the membrane itself (e.g. Hg and B), and large carry over effects for certain elements (e.g. Os – Durham Conference 2000). As far as routine optimization goes, our work comparing two different USN-ICPMS systems suggests that the ability to control the physical positioning of the plasma in all three dimensions, while other parameters (e.g nebuliser flow) remain unchanged, is critical to obtaining maximum signal enhancement with this technique.

5 REFERENCES

1.   S. H. Tan and G. Horlick, *Applied Spectroscopy*, 1986, **40**, 445.
2.   A. A. van Heuzen and N. M. M. Nibbering, *Spectrochimica Acta*, 1993, **48B**,1013.
3.   N. Jakubowski and L. Moens, *Spectrochimica Acta*, 1998, **53**, 1739
4.   S. D. Tanner, M. Paul and E. R. Denoyer, *Atomic Spectroscopy*, 1995, **16**, 16.
5.   J. Bowman, B. Fairman and T. Catterick, J. Anal. Atom. Spectrom., 1997, **12**, 313.
6.   I. Feldmann, N. Jakubowski and D. Stuewer, Fres. J. Anal. Chem., 1999, **365**, 422.
7.   S. E. Campana et al., *Can. J. Fish. Aquat. Sci.*, 1997, **54**, 2068.
8.   Geigy Scientific Tables, Volume 3, P78, Publ. Ciba-Geigy, Switzerland, 1984.
9.   Handbook Of Chemistry & Physics, 80th edition, Publ. CRC Press, USA.

THE CHARACTERISTICS OF AN OFF-LINE MATRIX REMOVAL/PRE-CONCENTRATION SYSTEM WHICH USES A CONTROLLED PORE GLASS IMINODIACETATE AS A REAGENT FOR INDUCTIVELY COUPLED PLASMA MASS SPECTROMETRY

Ted McGowan and Noel Casey

Institute of Technology Sligo, Ballinode, Sligo, Republic of Ireland.

1 INTRODUCTION

The technique of Inductively Coupled Plasma Mass Spectrometry (ICP-MS) provides superior detection capability for trace element analysis. Most elements can therefore be determined at sub parts per billion (ppb) levels or lower in many types of samples. However the technique suffers from some well known interferences (matrix, spectral, etc.) which have the potential to both degrade detection limits and contribute to inaccuracy in analytical data[1-3]. For example, when attempting the trace element analysis of biological and seawater samples the formation of polyatomic spectral interferences from high concentrations of alkali/alkaline earth metals and anions cause serious problems. Possible solutions include prior matrix removal, application of interference correction factors and instrument modifications including the incorporation of reaction cells.

ICP-MS also exhibits a low tolerance to dissolved solids (<0.2% w/v). A high dissolved solids content leads to the physical deposition of salts on the sampler and skimmer cones thereby restricting the orifices and causing a subsequent loss of sensitivity. The dissolved solids content of a sample can be reduced by simple dilution of the sample. This is feasible provided the trace elements are present at levels well above the detection capability of the instrument. Here also matrix removal with preconcentration provides an alternative solution.

Matrix removal/pre-concentration techniques[4] include the use of coprecipitation[5], solvent extraction[6] and chelation ion exchange[7]. By using chelating ion exchange resins for matrix removal and pre-concentration ICP-MS detection limits can be lowered by two orders of magnitude or greater for samples having complex matrices. A simultaneous reduction in overall spectral interferences and dissolved solids content of the sample can also be achieved.

Fang et al.[8-10] provides a comprehensive overview of the design and characteristics of chelating ion exchange resin based matrix removal/pre-concentration systems. The availability of several commercial chelating ion exchange resins based on the iminodiacetate functional group has facilitated the development of matrix removal/pre-concentration systems incorporating these resins. These resins have been incorporated into automated systems for either online or offline applications and therefore have a distinct advantage when compared to techniques based on coprecipitation and solvent extraction. Iminodiacetate based chelating ion exchange resins have also been successfully used for ICP-MS applications[11].

Several forms of iminodiacetate based chelating ion exchange resins are commercially available including Metpac CC-1 (Dionex Corp, Sunneyvale, CA, USA), Chelex 100 (Biorad Lab., Richmond, VA, USA), Prosep (Bioprocessing, Consett, Durham, UK) and AF 650M Chelate (Tosohaas, Montomeryville, PA, USA). Each of these resins use the same iminodiacetate chelating ligand immobilised onto the surface of a polymer or controlled pore glass substrate. The physical/chemical properties of the substrate can affect the operating performance of the resin. Volume instability (swelling) of the polymer substrate (polystyrene divinylbenzene) with changes in pH and ionic strength adversely affects their operating performances. Conditioning of the resin using buffer solution is required to overcome the swelling. This extra step in the analytical method increases both analysis time and method blank levels.

For example, recent studies indicate that Prosep[12] and AF-650M[13] have some advantages over the Metpac CC-1[14] and Chelex 100[15] resins in terms of volume stability over a range of pH and ionic strength.

A number of recent publications have provided excellent descriptions of iminodiacetate based matrix removal/preconcentration applications which facilitate better ICP-MS measurements. Nicolai et al.[14] describe a hyphenated method using ion chelation chromatography coupled online to an ICP-MS instrument. This matrix removal/pre-concentration system utilises the Metpac CC-1 resin. The retention behaviour of 19 elements on the resin is discussed. Yabutani et al.[15] describes the ICP-MS multi-element determination of 29 trace elements in coastal seawater using Chelex 100 with a batch methodology. Willie et al.[13] describes a matrix removal/pre-concentration system using the AF 650M chelate for the determination of trace elements in seawater by ICP-MS.

The behaviour of the common matrix elements i.e. calcium, magnesium, sodium and potassium on the iminodiacetate chelating ion exchange resin is important in order to develop successful matrix removal/pre-concentration systems and strategies. A study of the fundamental parameters governing the retention of matrix elements on an iminodiacetate resin and experimental conditions required for optimal matrix removal has been carried out by Luttrell et al.[16]. A parameter study of how to efficiently carry out a matrix removal/pre-concentration step using an iminodiacetate chelating ion exchange resin has also been investigated by Nickson et al.[17].

Novel uses of iminodiacetate based chelating ion exchange including speciation and anion exchange have also been reported. The speciation of Selenium[18-20], Thallium[21, 22] and Vanadium[23] has been carried out using either a chemically modified form of the resin[24] or using unique experimental procedures. In acidic solutions the iminodiacetate chelating ion exchange resin has been used as an anion exchanger with the ability to separate anionic species of various elements[25]. These studies demonstrate the potential of iminodiacetate based chelating ion exchange resins particularly with regard to speciation and much of this potential has yet to be researched.

This paper describes a laboratory based matrix removal/pre-concentration system using the Prosep controlled pore glass iminodiacetate reagent. The basic analytical features of the system are described including the behaviour of a number of elements to variations in pH and other operating parameters.

2 EXPERIMENTAL

2.1 Reagents

All solutions were prepared with ultrapure water (18 MΩ cm) from a Millipore 2 analytical grade water purification system (Millipore, Badford, MA). A multielement stock solution (100 µg ml⁻¹ Spex chemical) was used to prepare calibration and standard solutions. Ammonium acetate (Merck, Poole, Dorset, UK) solutions were prepared from the solid. Adjustments in pH were made using acetic acid or ammonium hydroxide as appropriate. 0.5 M Nitric acid was prepared by diluting concentrated aristar grade Nitric acid (Merck, Poole, Dorset, UK).

The iminodiacetate reagent (Prosep, Bioprocessing, Consett, Durham, UK) was used as supplied.

2.2 Instrumentation

ICP-MS measurements were made using a VG Elemental PlasmaQuad Inductively Coupled Plasma Mass Spectrometer (VG Elemental, Winford, Chesire, UK). The instrument was calibrated and optimised prior to operation using a solution containing the elements Be, Mg, Co, Y, Bi at 10 ng ml⁻¹ in a 2% nitric acid matrix. The analyte peaks were monitored in the peak-jumping mode using the instrument conditions described in Table 1.

Table 1   *ICP-MS instrument conditions*

Cool flow:	14.5 l min⁻¹
Auxiliary flow:	0.9 l min⁻¹
Nebuliser flow:	0.843 ml min⁻¹
Incident power:	1500 w
Reflected power	5 w
Channels/amu:	20
Detector mode	Pulse counting
Points per peak	3

2.3 Matrix Removal/Pre-concentration System

A schematic diagram of the matrix removal/pre-concentration system is shown in Figure 1. This was constructed from PTFE tubing and the column was packed with 100 mg of Prosep. The valve used in this system is a four-stream selector valve (Model C25F-3184-EMH, Valco Europe).

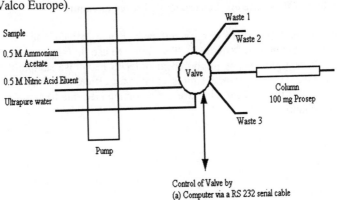

Figure 1 *Schematic Diagram of Matrix Removal/Pre-concentration system*

Control over stream selection is achieved either manually using a manual control unit or automatically by computer via an RS 232 serial cable using an MS-DOS© based timetable program. The reagents were pumped through the column at 4.0 ml min^{-1} per channel using 1 mm i.d. Tygon® peristaltic pump tubing (VG Elemental, Winford, Chesire, UK) via a 4-channel peristaltic pump (Gilson Minipuls 2). A dual electrode pH meter (Model 691, Metrohm) was used throughout.

2.4 Matrix Removal/Pre-concentration Operating Procedure

The steps involved in the matrix removal/pre-concentration procedure are listed in Table 2.

Table 2　　　　*Matrix Removal/Pre-concentration Operating Procedure*

Step	Description	Solution	Time (min)
1	Column regeneration	Nitric Acid	1.5
2	Conditioning of column	Buffer Wash	1.5
3	Loading of sample	Sample	12.5
4	Matrix Removal	Buffer Wash	1.5
5	Rinse	Water	1.5
6	Elution start	Nitric Acid	2.5

Memory effects from a previous sample are removed in the initial acid washing step (step1). The column is then conditioned by washing with the buffer solution (step 2) followed by loading of the sample for pre-concentration and matrix removal (step 3). Another buffer washing step (step 4) removes the sample matrix. The buffer itself is then washed from the system with a water rinse step (step 5). Finally the concentrated trace elements are eluted from the column with a nitric acid wash (step 6). Multielement analysis of the eluent solution is then performed by ICP-MS. A flow-rate of 4 ml/min^{-1} was chosen for most studies for the reasons outlined in section 3.1 below.

3 RESULTS AND DISCUSSION

3.1 The effect of flow-rate on the recovery of trace elements

The results indicate that there is no significant change in recoveries over the range of flow-rates from 1-4 ml min^{-1} (figure 2). The ability to use flow-rates as high as 4 ml min^{-1} without sacrificing recovery is important in order to maximise sample throughput. This indicates that the current system design is suitable for efficient element pre-concentration. Flow-rates greater than 4.5 ml min^{-1} are not recommended due to an unacceptable build-up of back pressure.

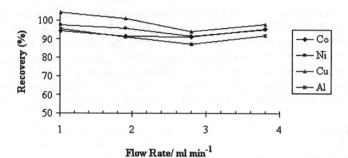

Figure 2 *The effect of flow rate on the recovery of trace elements, Relative Standard Deviation (RSD) <5%*

3.2 pH dependence of Matrix Removal and Pre-concentration

The pH dependence (pH 0.5- 10) of the system incorporating the Prosep reagent was evaluated for a range of elements. A 50 ppb multi-element solution was prepared in an ammonium acetate buffer system. The initial buffer (0.2 M ammonium acetate/acetic acid) was adjusted to the desired pH using either acetic acid or ammonium hydroxide.

Recoveries for most elements vary across the pH range with generally low recoveries at pH values less than 5 and higher recoveries as the pH is increased. Recoveries for Cd, Al, Fe and Ti reach greater than 79% at pH 5.5 (Figure 3(a)) while recoveries for Co, Ni, Cu and Zn are greater than 95% at this pH (Figure 3(b)).

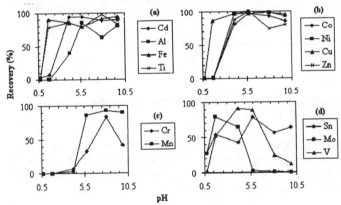

Figure 3 *The effect of ammonium acetate buffer pH on the recovery of trace elements, RSD <5%*

The effect of pH on Cr and Mn is shown in figure 3(c). The recovery of both elements peaked at pH 8. It has been previously reported that with proper pH control Cr and Mn show an optimum recovery at this pH[11]. Working at pH 8 may quantitatively retain these elements but matrix removal of alkaline earth metals from the sample will not be achieved. The almost quantitative retention of Cr is interesting in that there are few reports of this element having been successfully recovered using an iminodiacetate reagent by the same experimental procedure[15].

The pH dependence of Sn, V and Mo is shown in figure 3(d). The recoveries of Sn and V are greater than 79% at pH 5.5. Recoveries for Mo reached a maximum at pH 2. This element forms an oxoanion at low pH. Since the iminodiacetate reagent can behave as an anion, cation and chelate exchanger therefore at pH 2 the iminodiacetate anion exchange properties are most dominant, which may explain the high recovery of Mo at this pH. As the pH increases the iminodiacetate reagent loses its anion exchange properties and acts as a chelate exchanger and therefore the reagent no longer retains the Mo oxoanion.

The pH dependence of Tl is shown in figure 4. Recoveries for thallium over the entire pH range are poor but do increase with increasing pH. It has been previously reported that thallium has not been quantitatively retained on the iminodiacetate reagent[26]. Speciation procedures for the determination of both Cr^{27} and $Tl^{21, 22}$ using an iminodiacetate resin by a different experimental procedure have also been reported. Both of these procedures do not use ammonium acetate to control the pH of the sample. The presence of ammonium acetate may inhibit the uptake of these elements by the reagent. These conflicting reports indicate that more study is required on the behaviour of these elements on the iminodiacetate reagent under various experimental conditions.

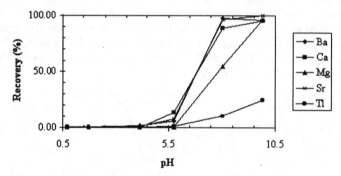

Figure 4 *The effect of ammonium acetate buffer pH on the recovery of alkaline earth metals and thallium, RSD <5%*

The pH dependence profile for the alkaline earth metals is also shown in figure 4. Quantitative recovery of these elements (95-100%) is achieved at pH 9.9. The optimum pH for the analysis of trace levels of these elements is therefore greater than pH 9. Recoveries at this pH are summarised in Table 3.

Table 3   *Recoveries of alkaline earth metals studied at pH 9.9*

Element	Recovery (%)	RSD* (%)
Ba	96.2	2.8
Ca	95.4	4.1
Mg	96.3	4.6
Sr	100.1	4.5

* The RSDs were calculated from three replicate preconcentration measurements.

For most efficient matrix removal of alkaline earth metals from the iminodiacetate reagent a pH of 5.5 or lower is required. The poor retention of these elements at pH 5.5

or below is in agreement with a previous retention study of the alkaline earth metal, calcium, on a similar iminodiacetate resin[16].

Overall, the best compromise pH for the quantitative retention of the majority of trace elements while achieving efficient matrix removal of alkaline earth metals is about pH 5.5. Recoveries are tabulated in Table 4.

4 CONCLUSIONS

A semiautomated matrix removal/preconcentration system has been constructed and its basic features evaluated. The retention behaviour of 18 elements have been evaluated over a pH range 0.5-10. Efficient preconcentration and matrix removal was achieved at pH 5.5. Preliminary results are reported for the behaviour of Cr, Tl and Mo.

Further work will involve the optimisation of the matrix removal/preconcentration experimental parameters such as eluent concentration, buffer concentration, buffer type etc. Method validation of the system using standard reference materials will also be carried out.

The iminodiacetate reagent can behave as an anion, cation and chelate exchanger. Difficult to analyse elements such as selenium and arsenic form oxoanions in solution and may potentially exhibit the same retention behaviour as the Mo oxoanion on the iminodiacetate reagent. Further study of the behaviour of these elements on the reagent is required.

Table 4   *Recoveries of trace elements at pH 5.5*

Element	Recovery (%)	RSD[*] (%)
Cd	94.6	2.8
Al	85.0	3.7
Fe	79.6	1.3
Ti	78.9	7.4
Co	96.4	2.7
Ni	99.9	3.4
Cu	96.6	2.3
Zn	99.8	17.3
Mn	7.8	6.3
Cr	3.0	8.5
Sn	79.0	n/a
Mo	2.2	2.5
V	88.9	0.8

[*] The RSDs were calculated from three replicate preconcentration measurements.
n/a not available

Conflicting results with other researchers for the retention of Cr and Tl indicate that more study is required on the behaviour of these elements on the reagent under various experimental conditions. The optimum pH for the recovery of alkaline earth metals is greater than pH 9. Further study of the retention behaviour of these elements on this reagent is also required. Possibilities also appear to exist for the analysis of trace levels of alkaline earth metals (Be, Sr, Ba) at a high pH and other species such as Mo, Cr, and Tl at a low pH.

5   REFERENCES

1.   S.H. Tan, G. Horlick, *Applied Spectroscopy*, **40**, 1986, 445.
2.   D. C Grégoire., *Spectrochimica Acta,* **42B**, 1987, 895.
3.   J.W. McLaren, *At. Spectroscopy*, 1993, **14**, 191.
4.   *Preconcentration Techniques for Trace Elements*, ed. Z.B Alfassi, Chien M. Wai, CRC Press, 1991.
5.   J. Wu, E.A. Boyle, *Anal. Chem.*, **69**, 1997, 2464.
6.   G.J. Batterham, N.C. Munksgaard, D.L. Parry, *J. Anal. Spectrom.*, **12**, 1997, 1277.
7.   H.M. Kingston, *Anal. Chem.*, 1978, **50**, 2058.
8.   Zl. Fang, *Analytica Chimica Acta*, 1984, **164**, 23.
9.   Zl. Fang, *Analytica Chimica Acta*, 1987, **200**, 35.
10.   Zl. Fang, *Flow Injection Separation and Preconcentration*, VCH, 1993.
11.   M. H. Yang, *Anal. Chem.*, 1997, **69**, 1185.
12.   S.N. Nelms, G.M. Greenway, *J. Anal. At. Spectrom.*, 1996, **10**, 907.
13.   S.N. Willie, *Atomic Spectroscopy*, 1998, **19**, 67.
14.   M. Nicolai, *Talanta*, 1999, **50**, 433.
15.   T. Yabutani, *Bull. Chem. Soc. Jpn.*, 1999, **72**, 2253.
16.   G.H Luttrell, *Anal. Chem.*, 1971, **43**, 1370.
17.   R.A. Nickson, *Analytica Chimica Acta*, 1997, **351**, 311.
18.   K. Pyrzynska, *Analytical Letters*, 1998, **31**, 1777.
19.   J.M. Besser, *Chemosphere*, 1994, **29**, 771.
20.   T. Ferri, *Analytica Chimica Acta*, 1996, **321**, 185.
21.   T.S. Lin, *Analytica Chimica Acta*, 1999, **395**, 301.
22.   T.S. Lin, *Environ. Sci. Technol.*, 1999, **33**, 3394.
23.   M. Pesavento, *Analytica Chimica Acta*, 1996, **323**, 27.
24.   A. Siegel, E. Degens, Science, 1966, **151**, 1098.
25.   F.H. Elsweify, *J. Radioanal. Nucl. Chem.*, 1997, **222**, 55.
26.   Technical Note 28, Dionex Corp., 1992.
27.   S. Hirata, Spectrochimica Acta Part B, 2000, **55**, 1089.

6   ACKNOWLEDGEMENTS

We wish to acknowledge the financial assistance of the Graduate Training Programme (GTP).

MOVING TO THE NEXT LEVEL: SAMPLE INTRODUCTION AND PLASMA INTERFACE DESIGN FOR IMPROVED PERFORMANCE IN ICP-MS

E. McCurdy, S. Wilbur, G. D. Woods and D. Potter

Agilent Technologies Ltd., Lakeside, Cheadle Royal Business Park, Stockport, Cheshire, SK8 3GR, United Kingdom

1 ABSTRACT

The technique of Inductively Coupled Plasma Mass Spectrometry (ICP-MS), though powerful, suffers from several well-documented limitations, principally with respect to its matrix tolerance and ability to decompose matrix-based interferences. Many different approaches have been used to overcome these limitations, some relying on hardware modification or software correction, whilst others have investigated changing operating conditions to minimise formation of the interfering species.

In this paper we will illustrate some of the key parameters that affect the fundamental performance of the instrument. We will also present results showing how sample introduction, plasma and spectrometer interface design and optimisation may be used to improve the performance of the ICP-MS with respect to sensitivity, sample throughput, tolerance to certain difficult matrices and specific interference removal. These subjects will be illustrated using 4 specific applications, covering a range of sample types and industry requirements.

2 INTRODUCTION

ICP-MS is widely accepted for a diverse range of multi-element measurements in a wide range of sample types. Virtually all elements can be measured, high sensitivity and low background signals combine to give very low detection limits (ng/L in most cases) and the rapid scanning quadrupole analyser means that measurement of a full suite of elements takes only about 4 minutes per sample.

Whilst ICP-MS is clearly enormously powerful, any analytical technique (ICP-MS included) has some limitations. Dissolved solids levels must be controlled carefully for ICP-MS analysis, typically no higher than 0.2%, to avoid matrix deposition on the spectrometer interface. Compared to ICP-OES, ICP-MS has a limited capability in the determination of very high analyte concentrations (> 100's mg/L), although this has been addressed with new detector technology in the latest generation of instruments. Finally, one of the strengths of ICP-MS, its spectral simplicity, can work against it by making spectral overlaps more difficult to avoid.

The question of spectral overlaps has been discussed since the very first papers on ICP-MS were written. Even Gray's 1975 paper on direct current plasma-mass spectrometry (DCP-MS), regarded by many as one of the papers that led to the development ICP-MS, contains references to the appearance of undissociated molecular species in the mass spectrum[1]. The key limitations of the DCP (low sensitivity for poorly ionised elements and severe matrix effects) were related to its low temperature (about 5000K), which is of relevance even with modern ICP-MS instruments.

The early ICP-MS instruments operated at a much higher plasma temperature (around 7500K) than the DCP, so many of these limitations were overcome[2], but reports of performance limitations of ICP-MS, due to high salt matrices or specific spectral overlaps, still appeared in the literature[3,4]. These two papers identified two of the key performance parameters by which ICP-MS is still measured today, namely tolerance to high levels of easily ionised salt matrix and the formation of refractory oxide interferences, based on components of the sample matrix.

In considering the performance of an ICP-MS instrument, it is beneficial to assess which limitations are inherent to the technique as a whole and which can be alleviated through incorporation of suitable hardware and the use of appropriate operating conditions. An extreme example of this can be seen in the performance comparison between the DCP and the ICP as ion sources, which serves to highlight the importance of plasma conditions in the overall performance of the instrument. Performance variations can still be identified amongst the various plasma sources used in ICP-MS instruments today and selection of the most suitable conditions will have a dramatic effect on the instrument performance in certain applications.

3 FUNDAMENTAL PERFORMANCE

The processes that occur as a sample aerosol droplet passes through the ICP can be simplified to drying, decomposition, dissociation, atomisation and ionisation. The sample aerosol is travelling very quickly through the plasma and has a residence time of only a few milliseconds. In this time, the plasma must transfer sufficient heat energy into the sample to give, as far as possible, 100% conversion of the sample analyte components into singly charged positive ions. There are several factors that will affect the efficiency of this process and these factors will have a direct effect on the performance of the ICP-MS instrument. They are plasma temperature, sample residence time, sample aerosol density and water vapour content of the sample aerosol.

Clearly, some of these parameters are under the control of the instrument operator and others are dependent on the selection of suitable hardware in the design of the ICP-MS, but the fundamental requirement is for a system and optimisation that maintain a high plasma temperature. Although it is difficult to measure directly, the important parameter is the temperature of the central channel of the plasma, since this is where the sample aerosol is travelling. Effective transfer of energy from the outer part of the plasma into the central channel is therefore beneficial. The influence of plasma temperature on sensitivity is highlighted in Table 1, which shows the degree of ionisation for a series of elements, based on different plasma temperatures. It is evident from the data that reducing plasma temperature can have a dramatic effect on the ionisation (and therefore sensitivity) of many elements, particularly those with high first ionisation potential.

Considering design parameters first, there are 4 areas where performance of the sample introduction and plasma can be influenced: Plasma RF generator design, ion lens design (to ensure high sensitivity and thereby allow operation at low solution flow rate), plasma torch design and spraychamber temperature control.

The RF generator used in most ICP-MS instruments operates at a frequency of 27.12MHz, which gives a higher central channel temperature than an equivalent

Table 1 *Ion Population (% ionisation) as a Function of Plasma Temperature and Ionisation Potential, Calculated Using the Saha Equation*

Element	Ip (eV)	plasma temperature			
		5000 K	6000 K	7000 K	8000 K
Cs	3.89	99.4%	99.9%	100.0%	100.0%
Na	5.14	90.0%	98.9%	99.8%	99.9%
Ba	5.21	88.4%	98.7%	99.8%	99.9%
Li	5.39	83.4%	98.2%	99.7%	99.9%
Sr	5.69	71.5%	96.8%	99.5%	99.9%
Al	5.98	56.2%	94.5%	99.1%	99.8%
Pb	7.42	4.3%	51.2%	91.1%	98.3%
Mg	7.64	2.6%	40.7%	87.7%	97.7%
Co	7.86	1.6%	31.0%	83.2%	96.9%
Sb	8.64	0.3%	9.0%	57.6%	90.9%
Cd	8.99	0.1%	4.8%	43.2%	85.7%
Be	9.32	0.1%	2.6%	30.6%	78.8%
Se	9.75	0.0%	1.1%	17.8%	66.6%
As	9.81	0.0%	1.0%	16.4%	64.6%
Hg	10.43	0.0%	0.3%	6.5%	42.6%

powered generator operating at 40.68MHz, the frequency commonly used in ICP-OES instruments, as a result of the greater depth to which the lower frequency energy penetrates into the bulk plasma from the outer surface[5,6]. The role of the RF generator may seem to be identical in ICP-OES and ICP-MS, but in fact very different priorities exist. In ICP-OES, many of the analytical lines used are atomic lines, so a high atom population is advantageous. By contrast, only ions are measured in ICP-MS, so conversion of atoms to ions, which requires extra energy, must be maximised.

The ion lens design may seem only indirectly connected with plasma efficiency, but in fact has a crucial role. Improved ion transmission through the ion lens allows the ICP-MS sample introduction design to be optimised for matrix tolerance, rather than maximum sensitivity. Without an efficient ion lens design, changes to the sample introduction area would result in an unacceptable loss of sensitivity.

The plasma torch may also seem a basic and consistent part of the ICP, but a wide internal diameter (ID) torch injector will give a slower carrier gas velocity and a longer sample residence time. This gives an effective increase in plasma temperature, since the longer residence time leads to improved energy transfer from the plasma to the aerosol. The wide ID injector also ensures a more diffuse sample aerosol, so improving drying and decomposition efficiency.

Water vapour is produced during the nebulisation process and will absorb plasma energy, unless it is removed by chilling of the spraychamber. Use of a peltier device to cool the spraychamber to below 5°C offers significant performance benefits, as removal of the water vapour reduces the solvent load on the plasma, leaving more energy available for matrix decomposition and analyte ionisation[7].

Combined with these hardware design criteria are several optimisation parameters that influence the matrix decomposition and interference removal capability of the ICP. Again, the main aim of these optimisation parameters is to maintain a high temperature in the plasma central channel. The key parameters are plasma RF forward power, sampling depth, carrier gas flow rate and sample uptake rate. Simple adjustment of all these parameters is essential and modern ICP-MS instruments typically allow computer control and auto-tuning, to ensure consistency of setup.

Increasing the plasma forward power and the sampling depth will both contribute to an increase in effective plasma temperature. Reducing the carrier gas flow rate works in combination with a wide torch injector internal diameter, while a reduced sample uptake rate (typically <400uL/min) is particularly important as the sample matrix level is increased.

With all of these optimisation parameters, it is important to consider the impact of the optimisation on the other performance characteristics of the instrument. For example, a system that is not designed to operate at low sample flow rate may show an unacceptable loss of sensitivity if the flow rate is reduced in pursuit of better matrix tolerance.

The plasma robustness criterion that has been widely adopted to monitor ICP-MS performance is the CeO/Ce ratio. This ratio indicates the efficiency with which the plasma can decompose the strong Ce-O bond and acts as an indicator for many other matrix-based interferences, such as those derived from chloride or sulphate in the sample. Furthermore, the CeO/Ce ratio acts as an indicator of general plasma performance, since a plasma that is unable to decompose CeO efficiently will have insufficient residual energy to give efficient ionisation of the elements with high first ionisation potentials. On ICP-MS instruments where matrix tolerance is not a high priority, the CeO/Ce ratio may be as high as 3%[8], whereas good design and appropriate optimisation can reduce this figure to around 0.3% - a factor of 10 improvement[9].

The impact of adjusting two of these parameters (sample flow rate and plasma to sample cone distance or sampling depth) is shown in Figures 1 and 2. These Figures illustrate that an increase in sensitivity can easily be achieved by sampling at a shorter coil to sample cone distance, or by increasing the sample flow rate. In each case, however, the sensitivity increase is at the expense of matrix tolerance, represented by the CeO/Ce ratio.

It should also be noted that the worst CeO/Ce ratio achieved in these single parameter optimisation studies was less than 0.45%, which is still exceptionally low for ICP-MS, for which values of around 1% to 3% are typically reported [e.g. 8].

The second aspect of the fundamental performance of the ICP-MS sample introduction and plasma investigated was the relationship between the CeO/Ce ratio and other matrix derived polyatomic interferences. Several matrices were evaluated and several commonly-reported polyatomic interferences were measured. Data are presented for just one, the $^{40}Ar^{35}Cl$ polyatomic overlap on mono-isotopic arsenic at mass 75. This interference has been the subject of many recent publications [e.g. 10, 11], some of which have given the impression that the interference is intractable and an inevitable consequence of normal ICP-MS operation. Our work has indicated, by contrast, that plasma optimisation, coupled with a sample introduction and plasma configuration designed to maintain a high and stable plasma temperature, can attenuate the ArCl interference to such an extent that determination of As at sub-10ug/L (ppb) levels in high salt matrices becomes possible. This interference attenuation requires no special operating conditions and is simply a consequence of the normal optimisation process, if matrix tolerance (low CeO/Ce ratio) is considered a priority. The relationship between the CeO/Ce ratio and ArCl formation in a high NaCl matrix is illustrated in Figure 3. This Figure shows the measured As from a series of standards, all of which contained 10ug/L As. The samples

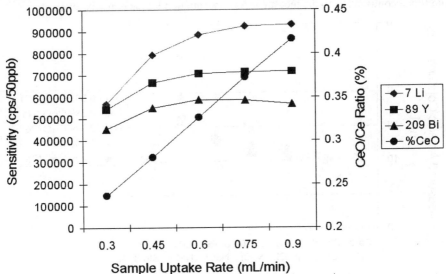

Figure 1 *Influence of Sample Uptake Rate on Sensitivity and CeO/Ce ratio*

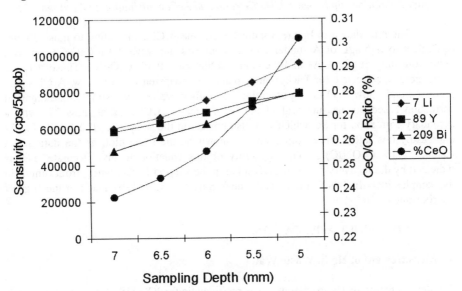

Figure 2 *Influence of Sampling Depth on Sensitivity and CeO/Ce ratio*

were spiked with an increasing concentration of a NaCl matrix at concentrations from 0mg/L to 10,000mg/L (1%), to represent the extremes of high matrix analysis normally tolerated by an ICP-MS. The calibration was in a 2% HNO_3 matrix and no matrix-specific optimisation or interface conditioning was carried out prior to the analysis. Each matrix sample was analysed under 4 different tuning conditions, characterised by CeO/Ce ratios of 0.3%, 0.5%, 1.0% and 3.0%. The level of ArCl formed was assessed by the additional contribution to the measured As concentration for each matrix level.

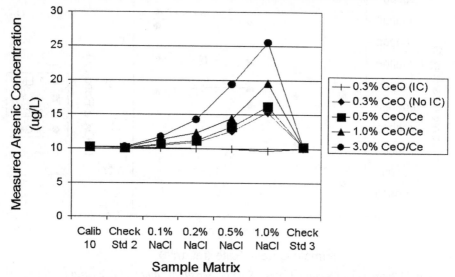

Figure 3 *Relationship between CeO/Ce ratio and ArCl contribution to As at mass 75*

The plot shown in Figure 3 indicates that the ArCl contribution to mass 75 was equivalent to >15ug/L of As under tuning conditions that gave 3.0% CeO/Ce, but was reduced to only 5ug/L of As under conditions that gave 0.3% CeO/Ce. These data are completely uncorrected for background, blank or interference. The low ArCl signal produced under low CeO tuning conditions is easily corrected by a simple mathematical correction based on measurement of the second major ArCl peak at mass 77 (with a secondary correction for Se, which was also present in the samples). The results for As after using this interference correction are shown on the same plot, as the data series labeled "0.3% CeO (IC)". The capability of this equation to give accurate data is indicated by the recovery of the 10ug/L spike in the varying NaCl matrix, from 0mg/L in the samples identified as Calib 10 and Check Std2 and 3, to 1.0% NaCl in the highest matrix sample.

4 DIFFICULT APPLICATIONS

4.1 Measurement of Hg in Waste Water

The determination of Hg in wastewater is problematic for ICP-MS, due to the low levels of Hg typically present, the variation in concentrations and the common requirement to include Hg analysis in a large suite of elements routinely measured in this matrix. The actual measurement of Hg is difficult by ICP-MS, due to the relatively low degree of ionisation of the element, the fact that the largest Hg isotope (mass 202) is less than 30% abundant and the long uptake and washout times required to flush Hg from the sample introduction system.

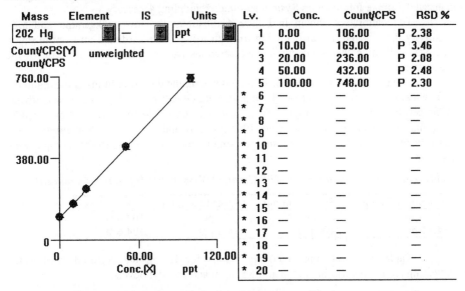

Figure 4 *Calibration for Hg in HPS Certified Wastewater (Hg isotopes summed)*

The degree of ionisation of high ionisation potential elements is greatly improved by the maintenance of a high plasma temperature (as shown in Table 1), which improves the sensitivity of elements such as Hg. The isotopes of Hg can be summed, using a correction equation, to further improve the count rate measured. Hg stability and washout is greatly improved through the use of Au addition to the samples, standards, blanks and rinse solution. Although the actual mechanism for the Hg stabilisation is the subject of some debate[12], addition of a gold concentration of 200ug/L has been found to

Table 2 *Detection Limit for Hg in HPS Certified Wastewater (Hg isotopes summed)*

File:	Date/Time:	Mercury /199	Mercury /200	Mercury /201	Sum (199,200,201,202)
009SMPL.D#	2/17/2000 11:50	11.19	10.78	12.68	16.53
010SMPL.D#	2/17/2000 11:55	10.84	10.16	10.3	15.54
011SMPL.D#	2/17/2000 11:59	10.34	10.9	12.48	16.23
012SMPL.D#	2/17/2000 12:04	11.82	11.09	12.93	16.41
013SMPL.D#	2/17/2000 12:09	10.54	10.68	11.53	15.88
014SMPL.D#	2/17/2000 12:13	10.62	9.245	10.43	15.72
015SMPL.D#	2/17/2000 12:18	10.3	11.13	11.42	16.09
	3 Sigma MDL (ppt)	1.676	2.077	3.040	0.938

be successful at concentrations up to around 20ug/L of Hg[13]. Finally, and most importantly, the determination of low level Hg is vastly improved if the overall Hg load on the ICP-MS sample introduction system is reduced. On the Agilent 7500, this is done through the use of a low-flow nebuliser (400uL/min is typical for this instrument) and this can be further reduced by using a method of constant-flow nebulisation, utilising the Integrated Sample Introduction System automated sampling accessory.

The calibration shown in Figure 4 illustrates the low background and high sensitivity that can be obtained for Hg through using these hardware and optimisation methods, indicating that single ng/L concentrations can be measured with ease in the wastewater matrix.

The results for Hg shown in Table 2 illustrate the detection limits achieved in the wastewater matrix, calculated from 3 x the standard deviation of 7 separate measurements of the certified wastewater sample. These results illustrate the improvement obtained by summing the 4 main Hg isotopes (m/z 199, 200, 201 and 202). Detection limits below 5ng/L were obtained for the single isotopes, improving to <1ng/L when the isotopes were summed.

Table 3 *Spike Recovery for Hg in HPS Certified Wastewater (Hg isotopes summed)*

[Unspiked Sample]	Spike Amount	[Spiked Sample]	% Recovery
14.11 ppt	50 ppt	65.9 ppt	103.6 %
14.11 ppt	100 ppt	118.8 ppt	104.6 %

Table 3 shows spike recovery results for Hg in the certified wastewater, indicating recoveries within +/- 5% at the 50ng/L and 100ng/L levels.

4.2 Determination of Iron and Other Trace Elements in Drinking Water

ICP-MS is commonly applied to the determination of iron in drinking water, with the required limits typically being in the region of 15 to 20 ug/L[13]. Recent legislative changes

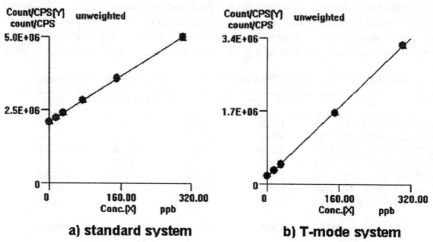

a) standard system   **b) T-mode system**

Figure 5 *Calibrations for ^{56}Fe, without (a) and with (b) T-Mode Interface*

in countries such as Japan have, however, reduced the detection limit required for Fe to 10ug/L or below, with the result that conventional ICP-MS is unable to guarantee detection at the required level. A newly developed ICP-MS interface option, called T-Mode, on the Agilent 7500 Series utilises ion/molecule interactions in the expansion stage to reduce the contribution of molecular species such as ArN and ArO to the ICP-MS spectrum. Reduction of these molecular species reduces the background for the 2 most abundant isotopes of iron, at mass 56 and 54 respectively, so reducing detection limits for Fe at these masses. Detection limits obtained using the T-Mode, for a series of trace and major elements in drinking water are shown in Table 4.

Table 4 *Detection Limits for Trace and Major Elements in Drinking Water, using the Agilent T-Mode Interface*

Isotope	3s LOD	Unit
11 B	0.2	ug/L
23 Na	0.0008	mg/L
24 Mg	0.0004	mg/L
27 Al	0.03	ug/L
39 K	0.0001	mg/L
43 Ca	0.01	mg/L
52 Cr	0.01	ug/L
53 Cr	0.04	ug/L
54 Fe	2	ug/L
56 Fe	0.5	ug/L
55 Mn	0.01	ug/L
60 Ni	0.02	ug/L
65 Cu	0.03	ug/L
66 Zn	0.03	ug/L
75 As	0.008	ug/L
82 Se	0.03	ug/L
95 Mo	0.01	ug/L
111 Cd	0.005	ug/L
121 Sb	0.001	ug/L
202 Hg	0.002	ug/L
208 Pb	0.004	ug/L
238 U	0.0003	ug/L

A comparison of the background signals and calibration curves obtained under standard and T-Mode configurations is shown in Figure 5, which highlights the reduced contribution from ArO at mass 56, leading to significantly lower blank signal and so improved signal to noise under T-Mode conditions (Figure 5 (b)). The data for Fe in Table 4 indicate that 3 sigma limits of detection for the commonly measured isotopes (^{54}Fe and ^{56}Fe) were both well below the required level of 10ug/L.

The T-Mode interface also has a positive impact in reducing several other commonly reported ICP-MS polyatomic interferences, giving reduced detection limits for chromium at mass 52 and selenium at mass 82 (3 sigma limits of detection of 10ng/L and 30ng/L, respectively).All required elements can be determined in drinking water, in a single analysis, under a single set of operating conditions. It should be noted that the major elements (Na, Mg, K and Ca) were all determined at the same time as the trace elements and that the trace elements included difficult analytes such as Hg (see Section 4.1). Uranium, which has been proposed as a controlled element in drinking water, was also measured.

4.3 Low Level Analysis of Trace Metals in Simple and Complex Matrices

ICP-MS is widely used in the semiconductor industry for the determination of a range of trace metallic impurities that would affect the integrity of semiconductor devices, should the trace metals be allowed to contaminate products during manufacture. Included in the trace elements routinely monitored are K, Ca and Fe, all of which were once considered difficult by quadrupole ICP-MS, due to the presence of plasma-based polyatomic interferences on their most abundant isotopes, ^{39}K, ^{40}Ca and ^{56}Fe, respectively.

The inherent problems of analysis by quadrupole ICP-MS led to these elements being measured using a separate technique, typically graphite furnace atomic absorption spectrometry (GFAAS) or high resolution ICP-MS, until the commercialisation of reliable cool plasma technology on the Agilent 4500 in 1994[14]. Since this time, quadrupole ICP-MS operating under cool plasma conditions has become accepted as the industry standard method for routine determination of trace metallic contaminants in

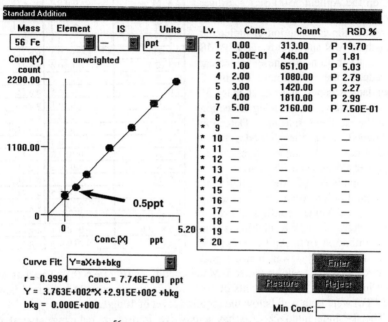

Figure 6 *Calibration for ^{56}Fe in Ultrapure Water, Showing Removal of ArO Overlap*

semiconductor process chemicals, such as ultrapure water, mineral acids and wafer cleaning baths [e.g. 15, 16]. An example of the performance achieved routinely by cool plasma ICP-MS, utilising the ShieldTorch System of the Agilent 7500, is shown in Figures 6. This Figure shows a standard addition calibration for ^{56}Fe in ultrapure water, with spike levels of 0.5, 1.0, 2.0, 3.0, 4.0 and 5.0ng/L (ppt).

Cool plasma ICP-MS has, however, been considered unsuitable for analysis of these same trace metals in more complex samples, due partly to reported problems of matrix tolerance and matrix decomposition efficiency with some cool plasma configurations [e.g. 17, 18]. In addition, it has been suggested that certain implementations of cool plasma conditions are unable to maintain sufficient plasma energy to give adequate ionisation of poorly ionised elements such as B, which is a critical element for monitoring the efficiency of water purification equipment in the semiconductor industry. Whilst it is certainly correct to say that refractory oxide species are less well decomposed under cool plasma conditions than under "normal" plasma conditions, it is not correct to conclude that cool plasma conditions are only suitable for "simple" or "clean" matrices. Furthermore, the cool plasma configuration used on the Agilent 7500 ICP-MS retains sufficient plasma energy to allow determination of elements such as B at analytically useful levels, as illustrated by the calibration shown in Figure 7.

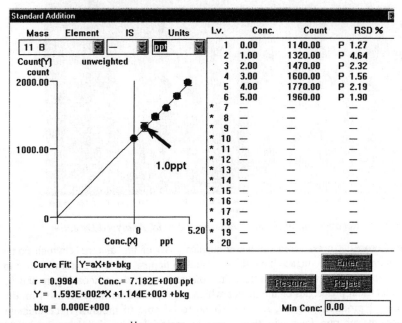

Figure 7 *Calibration for ^{11}B in Ultrapure Water, Showing Good Ionisation Efficiency*

Operating cool plasma conditions at very low power (<750W), may result in some degradation in matrix tolerance and interference levels and such effects have been reported by several workers. However, it should be noted that these effects are very much less problematic when the cool plasma is operated at powers in excess of 900W, as is typically the case when the Agilent 7500 is used for analysis of high matrix semiconductor samples. Under these higher power cool plasma conditions, ppt levels of K, Ca and Fe, together with the other critical contaminant trace elements, can be monitored and controlled routinely in relatively complex sample matrices such as sulphuric acid and organic process chemicals. The use of high power cool plasma conditions reduces the need to carry out sample preparation (evaporation or digestion) or to use sample desolvation with its disadvantages of complexity, extended sample path and loss of volatile analytes. The detection capability of the Agilent ShieldTorch System in the determination of trace elements in sulphuric acid is illustrated in Figure 8, which shows cool plasma spike recovery results at 5ng/L (ppt) obtained from the analysis of 10% sulphuric acid, analysed directly. All analytes were recovered within the required 25%.

A further group of samples of increasing interest in the semiconductor industry is those organic process chemicals used at various stages of the wafer production process. In addition to cleaning solvents such as methanol, n-methyl pyrrolidone (NMP) and tetra-methyl ammonium hydroxide (TMAH), organic materials are used to mask regions of the wafer during the etching process, whilst further organic solvent mixtures are used to

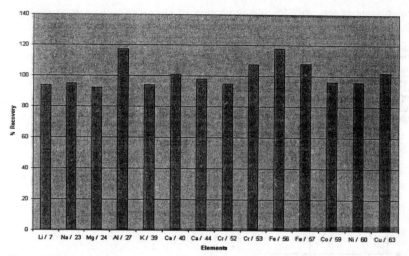

Figure 8 *Spike Recovery at 5ng/L in 10% Sulphuric Acid, Analysed Directly*

remove the photoresist mask from the wafer surface. All of the materials which come into contact with the wafer surface have the potential to deposit surface contamination in the form of trace metals, which will affect the conductivity of the silicon layers and so damage the electrical properties of the device which is fabricated on the wafer. From the point of view of product integrity, it is beneficial to be able to monitor trace element contamination at sub-ppb levels in these organic materials, but this has previously required analysis using GFAAS (essentially a single-element technique) or a pre-digestion using wet ashing with concentrated acid.

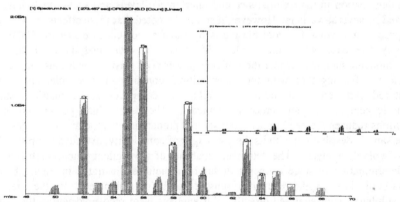

Figure 9 *Spectrum of Organic Post-Etch Cleaner, With and Without 1ug/L Spike of Cr, Mn, Fe, Ni, Co, Cu and 5ug/L Spike of Zn*

The requirements for rapid reporting of results on a sample, which may affect the current wafer production line, means that GFAAS analysis is becoming obsolete in the semiconductor industry and many organic solvent mixtures respond badly to acid digestion, resulting in violent reactions or explosion. In addition, acid digestion may require a relatively large volume of acid compared to the sample volume, so the accuracy of the analysis is compromised by trace contamination in the acid itself. A preferable approach would be the direct analysis of the organic solvent by a multi-element technique such as ICP-MS, after simple dilution in an appropriate aqueous or organic solvent, such as deionised water, dilute acid, ethyl lactate or xylene.

As with the analysis of sulphuric acid, low-power cool plasma ICP-MS has limited applicability for the analysis of organic samples, due to its reduced ability to decompose the sample matrix. However, as with the analysis of sulphuric acid, the operation of the cool plasma at higher power (900W or above) allows complex organic solvent mixtures to be analysed successfully. The cool plasma conditions lead to the removal of polyatomic ions derived from carbon, as well as reducing the ionisation of C itself. Several difficult elements are therefore accessible under cool plasma conditions, as shown in the spectra in Figure 9.

Table 5 *Detection Limits and Spike Recoveries for Trace Metals in Isopropyl Alcohol, Analysed Directly*

Element	LOD (ppt)	20 ppt Spike Recovery	Plasma Mode
Li	0.08	98	Cool
Na	1.14	101	Cool
Mg	4.27	101	Cool
Al	1.20	100	Cool
K	1.23	102	Cool
Ca	3.03	112	Cool
Ti	0.50	97	Normal
V	1.43	104	Normal
Cr	4.62	99	Cool
Mn	0.32	103	Cool
Fe	1.73	113	Cool
Co	1.22	106	Cool
Ni	2.72	100	Cool
Cu	3.44	107	Cool
Zn	7.81	98	Normal
As	6.93	105	Normal
Sb	0.21	109	Normal
Ba	0.02	107	Normal
Pb	0.37	103	Normal

The two spectra show the same organic post-etch cleaner, both unspiked (inset) and with the addition of a 1ug/L spike of the trace metals Cr, Mn, Fe, Ni, Co and Cu and 5ug/L of Zn (main spectrum). Zn was added at a higher concentration to compensate for its reduced sensitivity, due to its high first ionisation potential. The inset spectrum shows the near-complete removal of polyatomic ion interferences on Cr at mass 52 and Fe at mass 56. A trace of Zn contamination can also be observed in the unspiked sample.

While the post-etch cleaner shown in Figure 9 was water soluble, high power cool plasma is also applicable to the measurement of ng/L concentration contaminants in non water soluble semiconductor matrices, such as methanol, isopropyl alcohol, xylene, toluene, etc. The data shown in Table 5 illustrate the limit of detection and spike recovery performance of the Agilent cool plasma configuration, in the direct analysis of isopropyl alcohol. For this analysis, automatic switching between cool and normal plasma conditions was used, to allow measurement of all semiconductor elements in a single analysis.

4.4 Ultra-High Sensitivity Measurement Using the Agilent ShieldTorch System

Whilst the capability of the Agilent ShieldTorch System for the removal of plasma and matrix-based polyatomic interferences is well documented and used in routine laboratories, it is less well known that the same ShieldTorch hardware gives a dramatic increase in ICP-MS sensitivity, when used under normal, high-power plasma conditions. The reason for this sensitivity increase is that the ion energy spread is reduced as a result of the effective grounding of the plasma, with respect to the spectrometer interface. Since the ions are extracted at closer to the same ion energy, they all follow more closely the same trajectory, under the influence of the ion lens system, resulting in increased transmission through the spectrometer interface.

The are many applications for which increased sensitivity is of interest, including determination of low level elements in natural waters, trace element determination in geochemical matrices and determination of trace actinides in environmental and clinical samples. The latter application is illustrated in the spectrum in Figure 10, which shows the calibration for ^{238}U in a matrix of 10x diluted urine. The sensitivity improvement achieved using the Agilent ShieldTorch is around 5 fold, compared to the standard instrument, without any compromise in matrix tolerance or other analytical figures of merit, such as random background.

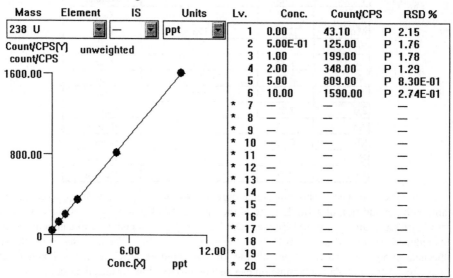

Figure 10 *Calibration for ^{238}U in 10x Diluted Urine*

Despite the presence of trace amounts of uranium in the sample, the detection limit (3 sigma) was 17.9pg/L (ppq) and the stability at the 500pg/L level was below 2%SD. Routine analysis at these levels is possible, since the matrix tolerance and robustness of the sample introduction system is maintained, in contrast with the use of secondary sample introduction devices such as ultrasonic or desolvating nebulisers.

5 CONCLUSIONS

The Agilent 7500 is designed and generally optimised with matrix tolerance as the highest priority. As a result, the sample introduction system and plasma give very efficient decomposition of matrix derived interferences, to levels around 10x lower than reported on other ICP-MS designs.

The improved matrix tolerance is derived principally from the maintenance of a high temperature in the plasma central channel. As a result of this high plasma temperature, several difficult applications can be addressed, including the determination of poorly ionised elements, such as Hg, at very low levels in high matrix samples.

Sample introduction options and accessories extend the range of ICP-MS applications, to include measurement of Fe at below 10ug/L in drinking water and determination of contaminant elements in complex semiconductor matrices such as sulphuric acid and organic solvents. The ShieldTorch System, as well as allowing high power cool plasma analysis, also results in increased sensitivity when operated under normal plasma conditions, so allowing pg/L level measurements of trace actinides in biological and environmental samples.

6 REFERENCES

1.   Gray, A. L., Analyst, 1975, **100**, 289.
2.   Houk, R. S., Fassel, V. A., Svec, H. J., Gray, A. L. and Taylor, C. E., Anal. Chem., 1980, **52**, 2283
3.   Olivares, J., A. and Houk, R., S., Anal. Chem., 1986, **58**, 20
4.   Gray, A., L., Spectrochimica Acta, 1985, **408**, 1525
5.   Abdalla, M., H., Diemiastzonek, R., Jarosz, J., Mermet, J-M. and Robin, J., Anal Chim Acta, 1978, **33B**, 55
6.   Walters, P., E., Gunter, W., H. and Zeeman, P., B., Spectrochim Acta, 1983, **41B** 133-141
7.   Hutton, R., C. and Eaton, A., N., J. Anal. At. Spectrom., 1987, **2**, 595
8.   Tromp, J., W., Tremblay, R., T., Mermet, J-M. and Salin, E., D., J. Anal. At. Spectrom., 2000, **15**, 617
9.   De Boer, J., L., M., in Plasma Source Mass Spectrometry , ed. G. Holland and S. D. Tanner, Royal Society of Chemistry, Cambridge, 1999, 27
10.   Vollkopf, U., Baranov, V. and Tanner, S., in Plasma Source Mass Spectrometry , ed. G. Holland and S. D. Tanner, Royal Society of Chemistry, Cambridge, 1999, 63
11.   Neubauer, K. and Vollkopf, U., At. Spectrosc., 1999, **20 (2)**, 64
12.   Sturman, B., T., J. Anal. At. Spectrom., 2000, **15**, 1512

13.   Woods, G., D. and McCurdy, E., J., in Plasma Source Mass Spectrometry , ed.
        G. Holland and S. D. Tanner, Royal Society of Chemistry, Cambridge, 1999, 108
14.   Sakata K and Kawabata K, Spectrochim. Acta, 1994, **49B**, 1027
15.   Hoelzl, R., Fabry, L., Kotz, L. and Pahlke, S., Fresenius J. Anal. Chem., 2000,
        366, 64
16.   Shive, L., Ruth, K. and Schmidt, P., Micro, 1999, **17 (2),** 27
17.   Bollinger, D., S. and Schleisman, A., J., in Plasma Source Mass Spectrometry ,
        ed. G. Holland and S. D. Tanner, Royal Society of Chemistry, Cambridge, 1999,
        80
18.   Settembre, G. and Debra, E., Micro, 1998, **16 (6),** 79

Section 2

Mass Analyser Instrumentation

QMF OPERATION WITH QUADRUPOLE EXCITATION

V.I. Baranov[1], N.V. Konenkov[2] and S.D.Tanner[1]

[1]MDS SCIEX 71 Four Valley Dr., Concord, Ontario, L4K 4V8, Canada
[2]Department of General Physics, Ryazan State Pedagogical University, Svoboda Street 46, Ryazan, Russia

1   INTRODUCTION

Since the discovery in 1953 by Paul and Steinwedel [1,2] of the usefulness of radio frequency (rf) quadrupole field for trapping charged particles, a large body of theoretical and experimental work [3-12] has been aimed at improving ion trap and quadrupole mass filter (QMF) operations. Precise electrode manufacturing, discovery of mass selective axial ejection [13,14], a variety of ion sources, fast electronics and ion detectors and different excitation methods, transformed the quadrupole devices into state of the art scientific instrumentation.

The QMF has been applied successfully to the ICP MS technique, where the requirements to sustain high ion current from plasma while maintaining outstanding abundance sensitivity, signal stability and sensitivity are extreme. There are several contradictory constraints which this analytical method superimposes on the vacuum interface, QMF, and ion focusing optics. In order to achieve high abundance sensitivity, the ion beam should have low axial translational energy with narrow distribution and to be well collimated. This is difficult to fulfill since the ion source is space charge limited. Also, it requires a long residence time of ions in QMF and, as a result, high quality of long quadrupole assembly. After many experimental adjustments, an acceptable arrangement can be achieved, however it is destined to be very inflexible. In order to achieve significant improvement in the essential characteristics of the QMF, one has to step outside of the existing "equilibrium" and consider a radical change. Quadrupole excitation offers an opportunity to modify the Mathieu stability diagram, dramatically changing the mass selecting property of QMF.

The resonant excitation of ion vibrations by application of an auxiliary rf signal has been demonstrated to be useful in controlling the motion of trapped ions. Its use enhanced the ion trap performance [15] remarkably. Different methods of resonant excitation are used now for promoting endothermic reactions with trapped ions, for mass range broadening and mass selective ejection in the Paul trap. The theory of resonant excitation is well developed [8,15-16].

In a recent paper [17], quadrupolar excitation with an additional rf signal was shown to have a parametric nature. The higher order resonance conditions were derived theoretically. Parametric resonance at fractional secular frequencies was observed in a 3D ion trap by M.A.N.Razvi et al [18] and recently was investigated in detail by Campbell et al [19] in the linear ion trap [20].

As was mentioned in [17], auxiliary quadrupolar excitation changes the stability conditions of an ion in the quadrupole field. Lines of instability appear on the Mathieu stability diagram due to the parametric resonance of the first and higher orders. As a result, the diagram splits into a number of stable areas presented schematically in [17].

In this work, we verified experimentally the splitting and investigated some basic characteristics of ion separation in selected stable "islands" close to the apex of the Mathieu stability diagram.

2 QUADRUPOLE EXCITATION

In the absence of excitation, the stable ion trajectory has a quasi-periodic character. Its frequency spectrum consists of an infinite number of harmonics that are not conventionally related to a fundamental one. Quadrupolar excitation has a parametric nature. The amplitude of radial displacement grows exponentially at the resonance leading to instability. It is well known [21], that parametric resonance appears in a band of frequencies near a resonant one. The strength of the resonance depends on the excitation parameters of the additional rf signal.

3 EXPERIMENTAL OBSERVATION OF THE DIAGRAM SPLITTING

An experimental study of quadrupolar excitation was performed on an ELAN 6100 (PerkinElmer SCIEX). An auxiliary rf signal from an arbitrary waveform generator (HP 33120A) was introduced into the main drive amplifier providing the possibility to support the ratio of auxiliary to fundamental signals constant during mass scanning. The magnitude and frequency of excitation was measured using a spectrum analyzer (Advantest R3361A). The input signal was created by an antenna, which was placed near the resonance coil. From measuring the relative signal amplitudes of two spectral components, the ratio was determined at a given mass number. The quadrupole mass filter used operates in the mass region 3 – 250 Da. The main drive frequency is 2.5 MHz.

Under normal ELAN6100 operating conditions, the ICP source, vacuum interface, and ion focusing optics produce an ion beam normally with initial ion energy of 3-5 eV having an energy spread of several eV (typically half the initial energy) at the 10% level.

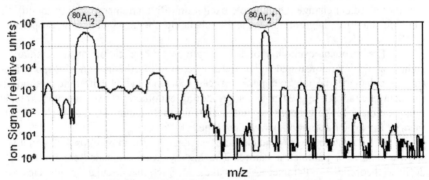

Figure 1a *Mass spectrum obtained from two excited stable regions. Translational ion energy is ≈5 eV*

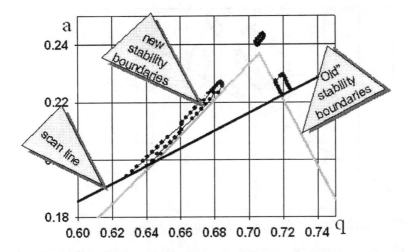

Figure 1b *The experimentally determined stable regions superimposed on the Mathieu stability diagram. The scan line corresponds to the spectrum in Figure 1a*

The observed mass spectrum is shown in Figure 1a in the case when the scan line crosses two stable regions (see Figure 1b). Superposition of two similar mass spectra partially overlapped, with different resolution was observed. The picture would be more apparent if only a single population of ions were produced by the ion source. In that case, the number of peaks would be equal to the number of stable regions, which were crossed by the scan line. The ICP source produces several abundant argon containing ions, and the region near the Ar_2^+ ion signal was selected for the example. As can be seen in Figure 1a, the overlapping leads to the interference of simultaneously transmitted ions on the same apparent mass. For instance, 80m/z from the first stable region interferes with 70m/z from the second region, and 80m/z from the second region interferes with 90m/z from the first one.

A more complex, but similar picture is observed when a scan line crosses many stable regions, which was observed under different excitation conditions.

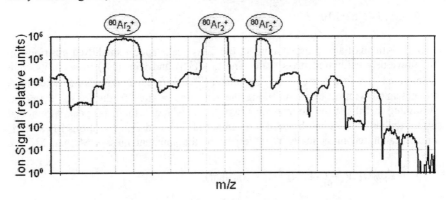

Figure 2a *Mass spectrum obtained from three excited stable regions*

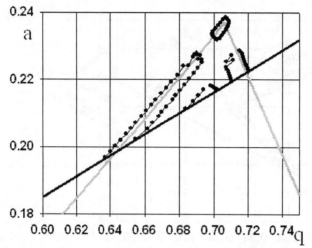

Figure 2b *The experimentally determined stable regions superimposed on the Mathieu stability diagram. The scan line corresponds to the spectrum in Figure 2a*

The procedure for mapping the excited stability diagram is as follows. The mass spectrum in the region of a reference mass peak (in this case Ar_2^+) was recorded in the absence of auxiliary rf. In practice, it is difficult to measure the scan parameter $\lambda = U/V$ (U is a direct voltage and V is a sinusoidal (zero to peak) voltage of the main resolving rf signal) due to the influence of the measurement circuit on the operation of the mass filter. Therefore, the relative number L, which is used as a resolution control parameter in the ELAN software, was employed to change the mass peak width ΔM. The resolution (at 10% of the peak height, $\Delta m_{0.1}$) $R_{0.1} = M/\Delta m_{0.1}$, as a function of the control parameter L was measured. Knowing $R_{0.1}$, the $a_{0.706}$ value (intersection of the scan line and $q_0 = 0.7060$ [12]) was determined as:

$$a_{0.706} = 0.23699 - 0.178/R_{0.1}$$

The scan parameter λ is equal to $a_{0.706}/2q_0$. The resulting experimental dependence of λ on L was:

$$\lambda = 0.154 + 6.00\ 10^{-6}\ L$$

With auxiliary rf signal applied, the mass peaks appear in shifted positions on the previously calibrated mass scale. The low (M_1) and high (M_2) mass shifts were determined as a function of the L parameter. The difference $\Delta M = M_2 - M_1$ determines the passband width Δa and Δq along the scan line relative to the ($a_{0.706}$, q_0) point. Assuming that the scan line intersects two stable "islands", the mass peak with actual mass number M_0 will be observed at rf amplitudes V_1 and V_2 in positions M_1 and M_2 on the calibrated mass scale. With respect to the apparent masses M_1 and M_2, we have:

$$V_1 = C_1 q_0 M_1,\ V_2 = C_1\ q_0 M_2\ \ (C_1 = \text{constant}),$$

or with respect to the actual mass M_0:

$$V_1 = C_1 q_1 M_0,\ V_2 = C_1\ q_2 M_0$$

From the above equations, one can determine the boundary points a_1, q_1 and a_2, q_2 which are intersected by given scan line $a=q \cdot a_{0.706}/q_0$:

$$a_1 = 2\lambda q_0 \frac{M_1}{M_0}, q_1 = q_0 \frac{M_1}{M_0}; a_2 = 2\lambda q_0 \frac{M_2}{M_0}, q_2 = q_0 \frac{M_2}{M_0}$$

Thus, by changing the slope of the scan line, the boundaries of the stable regions close to the apex of the Mathieu stability diagram can be determined.

Interestingly, a relatively small auxiliary signal leads to dramatic changes in the stability diagram. The first order quadrupole resonance is created by powerful instability bands which follow iso-β lines $\beta_x=0.9$ and $\beta_y=0.1$. The stable regions (outside of these instability bands) move beyond of the unexcited stability diagram. The unstable band which follows $\beta_x=0.8$ (the second order resonance) is weak but clearly recognizable. Bands of higher order resonances are yet weaker. Stable regions

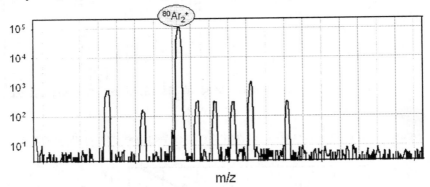

Figure 3a *Mass spectrum obtained from the tip of one stable region*

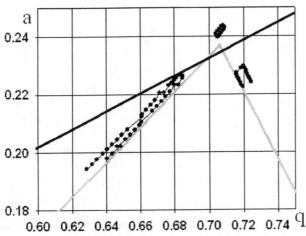

Figure 3b *The experimentally determined stable regions superimposed on the Mathieu stability diagram. The scan line corresponds to the spectrum in Figure 3a*

have boundaries, which follow iso-β_y and iso-β_x lines. Hence, on these boundaries one-dimensional ion separation in the *y*-direction or in the *x*-direction occurs. Under certain conditions, the apex of the internal region can be observed (see Figure 2b). With increasing amplitude of excitation, the instability bands are widened and the width of stable regions are narrowed. New stable regions move out from the conventional Mathieu stability boundaries.

Resolution control might be achieved either by changing the slope of the scan line at the apexes of the new stable region or by changing the amplitude of the auxiliary rf signal. The first approach is demonstrated in Figures 3 and 4.

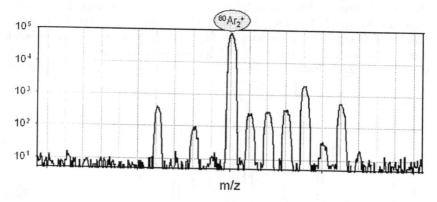

Figure 4a *Mass spectrum obtained from the tip of the upper most stability region*

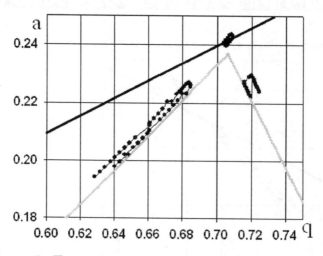

Figure 4b *The experimentally determined stable regions superimposed on the Mathieu stability diagram. The scan line corresponds to the spectrum in Figure 4a.*

The second approach is preferable because in that case the scan line crosses the whole stable region where the QMF acceptance is maximized. This method of operation is based on the observation that the width of the stable region decreases with increasing excitation level. In comparison to operation at the apex of the Mathieu

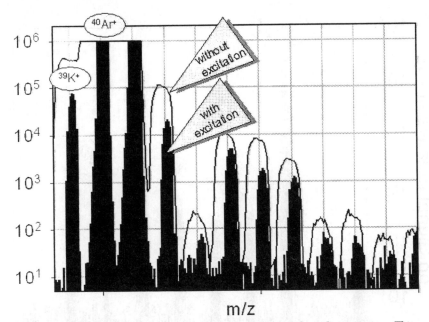

Figure 5 *Resolving of m/z 39, 40 and 41 using the quadrupole excitation. This result is achieved primary due to the exceptional improvement in the abundance sensitivity rather than in resolution*

stability diagram, the ion transmission dependence on resolution is weaker in this case. Using this approach, m/z 39, 40 and 41 can be resolved successfully with high abundance sensitivity.

In practice it would be useful to reduce the ion residence time in the QMF if one could retain the resolving power and abundance sensitivity intact. Experimentally this can be achieved by variation of the QMF offset voltage. The initial kinetic energy of ions from the ICP source is of the order of 3-5eV. As can be seen in Figure 6, mass peak broadening with increasing transport ion energy up to 13-15eV is insignificant when quadrupole excitation is applied. Similar results are not achievable when operating the QMF in the proximity of the apex of the Mathieu stability diagram under otherwise similar experimental conditions.

In addition, the increase of the ion transport energy up to 13-15eV enhances the sensitivity. It is our observation that the limiting resolution in comparison with that achieved for conventional operation is not degraded. The resolution is generally predetermined by the properties of the first stability region, which are not changed after application of the auxiliary signal, although its geometry is dramatically affected. Auxiliary excitation improves the abundance sensitivity by efficient removal of the mass peak tails as can be seen in Figures 5 and 6.

A weak dependence of the mass peak width on the ion residence time in the QMF directly translates into an increase in the abundance sensitivity. High abundance sensitivity is useful when an intense peak overlaps a neighboring low abundance mass peak. This can be seen in Figure 7 where the mass spectrum of a 10 ppm mixture of

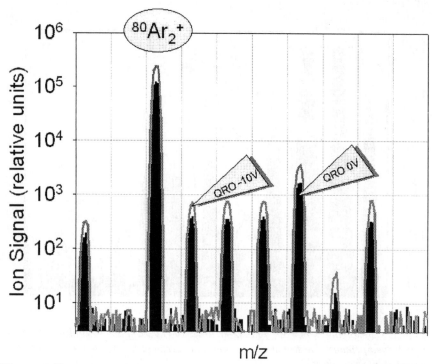

Figure 6 *The mass peak broadening due to increase in the ion axial energy. No evidence of an additional defocusing of the ion beam in the QMF fringing field due to the quadrupole excitation was observed*

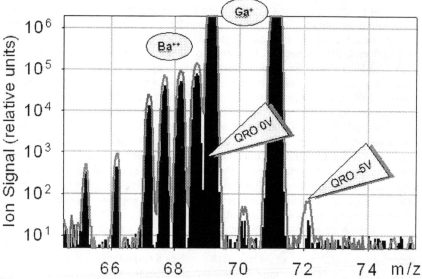

Figure 7 *Mass spectrum of 10 ppm mixture of Ga and Ba employing the quadrupole resonant excitation. Two different ion beam energies were used to explore dependence of the abundance sensitivity on the ion energy*

Ga and Ba obtained with quadrupolar excitation is presented. It is apparent that the mass peak tails vanish due to the action of the resonant quadrupolar excitation.

Removing the peak tails on the $^{69}Ga^+$ ion signal using auxiliary rf excitation allows resolution of the $^{137}Ba^{++}$ signal. Although the resolution at 10% of the peak height is not significantly improved, the outstanding abundance sensitivity allows distinction between ion signals separated only by 0.5amu and different in intensity by 10^5-10^6 times.

4 CONCLUSIONS

Auxiliary quadrupole excitation is useful for the creation of new stable regions with new ion separating properties. At low excitation levels, the increase in the abundance sensitivity in comparison with conventional mode of the QMF operation is substantial. Using the excitation mode of operation does not considerably increase the limiting resolution at 10% of the peak height employing the ICP ion source, however, it significantly diminishes dependence of resolution and abundance sensitivity on the ion energy.

5 REFERENCES

1. W. Paul, H. Steinvedel, *Z. Naturforschung.* 1953, **8a**, 448.
2. W. Paul, M. Raether, *Z. Phys.* 1955, **140**, 162.
3. E.W. Blauth, *Dynamic Mass Spectrometers*, Elsevier, Amsterdam, 1966, pp. 119-137.
4. F.A. White, *Mass Spectrometry in Science and Technology*, Wiley, New York, 1968, pp.66-107.
5. P.H. Dowson. *Advances in Electronics and Electron Physics.* Academ. Press. Inc. 1980, **53**,153.
6. P.H. Dawson (Ed.), *Quadrupole Mass Spectrometry and Its Applications*, Elsevier, Amsterdam, 1976, reissued by AIP, Woodbury, New York, 1995.
7. R.E. March, R.J. Hughes, J.F.J. Todd. *Quadrupole Storage Mass Spectrometry*, Wiley, New York, 1989, pp.31-110
8. R.L. Alfred, F.A. Londry, R.E. March, *Int .J. Mass Spectrom. Ion Process*, 1993, **125**, 171.
9. Y. Wang, J. Franzen, K.P. Wanczek, *Int. J. Mass Spectrom. Ion Processes*, 1993, **124,** 125.
10. J. Franzen, *Int. J. Mass Spectrom. Ion processes*, 1993, **125**, 165.
11. Y. Wang, J. Franzen, *Int. J. Mass Spectrom. Ion processes*, 1994, **132**, 155.
12. V.V. Titov, *J. Am. Soc. Mass Spectrom.*, 1998, **9**, 50.
13. G.C. Stafford, Jr., P.E. Kelley, J.E.P. Suka, W.E. Reinolds, and J.F.J. Todd, *Int. J. Mass Spectrom. Ion Processes*, 1984, **60**, 85.
14. G.C. Stafford, P.E. Kelley, D.R. Stephens, U.S. Patent 4,540,884
15. R.E. March. *Advances in quadrupole ion trap mass spectrometry: Instrumentation development and applications/* In Advances in Mass Spectrometry, Vol. 14; Elsevier: Amsterdam, 1998.
16. R.E. March, A.W. McMahon, E.T. Allinson, F.A. Londry, R.L. Alfred, J.F.J. Todd, F. Vedel, *Int. J. Mass Spectrom. Ion Processes*, 1990, **99**, 109.

17.   M. Sudakov, N. Konenkov, D.J. Douglas and T. Glebova, *J. Am. Soc. Mass Spectrom.*, 2000, **11**, 10.

18.   M.A.N. Razvi, X.Z. Chu, R. Alheit, G. Werth, R. Blumel, *Phys. Rev. A*, 1998, **58**, R34,

19.   B.A. Collings, D.J. Douglas, *J. Am. Soc. Mass Spectrom.*, 2000 (accepted)

20.   J.M. Campbell, B.A. Collings and D.J. Douglas, *Rapid Commun. Mass Spectrom.* 1998, **12**, 1463.

21.   L.D. Landau, E.M. Lifshitz, *Mechanics*, third edition, Pergamon Press, Oxford, 1960

ALL THE IONS ALL THE TIME: DREAM OR REALITY?

Gary M. Hieftje, Steven J. Ray, John P. Guzowski, Jr., Andrew M. Leach,
Denise M. McClenathan, David A. Solyom, William C. Wetzel, and Øle A. Gron

Department of Chemistry, Indiana University, Bloomington, IN 47405 U.S.A.

1 INTRODUCTION

Plasma source mass spectrometry has earned a justifiable reputation for extraordinarily low detection limits, on the order of one part per trillion (ppt) for quadrupole-based systems and in the range of parts per quadrillion (ppq) for sector-field systems. Yet, even in these tremendously sensitive instruments, most ions originally present in the sample solution are never measured. Instead, they are lost in the sample-introduction process, in extracting the ions into the mass spectrometer, through losses within the spectrometer itself, or by inefficient detection arrangements. It is justifiable to question whether such losses should be of concern; after all, most routine determinations are already limited by contamination in the sample, in the blank, or in the laboratory environment. Yet, in nanotechnology, biotechnology, and in any other situation where sample size or volume is limited, such losses might prove intolerable. In those circumstances it is often not the concentration of the elements or isotopes in a sample that is important but rather their total number. With effective sample volumes approaching picoliters in many chip-based laboratory schemes, with modern high-speed separations or chromatographic techniques, or in the examination of microscopic structures in solids, every sample atom is important and diligence must be exercised to avoid waste.

In accordance with this realization, the goal of the present manuscript is to examine one aspect of sample-utilization efficiency, specifically the one that arises within the spectrometer itself. Alternative mass-spectrometer configurations will be compared on the basis of the duty factor they offer; that is, the fraction of the ions that are presented to the analyzer in a continuance basis that are capable actually of being registered by the spectrometer. Depending on whether a particular spectrometer employs ion storage, time-of-flight technology, scanning, or continuous detection, its duty factor will be dramatically different. Schemes for improving the duty factor in several configurations will then be offered, and results for a particular system, a time-of-flight mass spectrometer (TOFMS) will be emphasized. Lastly, the importance of improved duty factor in TOFMS will be illustrated by means of a recent experiment involving laser-ablation sampling.

2 DUTY FACTOR - A USEFUL COMPARATIVE FIGURE OF MERIT

Let us begin by analyzing briefly the efficiency of an ICP-MS instrument that employs a quadrupole mass filter. Because this is the most popular current configuration, it seems an appropriate place to start. If the instrument consumes one mL/min of sample solution containing one ppm of analyte, it can be calculated that roughly 10^{14} ions/s are being aspirated. If ultrasonic nebulization is being employed, the sample-introduction efficiency is usually on the order of 10%, so about 10^{13} ions/s are being transported into the ICP. Most of those ions will then travel in the central channel gas stream and most will also be drawn into the sampling orifice to form a supersonic beam. However, because of geometric considerations, only one percent or so of those sampled ions pass the skimmer orifice, suggesting that 10^{11} ions/s enter the ion optics of the mass spectrometer. Modern instruments will then yield as many as 10^8 counts per second for these ions, indicating that losses of another factor of one thousand occur in the ion optics, in transmission, and in ion detection. Of course, efforts are ongoing to reduce these losses. Yet, they become even more worrisome when it is recognized that this analysis pertains to only one isotope of a single element; because the spectrometer must scan from one mass to another, it is forced to discard all ions at other masses while a particular one is being monitored. Such losses not only lengthen analysis time, but they necessarily sacrifice signal-to-noise ratios and thereby compromise detection limits and precision.

Although this brief analysis was made for a quadrupole mass filter, it will be shown in the paragraphs below that most other spectrometers also suffer substantial ion losses because of what we will define here as a duty factor. The spectrometers that will be compared are scanning units (quadrupole mass filters and sector-field systems) and a number of approaches that are viewed by many as being simultaneous. These "simultaneous" instruments include TOFMS, a quadrupole ion trap, Fourier Transform ion cyclotron resonance mass spectrometry, multi-collector sector-field instruments, and sector-field devices equipped with focal-plane detector arrays. Of these alternatives, it will be shown that only the last truly has the capability to measure all the ions all the time.

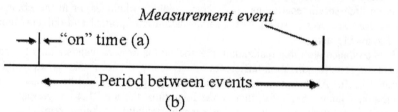

Figure 1   *Representation and definition of "duty factor" for a time-domain waveform*

Before we proceed, it is important to define what is meant by duty factor. As illustrated in Figure 1, duty factor is commonly considered to be the "on" time of a repetitive waveform. In Figure 1, a brief pulse occurs repetitively; between those pulses, the signal is "off" or simply ignored. If the period between pulses is b and the duration of the pulse is a, the duty factor will be defined as a/b, often expressed in the form of a percentage. It should be immediately obvious that this definition of duty factor applies directly to TOFMS, in which each sample-input event is represented by the "on" time of the pulse (a) and the interval between TOFMS spectra is given by the interval b. However, the same definition can be applied to scanned spectrometers, ion-storage devices, and others.

2.1 Scanned Spectrometers

As shown in Figure 2, the definition of duty factor for a scanned spectrometer is similar conceptually to that for a time-dependent waveform. In the case of a scanned spectrometer, the "on" time (a) is the portion of the total spectral range that is examined at once. Ordinarily, this involves only a single spectral peak. The period between events (b) indicated in Figure 1 is in the case of a scanning system the full spectral range that must be examined. Of course, this last parameter will change from one analytical

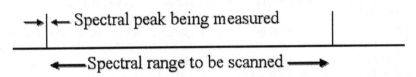

Figure 2 *Representation of duty factor for a scanned mass spectrometer*

situation to another. In the most advantageous case, when only a single isotope of a particular element must be recorded, the spectral peak to be measured is the same as the entire spectral interval of interest, so the duty factor is unity (or 100%). However, in the more likely event that a range of elements or isotopes must be determined, the duty factor becomes correspondingly less. In the worst-case situation, when the full atomic mass range is of interest, everything from ^{6}Li to ^{238}U will be required, in which case 207 isotopes must be measured. Thus, if the same time is allocated to the observation of each isotope, the duty factor is $1/207 = 0.005$ (0.5%).

On the other hand, if is only elemental information that is desired and when the isotopic composition of each element is of less interest, only about 85 elements need to be recorded. In this case, the duty factor is $1/85 = 0.012$ (1.2%).

In general, then, the duty factor in a scanning system is just the reciprocal of the number of peaks that must be recorded. However, this simple analysis assumes that the spectrometer is capable of being operated in a peak-hopping mode and that the time interval between peaks is zero. It also neglects any settling time required by the measuring equipment after a peak hop takes place.

2.2 Ion Storage Mass Spectrometers

There are two kinds of ion-storage spectrometers that have so far been applied to analytical atomic spectrometry: the quadrupole (or Paul) ion trap (QIT) and the Fourier Transform ion cyclotron resonance mass spectrometer (FTMS). Although the two instruments operate in a rather different manner, the duty factor of each is governed by the same factors, as illustrated in Figure 3. In both systems, the trap is filled by a flux of incoming ions. The time required for this filling process will be dictated by the incoming flux and by the capacity of the trap. After the storage device has been filled,

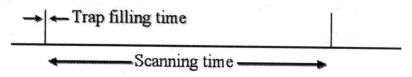

Figure 3   *Representation of duty factor for an ion-storage mass spectrometer*

the mass spectrum is recorded either by observing the motion of ions in the trap (with the FTMS and sometimes with the QIT) or by sequentially ejecting mass-selected ions out of the device (commonly used in the QIT). The duty factor of both will then be the time that is consumed by filling the trap divided by the period required to scan or encode the spectrum.

Let us first calculate roughly the time required to fill an ion-storage instrument. Interestingly, this time will be about the same for the both the FTMS or the QIT, since each has a capacity of roughly 10^5 ions before space-charge problems begin to arise and either degrade resolution or encourage ion loss. For the greatest duty factor, it is desirable to have the filling time as long as possible compared to the period needed to acquire a spectrum. Consequently, let us assume for the best-case scenario that argon ions in the incoming beam (from, say, an ICP) have been mostly eliminated by collision with a charge-transfer gas such as molecular hydrogen. Further, let us assume that ions from the solvent (probably water) will be eliminated either by similar charge-transfer processes or by means of resonant ejection from the trap. Thus, the full capacity of the ion-storage device will be able to be utilized for ions of the sample itself. Of course, in any real analytical situation, it will be necessary to prepare for the possibility of high sample concentrations, 1% dissolved solutes being a commonly accepted maximum in ICP-MS. If the efficiency of the ion production and delivery system are comparable to what is found in modern quadrupole-based mass spectrometers, it would be expected that a 1-ppm solution will yield at least 10^7 ions per second. Thus, a solution concentration of 1% (10^4 ppm) would produce 10^{11} ions per second. The filling time of the trap will then be 10^5 ion capacity/10^{11} ions/s, or 10^{-6}s.

In contrast, if the flux of argon ions into the trap is not quenched, as is assumed in the analysis above, an additional 10^{12} ions/s will flow into the trap. This will reduce the filling time to about 10^5 ions/10^{12} ions/s, or 10^{-7}s.

The time necessary to acquire a spectrum is different for an FTMS and a QIT. In FTMS, the scan time depends upon the desired resolution, with longer scan times being expected in order to generate the outstanding resolution for which FTMS is justifiably well known. If the scan time is taken as somewhere between 1 and 100s the duty factor can be computed from the filling time calculated above divided by this interval. Taking the best case of 10^{-6}s for the filling time, we would then obtain a duty factor of 10^{-6}s/scan time = 10^{-6}-10^{-8}.

If the QIT is operated in its conventional mode, in which ions are ejected from the trap sequentially, the time required to scan a spectrum will depend on the number of spectral peaks that must be acquired, just as with the scanning instruments discussed earlier. Typically, the QIT will require roughly 1ms/amu during a scan. Therefore, the best duty factor that can be obtained will occur when a single isotope is to be measured, which will provide a duty factor of $10^{-6}/10^{-3}$ or 10^{-3} (0.1%). More generally, the duty factor of a QIT will be $10^{-6}/(10^{-3})$(number of peaks to be measured). Although this number will depend upon the number of isotopic peaks to be examined, it will generally be on the order of 10^{-5} (0.001%).

2.3 Duty Factor in TOFMS

Because TOFMS produces a spectrum that follows very closely the scheme outlined in Figure 1, duty factor in the simplest instruments is straightforward to calculate. The "on" time is the period during which a packet of sample ions is injected into the TOFMS flight tube, while the period between events is the time required to accumulate a single mass spectrum. TOFMS instruments designed for use in atomic spectrometry ordinarily employ accelerating voltages on the order of 2000V and a flight tube approximately 1m in length. The time to record a full atomic mass spectrum is therefore on the order of 50μs. With the time for injection of an ion pack at 5ns, the duty factor then becomes 5ns/50μs = $(5\times10^{-9})/(50\times10^{-6}) = 10^{-4}$.

Fortunately, this extremely low duty factor, on the order of what is to be expected for a full spectral scan with a QIT, can be improved dramatically in TOFMS through use of a technique discovered over 35 years ago [1] In this scheme, portrayed in Figure 4, the ions sampled from an ICP or other plasma form a beam that travels at the thermal velocity of the expanding gas. In such a supersonic expansion, all species within the stream are expected to achieve the same velocity as that of the bath gas, in this case Ar. Because the ICP is rather hot, the velocity of this gas is quite high, on the order of thousands of meters per second. Compared to energies in TOFMS, however, the beam is relatively "cool". It can be calculated that one eV corresponds to a temperature of 11,600K. With typical ICP temperatures being measured in the range of 6,000K, the ICP therefore has a thermal energy corresponding to roughly 0.5eV.

This kinetic energy (0.5eV) is extremely low compared to the energy acquired by the ions when they are injected at a right angle into the flight tube of a TOFMS (2000eV). What occurs is that a packet of ions in the extraction zone shown in Figure 4 is accelerated into the TOFMS flight tube and the spectrum of that packet recorded. During that recording process, the slowly moving ion beam extracted from the ICP refills the extraction volume, so that the experiment can be repeated immediately after the mass spectrum of the preceding packet has been recorded.

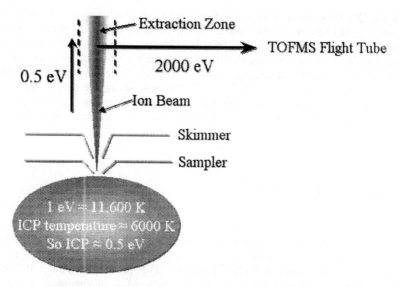

Figure 4   *Simplified diagram of the orthogonal-extraction geometry used to improve duty factor in time-of-flight mass spectrometry. See text for discussion*

A recent alternative to this classical orthogonal extraction geometry is an on-axis arrangement such as that shown in Figure 5. Here, the considerations are the same as in orthogonal extraction. That is, the slowly moving ion beam extracted from the ICP fills an extraction zone, from which ions are pulsed into a flight tube to produce a mass spectrum. While that spectrum is being recorded the extraction zone is refilled by the slowly moving input ion beam.

Each of these alternative arrangements has their own strengths and weaknesses. In both geometries, ions should flow into the extraction zone at the same velocity, at a speed corresponding to that of Ar. Consequently, all analyte ions are introduced into the extraction zone with the same efficiency, thereby minimizing mass bias. However, because all analyte ions have nearly the same velocity, they possess a mass-dependent kinetic energy. In the orthogonal extraction geometry, this kinetic energy variation causes ions of disparate mass to follow a different trajectory when they are deflected by a perpendicular field into the flight tube. That is, heavier ions tend to follow wider trajectories whereas lighter ions are deflected at closer to a right angle. Moreover, ions of any given mass will possess a range of kinetic energies, dictated by the kinetic-energy spread they attained in the plasma. As a result, even ions of a single mass will spread out laterally as they travel down the flight tube, so that a large detector or extraction zone is required in order to achieve an acceptable efficiency. Also, the detector or extraction zone must be enlarged even more if ions of all masses are to be accommodated, unless some kind of ion-steering device [2-4] is incorporated into the system. One of the greatest advantages of the on-axis arrangement is that the extracted ion packets possess a very small cross section in the direction perpendicular to the flight tube. Because the ion packets are narrow as they traverse the ion optics leading to the flight tube and as they pass through a reflectron, transmission is likely to be higher. Even more importantly, the small ion-beam cross section permits small detectors to be used. Particularly attractive in this regard is the use of discrete-dynode ion multipliers that are less prone to fatigue and saturation than their microchannel plate counterparts.

It should be recognized that the foregoing discussion is somewhat simplistic, in order to clarify duty-factor considerations in TOFMS. In truth, the duty factor in TOFMS can be shown to be:

$$Duty\ factor\ (TOFMS) = F \cdot d/V \qquad (1)$$

where F is the repetition rate of the TOFMS ($50\mu s$), d is the length of the extraction zone, and V is the velocity of the incoming primary ion beam. For a typical atomic TOFMS instrument, of either orthogonal-extraction or on-axis extraction design, the duty factor will therefore be $(20,000s^{-1})$ $(2.5\ cm)/(3\ X10^5)cm/s$, or about 0.16 (16%).

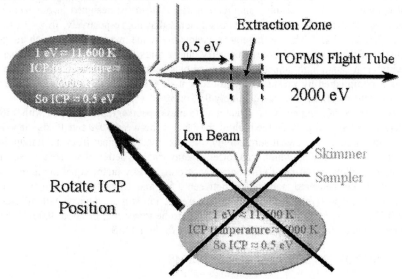

Figure 5 *Relationship between the orthogonal-extraction and on-axis extraction configurations used to raise duty factor in time-of-flight mass spectrometry*

2.4 Duty Factor in Array-Detector Sector-Field Mass Spectrometers

This type of mass-selection device is perhaps better called a mass spectrograph, since it should ideally be capable of recording an entire spectral range at once. In such a situation, the duty factor is obviously unity or 100%. Of course, this assertion neglects any time required for scanning the signal from a detector array, something that will become relatively more time-consuming when short signal-acquisition periods are employed. This situation will arise, for example, when transient samples such as those produced by chromatography, flow injection, laser ablation, or electrothermal vaporization are encountered.

Several focal-plane spectrographs have been designed in the past, two of which were intended specifically for atomic mass spectrometry[5, 6]. The instrument developed in our laboratory was designed for use with an argon plasma such as the ICP, so the mass range was intentionally broken into two segments to avoid the intense Ar^+ peak.

The concept behind both published atomic mass spectrographs is similar. Even if it is desirable to cover the entire mass range from hydrogen to uranium, only about 240 resolution elements are needed. Therefore, linear detector arrays should serve nicely. Further, because most commercial detector arrays are planar, it would be desirable to utilize a sector-field geometry with a flat focal plane. It is therefore not surprising that the Mattauch-Herzog configuration was selected for both. Unfortunately, this geometry, like most magnetic-sector instruments, produces a quadratic mass display, making it very difficult to satisfy simultaneously the requirements of broad spectral coverage and adequate spectral resolution. To avoid this complication, the instrument designed in our laboratory examines the atomic mass spectrum in two segments that go, respectively, from ^7Li to ^{39}K and from ^{41}K to ^{238}U. This arrangement yields the same mass ratio in each range, a requirement of the Mattauch-Herzog geometry, and avoids the problem of the intense Ar^+ peak, which can cause high detector background or saturation. Switching between the two ranges is accomplished by means of an adjustment in accelerating voltage. Obviously, in this sort of instrument, the duty factor is only 0.5, or 50%.

From the foregoing analysis, a sector-field mass spectrograph equipped with a focal-plane detector array will provide the highest duty factor, between 50% and 100%, depending on the configuration. Somewhat surprisingly, the next most efficient from the standpoint of duty factor is a TOFMS, which can offer a duty factor better than 10%. Scanning spectrometers follow, in terms of efficiency, by offering a duty factor equal to the reciprocal of the peaks to be measured, somewhere between 0.5% and 1.2%, for full elemental or full mass coverage, respectively. Last in efficiency are the ion-storage devices with efficiencies of only 0.1% for the QIT when a single isotope is to be measured, or about 0.001% for the QIT when full spectral coverage is desired, to 10^{-5}% for an FTMS.

3 IMPROVING DUTY FACTOR IN TOFMS

For all kinds of atomic mass spectrometer except for the sector-field systems equipped with a focal-plane detector, some improvement in duty factor seems possible. For example, parallel ion traps[7] are one possibility and multiple quadrupole systems have already been constructed[8]. Similarly, improving duty factor in TOFMS seems relatively straightforward and seems worthy of additional attention here.

From Equation 1, it is possible to improve duty factor in TOFMS by raising the repetition rate of spectral acquisition (F), by lengthening the extraction zone (d), or by reducing the velocity of the incoming ion beam (V). All three approaches are possible, but a number of tradeoffs arise.

For example, it is possible to raise the spectral-acquisition rate of a TOFMS (F) either by shortening the flight tube or by raising the accelerating voltage that sends ions down the tube. Unfortunately, reducing the length of the flight tube will likely lead to a loss in resolving power. Moreover, employing high accelerating voltages introduces instrumental complications, especially when it is realized that the spectral repetition rate is related to the square root of the accelerating voltage. Thus, unattractively high voltages must be employed to realize a very substantial gain in spectral acquisition time.

The duty factor in TOFMS can be improved also through use of a longer extraction region (d), shown in Figures 4 and 5. In the orthogonal acceleration geometry, such a change would mean enlarging an already wide ion packet as it sails down the drift tube. In turn, an even larger detector would be required and instrument dimensions might have to be increased accordingly. In the on-axis system, a longer ion packet would probably lead to a loss in resolution, since the packet has to be compressed by means of space focussing as it travels toward the detector. If the packet cannot be adequately compressed, leading ions will strike the detector earlier than those that follow, causing a wider ion pulse, a broader spectral peak, and poorer resolution.

Among these three options, then, the most attractive appears to be reducing the velocity of the incoming ion beam (V). Although the ion beam in Figure 4 or Figure 5 already travels at a relatively low velocity (energy), that energy corresponds to roughly 6,000K, about 20 times higher than room temperature. Consequently, lowering the effective kinetic temperature of the beam to room temperature should lead to an improvement in duty factor by at least 4.5-fold ($\sqrt{20}$). In fact, we shall see that an even greater gain is possible, enough to capture fully all the incoming ions.

Cooling the incoming beam to room temperature could be accomplished by several alternative means, the most attractive among which are through use of a collision cell [9-11] or by means of a drift tube such as used in ion mobility spectrometry[12].

Collision cells can be designed around a number of geometries, but the most common employed in atomic spectrometric instrumentation utilize quadrupoles, hexapoles, or octapoles. In each case, the principle is the same. Radiofrequency voltages applied to a symmetrically arranged radial series of metal rods produce an effective potential well in the center of the rod system, especially for species that are slowed by collisions with a relatively light gas such as helium or hydrogen. External ions directed into a collision cell therefore continue to drift down its center and are not rapidly lost by scattering. After a number of collisions the incoming ions are slowed to thermal velocities, those corresponding to the thermal temperature of the collision gas (He or H_2).

Such a collision cell has other well-documented uses also. For example, it can serve as a reaction chamber to aid in overcoming polyatomic interferences or to eliminate Ar^+ ions by means of charge exchange, as was mentioned earlier. Thermalization (cooling) of the incoming ion beam also serves to overcome the plasma offset potential that ordinarily arises through coupling of an ICP with a sampling interface. Lastly, the cell changes the incoming beam from an isokinetic one (with all ions moving at the same velocity) to an isoenergetic one (with all ions possessing the same kinetic energy). This last feature has the benefit of overcoming the mass dependent steering mentioned earlier that can afflict the orthogonal acceleration geometry. It can also boost resolving power in the on-axis geometry since the kinetic-energy variation of the incoming ion beam in the direction of the flight tube is reduced.

These effects are illustrated more quantitatively in Figures 6, 7, and 8. In Figure 6 are shown three curves, the top one taken from a set of experimental data that illustrates the kinetic-energy distribution of ions in a beam traversing the extraction zone of an ICP-TOFMS instrument. The data were obtained by incrementing the modulation grid bias on our instrument and monitoring the signal of the different elements to determine ion energies as a function of mass. The slope of the line corresponds to an incoming beam temperature of 4630K. However, an intercept is noted on the vertical axis and corresponds to a plasma offset potential of 2eV. The same curve but with the 2eV offset removed is also shown, as is the dramatically reduced slope of a hypothetical beam that had been cooled to 300K.

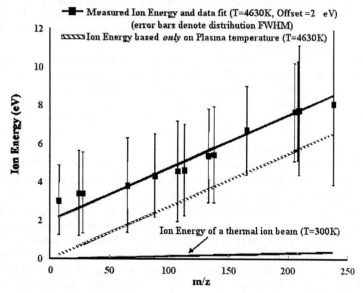

Figure 6   *Measured and calculated ion energies in the primary beam extracted from an ICP into an atomic mass spectrometer. Its slope indicates that the thermal temperature of the extracted beam is 4630 K; the intercept on the vertical axis indicates a plasma offset potential of about 2 eV. The cross-hatched line shows the energy distribution that would exist in the absence of the plasma offset potential. The bottom line indicates the low energies in a beam cooled to room temperature by, for example, a collision cell*

The ion velocities corresponding to these kinetic energies are shown in Figure 7. If there were no offset potential in the incoming ion beam, all ions would possess the same velocity, regardless of mass. However, the 2eV offset energy causes a rise in velocity for the lightest ions. This same mass-dependent velocity exists, of course, for the 300K cooled beam, although all velocities are considerably lower.

The impact of these changes on ICP-TOFMS duty cycle is illustrated in Figure 8. Values are shown for the three TOFMS instruments in our laboratory: a commercial instrument from LECO (the Renaissance), our original orthogonal-acceleration unit, and a later on-axis extraction instrument. Differences in duty factor among the three units are a result of variations in the dimensions of the extraction zone and of the flight tube; all three provide satisfactory analytical performance.

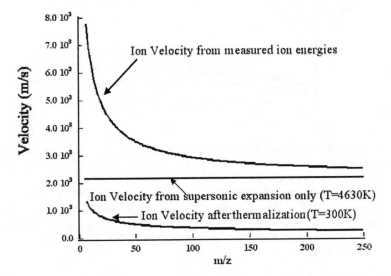

Figure 7 *Mass-dependent velocity distribution in a beam extracted from an ICP into a mass spectrometer interface. See caption to Figure 6 for discussion*

From these curves, the duty factor will be mass-dependent, not surprising in view of the mass-dependence in velocity seen in Figure 7. For both the LECO on-axis and our oldest orthogonal-acceleration geometry, the duty factor approaches 100% for even the transition elements. Further gains in duty factor would be possible by relatively minor changes in instrument dimensions.

Another highly attractive device for reducing the effective temperature of an incoming ion beam is through use of an ion mobility spectrometer[12]. Once considered low-resolution tools useful for airport drug screening and portable monitoring of atmospheric gases, modern laboratory instruments compete effectively with more traditional methods of separation, including gas chromatography, liquid chromatography, and capillary electrophoresis. Indeed, a modern ion mobility spectrometer is effectively a device for performing gas-phase electrophoresis.

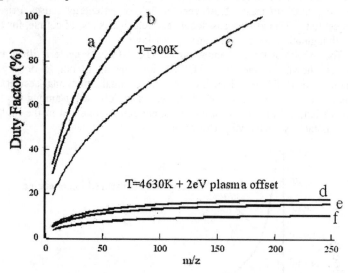

Figure 8   *Duty cycle that arises from the velocities shown in Figure 7 and for TOFMS geometries in use in the Hieftje group. Curves a, b, and c pertain to duty factors in a beam that has been cooled to room temperature whereas curves d, e, and f correspond to velocities in a beam extracted directly from an ICP. Curves a and d are calculated for the orthogonal-extraction instrument designed in our laboratory; curves b and e are for the geometry found in the Leco Renaissance instrument; curves c and f pertain to our original on-axis extraction system*

Expressed in terms familiar to separation scientists, such instruments are capable of delivering hundreds of thousands of theoretical plates and can be applied to either large or small molecules. Perhaps even more importantly, ion mobility spectrometers achieve separation of sample constituents in the gas phase rather than in condensed phase, so the separations proceed far more speedily. Just an in capillary electrophoresis, sample ions are injected into the front end of an ion mobility spectrometer in a small plug or packet and drift under the influence of an external electric field. Because the temperature, field gradient, and atmospheric pressure within the tube are ordinarily held constant, the ions achieve a drift velocity that depends on their charge state, and on their collisional cross section with the gas that fills the chamber in which they drift. Thus, sample constituents emerge from the ion mobility spectrometer at the temperature of the drift tube and in a sequence of pulses ideal for sending into a TOFMS. Coupling these two devices then seems to be a nearly ideal way to improve duty factor in TOFMS and simultaneously to perform separations or speciation on sample constituents.

3.1 Cross-Correlation TOFMS

A final way to elevate the duty factor in TOFMS is through a multiplexing approach. Several such approaches are possible, but all conceptually involve sharing the flight tube among different introduced ion packets. Some suggested approaches have employed sinusoidal or more complex waveforms to introduce ions continuously into the TOFMS flight tube. However, those approaches are not particularly attractive for atomic mass spectrometry, in which low backgrounds are essential if competitive detection limits are to be achieved.

A more attractive alternative is to introduce narrow, discrete ion packets as is done in conventional TOFMS, but to introduce those packets in a pseudo-random pattern that is tailored to avoid overlap of particular masses from one ion packet with those of another. These apparent restrictions are relaxed substantially when it is realized that atomic ions occur at discrete mass intervals, that only a relatively small number of such masses are of concern, and that the masses are far more widely spaced at the lower mass region that at higher masses. Because it is necessary to raise duty factor by only 10 fold, a computer-generated pulse sequence can be used to avoid overlap.

4 STUDIES WITH HIGH DUTY FACTOR ATOMIC SPECTROMETERS

In the author's laboratory are three plasma-source TOFMS instruments, two of which have been equipped with hexapole or octapole collision cells, and a sector-field mass spectrograph that is designed for coupling with a detector array. Recent progress with both kinds of systems will be briefly outlined in this section.

4.1 Mattauch-Herzog Mass Spectrograph

Our early work with a Mattauch-Herzog mass spectrograph[6, 13] explored the use of a conventional electro-optic ion detector (EOID). This device is comprised of a microchannel plate for spatially selective amplification, a following phosphor plate to convert the electron image to photons, and a CCD camera to capture the optical image. Similar devices have been used in the past by others but for organic mass spectrometry. Unfortunately, the conventional Mattauch-Herzog geometry requires the detector to be placed at a focal plane that is flush with the magnetic sector of the instrument. Because of this arrangement, fringing magnetic fields dramatically influence electron trajectories and amplification. The result is greatly degraded resolution, poor peak shapes, and non-linear detector response. Because of these limitations, our efforts turned to improving the mass spectrograph in a scanning mode, in order to test its ultimate capabilities, and to coupling it with alternative sources. Although the EOID was found to be clearly responsible for most of the original resolution loss, careful alignment and detailed iterative adjustment of ion optics led to a resolving power more than satisfactory for atomic mass spectrometry. Also, improvements in interface design and in ion optics raised instrument throughput and reduced noise to yield the figures of merit for a glow discharge shown in Table 1 and for an inductively coupled plasma shown in Table 2.

Table 1 *Figures of Merit for Glow Discharge Source with Array Detector Mass Spectrometer*

	Scanning Mode (SEM)	EOID Array
Limit of Detection	100-200 ppt	0.1-1 ppm
Isotope-Ratio Precision	0.15-0.5% rsd	1-3% rsd
Isotope-Ratio Accuracy	0.5% relative error	2-5% relative error
Resolving Power	800-900 (FWHM)	100-150 (FWHM)

Table 2 *Detection Limits in Array-Detector Sector Field ICP-MS*

Element	Isotope	Detection Limit (ppq
Li	7	10
Ti	49	30
V	51	45
Co	59	20
Ni	60	50
Cu	65	5
Zn	66	5
As	75	50
Sr	88	55
Mo	100	50
Cd	114	50
Sb	123	103
Ho	165	35
W	186	30
Ir	193	90
Au	197	160
Tl	203	85
Pb	208	10
U	238	20

With these instrumental improvements in hand, we have turned our attention back to coupling the instrument with an advanced multi-channel detector. The detector concept, being pursued jointly by a research team that includes Bonner Denton, David Koppenaal, and Philip Miller, is based on Faraday-cup detection. Although such an approach does not generate the multi-charge cascade that is found in micro-channel plates or discrete-dynode systems, it provides a signal that is independent of ion mass. Although this feature will be more important for macromolecular mass spectrometry than in elemental analysis, it is viewed overall as a benefit. Nevertheless, to achieve low limits of detection will then require an extremely low background from the detector. This low background will be realized by means of random-access integration and the flexibility of either destructive or non-destructive readout. These capabilities, common to what are found in modern charge-injection detector arrays, will enable each narrow Faraday Cup to be interrogated repeatedly in a non-destructive fashion. When a particular detector element has neared saturation, caused by a high incoming ion flux, that channel can be destructively read and integration can continue. During the same interval, channels that receive a lower ion flux can continue to integrate. Because many non-destructive reads of each channel can be carried out, the signal read noise can be averaged to an acceptably low level. It is anticipated that fewer than 10 atomic ions will be able to be detected in this way

4.2 Advances in Laser-Ablation ICP-TOFMS

The benefits of transient sampling are perhaps most evident when a TOFMS is employed. Because TOFMS systems sample all ions in a discrete packet as narrow as 5ns in temporal width, each ion packet represents a "snapshot" of the ion source at a particular moment in time. Consequently, errors in quantitation produced by spectral skew can be avoided. In this context, spectral skew refers to the quantitation errors that arise when sample concentration varies during the course of a spectral scan. This problem is particularly troublesome, of course, when transient sampling techniques are employed. In addition, TOFMS is capable of generating spectra at an astonishingly rapid rate approaching 20,000-30,000 per second.

Because of the absence of spectral skew and the ability to record spectra rapidly, it is attractive to consider the potential benefits of narrowing a sample pulse. In most quadrupole-based or sector-field ICP-MS instruments, laser ablation is carried out repetitively and the vaporized sample from each laser shot is stretched to produce a pseudo-steady state signal. In contrast, it is possible to examine each laser shot independently in TOFMS and, further, to narrow the vaporization pulse simply by modifying the ablation cell and its connections to the ICP. Let us imagine, for example, that it is possible to generate the same amount of sample vapor from an ablation event but to compress it into a pulse that is one hundred times narrower. Since the area of the pulse would not be changed, it should then become one hundred times taller, leading to a 100-fold peak signal enhancement. Alternatively, it might be preferable to record the area rather than height of the signal pulse. In this case, the narrower pulse will require examination of the vaporization event for a shorter period of time, so the total number of background counts that are collected should drop by a factor of 100. Because background noise is proportional to the square root of the background itself (because of Poisson statistics), the background noise will therefore drop by a factor of 10. Either way, with peak height or area measurement, a signal-to-noise gain is realized. We have demonstrated this level of improvement through modification of a commercial laser-ablation accessory, by employing a fast washout cell with a volume of only $0.7cm^3$. The signal pulse is produced by this arrangement exhibited a FWHM of only 40ms.

In another demonstration of the utility of TOFMS for laser-ablation sampling, we have found that it is possible to achieve standardless quantitation of metal-alloy samples with an accuracy between one and five percent. This achievement is possible because a TOFMS records an entire atomic mass spectrum for each ion packet introduced into the spectrometer. If ionization efficiency is not altered appreciably by changes in sample composition, the amount of sample material that is ablated should be related to the sum of all the ions that enter the spectrometer. In turn, the sum can be found by multiplying each mass-spectral peak by a mass-bias correction factor (fixed for a given spectrometer) and by a calibration factor that can be derived from a broad number of samples. The concentration of the given element is then simply equal to the ratio of its mass-spectral peak height, multiplied by its sensitivity factor and mass-bias correction, divided by the sum of the same products for every observed mass-spectral peak. With this scheme, we have found it possible to analyze metallic samples having matrices ranging from aluminum to brass, to copper and even to steel, all without additional standards.

5 CONCLUSION

There are many factors that affect the performance of a plasma source mass spectrometer. In this paper, only one of them has been addressed: the efficiency with which the spectrometer utilizes ions fed to it in a continuous fashion. For pulsed sources such as the laser-ablation system just discussed, these considerations will change. In addition, other instrument design features such as the interface arrangement, sample-introduction equipment, ion-optic design, detector choice, and signal-processing equipment will affect the signal-to-noise ratios that can be realized. Nevertheless, ion utilization efficiency is likely to become an increasingly important factor in the future and it is important that every effort be made to sacrifice as few sample ions as possible in their progress from sample to detector. It is suggested that duty factor is a useful figure of merit to gauge one aspect of this performance.

6   REFERENCES

1.   G. J. O'Halloran and L. W. Walker, Bendix Corporation, Technical Document Report No.ASD TDR 62-644, Parts I and II. Prepared under contract AF 33(657)-11018. (1964).
2.   M. Pellarin, B. Baguenard, M. Broyer, J. Lerme and J. L. Vialle, *J. Chem. Phys.*, 1993, **98**, 94.
3.   P. Milani and W. A. de Heer, *Rev. Sci. Instrum.*, 1991, **62**, 670.
4.   D. P. Myers, G. Li, P. P. Mahoney and G. M. Hieftje, *J. Am. Soc. Mass Spectrom.*, 1995, **6**, 411.
5.   E. F. Cromwell and P. Arrowsmith, *J. Am. Soc. Mass Spectrom.*, 1996, 7, 458.
6.   T. W. Burgoyne, G. M. Hieftje and R. A. Hites, *J. Am. Soc. Mass Spectrom.*, 1997, **8**, 307.
7.   E. R. Badman and R. G. Cooks, *Anal. Chem.*, 2000, **72**, 3291.

8.   A. R. Warren, L. A. Allen, H.-M. Pang, R. S. Houk and M. Janghorbani, *Appl. Spectrosc.*, 1994, **48**, 1360.

9.   D. J. Douglas and J. B. French, *J. Am. Soc. Mass Spectrom.*, 1992, **3**, 398.

10.   H. J. Xu, et al., *Nucl. Instr. and Meth. in Phys. Res. A*, 1993, **333**, 274.

11.   D. J. Douglas, *J. Am. Soc. Mass Spectrom.*, 1998, **9**, 101.

12.   C. A. Srebalus, J. Li, W. S. Marshall and D. E. Clemmer, *J. Am. Soc. Mass Spectrom.*, 2000, **11**, 352.

13.   D. Solyom, T. W. Burgoyne and G. M. Hieftje, *J. Anal. At. Spectrom.*, 1999, **14**, 1101.

7 ACKNOWLEDGEMENTS

This research was supported in part by the U.S. Department of Energy through grant DE-FG02-98ER14890, by ICI Technologies, and by Leco Corporation.

EVALUATION OF INDUCTIVELY COUPLED PLASMA – ION TRAP MASS SPECTROMETRY

Naoki Furuta, Akihiro Takeda and Jian Zheng

Department of Applied Chemistry
Chuo University, Tokyo Japan

Takayuki Nabeshima

Hitachi, Ltd., Central Research Laboratory, Tokyo, Japan

1 INTRODUCTION

Mass spectrometers were primarily applied to organic analysis such as GC-MS and LC-MS, and have been developed during the latter half of the century. At first, a quadrupole mass spectrometer was used for ICP-MS.[1] In the last two decades, the mass spectrometers which were developed for organic chemistry, were applied to inorganic analysis. The ICP high resolution mass spectrometer equipped with the double focusing MS, is commercially available for applications requiring precise isotope ratio measurements and ultra trace analysis.[2] The ICP-TOFMS has recently been developed, and has become commercially available also.[3] Another interesting combination with ICP is 3DQMS (Three Dimensional Quadrupole Mass Spectrometer), which is the ion trap type mass spectrometer.[4] Recently, Hitachi, Ltd. (Tokyo, Japan), commercialized the instrument, but it is now available only inside Japan. A laboratory-modified ICP-3DQMS based on this Hitachi instrument was used for this study.

The ICP-MS is widely used for its advantages of high sensitivity and capability of isotope ratio measurements. However, when the quadrupole mass spectrometer is used, spectral interferences due to polyatomic ions are severe problems for the determination of trace elements.[5,6] Several methods have been proposed to alleviate these spectral interferences such as operating the plasma under cool plasma conditions and installing a collision cell or a reaction cell in front of the quadrupole mass spectrometer.[7,8] The cool plasma condition which reduces plasma potential, is beneficial for minimizing Ar associated molecular ions for clean samples such as semiconductor samples, but does not have appropriate plasma power to decompose and ionize samples containing high matrix. On the other hand, the collision cell or the reaction cell is applicable for high matrix samples because the plasma is operated under hot plasma conditions, however, the polyatomic ions deriving from the sample matrix such as CaO can not be eliminated. The 3DQMS was used as the third approach to eliminate the polyatomic ion interference. The advantages of ICP-3DQMS, are that it uses high power to enable introduction of high matrix samples such as environmental samples, and has the potential to minimize the ion interference of both argon associated molecular ions and metal oxide ions.

The significant difference between the 3DQMS and the quadrupole mass spectrometer is the capability to trap the ions in the mass spectrometer. For quadrupole mass spectrometers, ions that are generated at the plasma are transiently transmitted through the ion optics and mass spectrometer. When the length of the quadrupole is 23 cm, it takes about 1-50 μs for the ions to pass through the mass spectrometer. However, the 3DQMS can trap the ions of interest for ms to seconds order, and it is possible to apply optimum conditions for the ions to collide, dissociate and react within the mass spectrometer, prior to detection.

2 INSTRUMENT OF 3DQMS

Ions produced in the ICP were introduced into the 3DQMS after passing through the Einzel lens, a double cylindrical electrostatic ion guide and a 90 degree deflector.[9] Pure helium was introduced in the 3DQMS as a buffer gas. The detector measured the trapped ions with a pulse counting mode.

The 3DQMS consists of one ring electrode and two endcap electrodes as shown in Figure 1. The ring electrode serves to trap the ions within the area surrounded by these electrodes, and the amount of ions that can be trapped in the area depends on the size of the ring electrode. The larger the radius (r_0), the more ions can be trapped, and r_0 and z_0 have the relationship of the following equation:

$$2z_0^2 = r_0^2 \tag{1}$$

The stability region can be obtained from Mathieu's equation:[10]

$$a_z = -16zeU \,/\, m(r_0^2+2z_0^2)\Omega_{ring}^2 \tag{2}$$

$$q_z = 8zeV_{ring} \,/\, m(r_0^2+2z_0^2)\Omega_{ring}^2 \tag{3}$$

The 3DQMS uses only AC, therefore only the q_z value should be considered. A radio frequency (Ω_{ring}) is constant, and the maximum amplitude for V_{ring} is limited (V_{ring} = a few kV). Unlike organic analysis which requires a wide mass range (up to 2000 amu), elemental analysis requires a mass range up to 300 amu, therefore, r_0 of the 3DQMS for inorganic analysis can be enlarged from that for organic analysis. By increasing the radius r_0, ions trapped in the area surrounded by the ring electrode and endcap electrodes could be increased by one order of magnitude (approximately 10^6 ions).

Ion trapping and ejection within the trapping area, are controlled by the V_{ring} amplitude. There is a corresponding V_{ring} amplitude for each m/z, and when a certain V_{ring} is applied, ions with m/z larger than that corresponding to the V_{ring}, are trapped in the trapping area. After trapping, the amplitude of V_{ring} is increased, and ions are ejected from the trapping area through the orifice of the exit side endcap electrode and detected.

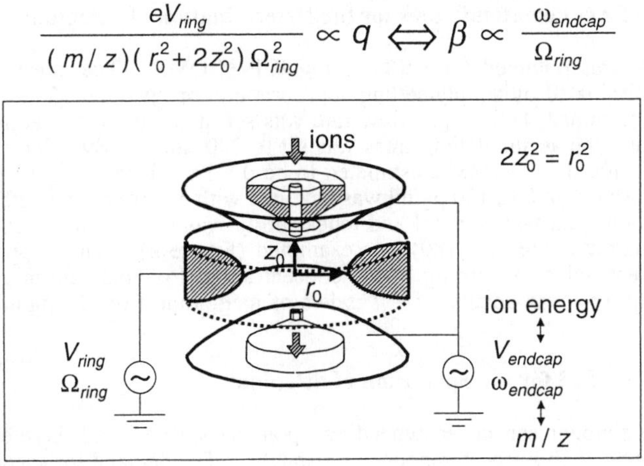

$$\frac{eV_{ring}}{(m/z)(r_0^2+2z_0^2)\Omega_{ring}^2} \propto q \Longleftrightarrow \beta \propto \frac{\omega_{endcap}}{\Omega_{ring}}$$

Figure 1 *Manipulation of 3DQMS*

The endcap electrode functions to resonate ions existing in the trapping area. Each m/z ion has a corresponding radio frequency ω_{endcap} to resonate itself, and this charateristic is used to collide the trapped ions with buffer gas molecules. Corresponding ω_{endcap} is applied to the endcap electrodes in order to resonate the ions of interest existing in the trapping area.   The collision can be controlled by the amplitude of V_{endcap} and the period of applying time ($V_{endcap} \le$ a few V).   By applying V_{endcap} and ω_{endcap}, the ions with the corresponding m/z collide with helium molecules, and the polyatomic ions with the same m/z will either break down or react to form different ions.

After the ions pass through the ion optics, they are introduced into the 3DQMS area. Meanwhile, the exit gate is closed, the entrance gate is opened and V_{ring} is applied to trap the ions.   FNF (filtered noise field) is applied to the endcap electrodes in order to isolate the ions with m/z of interest from other m/z ions.[11]   Once the analyte ions are trapped, the entrance gate is closed and appropriate V_{endcap} and ω_{endcap} can be applied to cause collision induced dissociation (CID) and reaction.   The condition for CID and reaction to occur, can be optimized by controlling the amplitude of V_{endcap}, the applying time of radio frequency ω_{endcap}, and the pressure inside the 3DQMS, which can be controlled by the flow rate of helium gas.   It is necessary to introduce helium buffer gas throughout the trapping process because it acts to decelerate the ions while trapping, and also plays an important role for CID and reaction.   After CID and reaction take place, the exit gate is opened, and the ions are measured by the detector.

FNF is used to isolate ions with m/z of interest.   If FNF is not applied, all ions trapped in the 3DQMS would eject to the detector, and a full spectrum will be obtained. However when FNF is applied to endcap electrodes, it is possible to eliminate unneeded ions and trap only ions with m/z of interest.   By isolating the ions of interest, the trapping efficiency is improved, and the obtained spectrum becomes simple.

As previously mentioned, each m/z has a corresponding radio frequency ω_{endcap} to resonate itself, and by applying that radio frequency at the optimum amplitude, collision occurs and polyatomic ions dissociate.   Collision in the trapping area not only dissociates the polyatomic ions, but also induces active reaction.   Under optimized conditions, polyatomic ions that interfere with the analyte, either dissociate or react to form other ions, and spectral interferences on the analyte are eliminated.

3   EVALUATION OF ICP-3DQMS

3.1   Elimination of Ar₂ Polyatomic Ions for the Determination of Selenium

A calibration curve was obtained for Se(80) using ICP-3DQMS.   The interference of $Ar_2(80)$ could be eliminated by simply setting appropriate trapping time.   The optimum conditions were determined; helium gas flow rate was set at 2.5 ml/min, trapping time was set at 50 ms, and the accumulation times was set to 200 times.   By using this flow rate, the pressure inside the 3DQMS is estimated to be 0.8 Pa.   Under this condition, a linear calibration curve from 0 to 100 pg/ml was obtained with a correlation coefficient of 0.999.   The detection limit was 6 pg/ml and repeatability was 4 - 5 % using 100 pg/ml solution.   The dynamic range of Se(80) was examined (Figure 2).   The trapping time was set at 4 different values according to the concentration, to avoid saturation at the trapping area.   Linearity was obtained for 7 orders of magnitude from 20 pg/ml to 100 µg/ml.

3.2   HPLC-ICP-3DQMS for the Speciation of Selenium

The toxicity of an element can differ depending upon the species, and therefore, it is important to be able to analyze the compounds separately.   The demand for speciation is great, and HPLC-ICP-MS is a powerful tool for these applications.[12]   There is a variety of liquid chromatography that can be used for speciation, and ion-pair reversed-phase chromatography was used for this experiment.

A LiChrosorb PR18 reversed-phase column was used.   Ion pair regents for

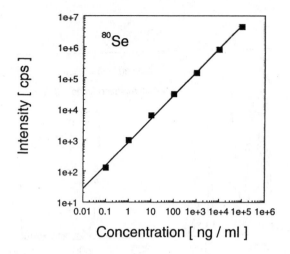

Figure 2 *Dynamic Range of Selenium (m/z =80)*

butanesulfonate and tetramethylammonium hydroxide were mixed and added to a mobile phase in order to simultaneously analyze cations, anions and neutral species. The sample injection volume was 100 µl, and the flow rate of the mobile phase was 1.0 ml/min. Table 1 describes the operating parameters for this experiment. Data was obtained for ICP-QMS and ICP-3DQMS for comparison purposes. Two inorganic species and six organic species containing selenium in different forms, were used for this experiment.

First, a 50 ng/ml mixture of 8 selenium standards was measured using HPLC-ICP-QMS.[13] Using the ICP-QMS, the optimum mass to measure selenium is m/z 82 (isotopic abundance: 9 %), because the major isotope of m/z 80 (isotopic abundance: 50 %) suffers from polyatomic ion interference of $Ar_2(80)$.

After optimizing the conditions, a chromatogram was obtained for a 100 ng/ml mixture of 8 selenium standards using HPLC-ICP-3DQMS (Figure 3). It was necessary

Table 1 *Operating Conditions for HPLC and ICP-MS Instruments*

HPLC

Column	LiChrosorb PR18 (250×4.6 mm I.D., 5 µm particle size)
Mobile phase	2.5 mM sodium 1-butanesulfonate –
	8 mM tetramethylammonuim hydroxide –
	4 mM malonic acid – 0.05 % methanol, pH 4.5
Flow rate	1.0 ml/min
Injection volume	100 µl
Column temperature	25 °C

ICP-MS

	QMS	3DQMS
Forward RF power	1300 W	1300 W
Plasma Ar flow	15.0 l/min	16.0 l/min
Auxiliary Ar flow	1.0 l/min	1.2 l/min
Nebulizer Ar flow	1.18 l/min	1.3 l/min
Integration time	100 ms	optimum

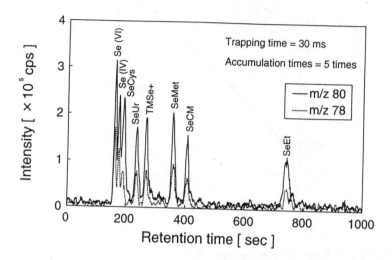

Figure 3 *HPLC-ICP-3DQMS Chromatogram obtained from a mixed Selenium Standard Solution (m/z =80)*

to shorten sampling interval because the HPLC produces transient signal, thus the trapping time and the accumulation times were optimized. When the concentration is high enough, it is possible to shorten sampling interval. Se(80) and Se(78) were measured alternatively during one injection of chromatographic separation, and the peaks for the two isotopes followed the natural isotopic abundance of selenium (Se(80) : Se(78) = 50 : 24). This result indicates that the peaks obtained in this measurement are due to selenium compounds, and polyatomic interferences are eliminated.

3.3 Elimination of ArO and CaO Polyatomic Ions for the Determination of Iron

The major isotope for ArO and CaO polyatomic ions are m/z 56, and these ions are known to interfere with the major isotope of Fe. Ca 100 µg/ml, Fe 100 ng/ml and blank solutions were measured using ICP-QMS. When the blank solution was measured, there was a peak on m/z 57, due to ArOH. When Ca 100 µg/ml was measured, a larger peak appeared on m/z 57, and this was due to the sum of ArOH and CaOH. The intensity of ArO on m/z 56 was estimated to be equivalent to approximately 125 ng/ml Fe, and the intensity of CaO was equivalent to approximately 175 ng/ml Fe.

The same solutions were measured using ICP-3DQMS. The sample was trapped for 50 ms, and CID was not applied. Significant difference between the chromatograms obtained by ICP-QMS and ICP-3DQMS were observed. The intensity of Fe was almost the same, but the intensity of ArO was smaller, and when Ca 100 µg/ml was measured, the peak on m/z 56 was smaller, and instead, a large peak appeared on m/z 57. The intensity of ArO was estimated to be equivalent to approximately 10 ng/ml Fe, and CaO was equivalent to approximately 80 ng/ml Fe.

The next attempt was to find the CID characteristics. CID time was varied to plot the analytical curves for Fe+ArO, CaO+ArO and ArO, using the same solutions. It was observed that Fe, ArO and CaO all decrease as the CID time gets longer. CaO decreased immediately after several hundred ms, and the CaO+ArO curve overlapped with the ArO curve, which suggests that CaO had disappeared. The CID effect on CaO was investigated in more detail between 0 and 400 ms, plotting CaO(56), CaOH(57) and the

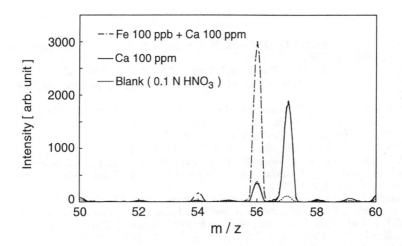

Figure 4 *ICP-3DQMS Mass Spectra using CID for the Determination of Iron*

sum of CaO+CaOH. From the curve, it was observed that CaO abundance becomes negligible at 300 ms. It was also observed that the intensity of CaOH increases as the intensity of CaO decreases, and the sum of CaO+CaOH was almost constant. This result indicates that CaO reacted with H to form CaOH, and thus the polyatomic ion interference due to Ca had disappeared from m/z 56.

Finally, chromatographic spectra were obtained for blank, Ca 100 μg/ml and the mixture of Fe 100 ng/ml and Ca 100 μg/ml (Figure 4). Trapping time was set at 50 ms, and the CID time was set at 300 ms. The peak for Ca 100 μg/ml on m/z 56 overlaps with the peak for blank on m/z 56, which indicates that all of CaO polyatomic ion intereference has disappeared from m/z 56. As mentioned, the CaO had reacted to form CaOH during CID, and a large peak was observed on m/z 57 to support the reasoning of reaction within the trapping area. Compared with the spectrum obtained for Fe 100 ng/ml without applying CID, the Fe signal intensity did not change when Ca 100 μg/ml matrix was added, and even if 300 ms of CID was applied. The peak for Ca 100 μg/ml and the peak for Fe 100 ng/ml+Ca 100 μg/ml completely overlap on m/z 57, and this result also supports there is no signal due to CaO on m/z 56.

4 CONCLUSION

The ICP-3DQMS instrument was evaluated for the analysis of elements that suffer from spectral interferences due to molecular ions.

The interference of $Ar_2(80)$ was eliminated using ICP-3DQMS by simply setting appropriate trapping time, and Se could be measured at m/z 80, the major isotope of Se. The detection limit of Se(80) was 6 pg/ml under optimum conditions, and the dynamic range of 7 orders of magnitude was achieved by using different trapping times according to the concentrations. The repeatability was 4 – 5 % by %RSD.

Ar_2 polyatomic ions were eliminated and Se ions could be detected at m/z 78 and 80 during chromatographic separation of Se compounds using HPLC-ICP-3DQMS.

The interference of CaO(56) was eliminated without losing the sensitivity of Fe(56), using ICP-3DQMS by keeping the ions trapped in 3DQMS for several hundred ms. CaO(56) ions did not dissociate, but simultaneously reacted to form CaOH(57) as ion trapping was taking place.

References

1. R. S. Houk, V. A. Fassel, G. D. Flesch, H. J. Svec, A. L. Gray and C. E. Taylor, *Anal. Chem.*, 1980, **52**, 2283.
2. F. Vanhaecke, L. Moens, R. Dams and P. Taylor, *Anal. Chem.*, 1996, **68**, 567.
3. D. P. Myers, G. Li, P. Yang and G. M. Heiftje, *J. Am. Soc. Mass Spectrom.*, 1994, **5**, 1008.
4. C. J. Barinaga and D. W. Koppenaal, *Rapid Commun. Mass Spectrom.*, 1994, **8**, 71.
5. M. A. Vaughan and G. Horlick, *Appl. Spectrosc.*, 1986, **40**, 434.
6. S. H. Tan and G. Horlick, *Appl. Spectrosc.*, 1986, **40**, 445.
7. K. Sakata and K. Kawabata, *Spectrochim. Acta*, 1994, **49B**, 1027.
8. S. D. Tanner and V. I. Baranov, in *Plasma Source Mass Spectrometry: New Developments and Applications*, eds. G. Holland and S. D. Tanner, Royal Society of Chemistry, Cambridge, 1999, p. 46.
9. Y. Takada, M. Sakairi and Y. Ose, *Rev. Sci. Instrum.*, 1996, **67**, 2139.
10. J. F. J. Todd, in *Practical Aspects of Ion Trap Mass Spectrometry*, eds. R. E. March and J. F. J. Todd, CRC Press, Inc., New York, 1995, Volume I, Chapter 1, p. 3.
11. D. V. Kenny, P. J. Callahan, S. M. Gordon and S. W. Stiller, *Rapid Commun. Mass Spectrom.*, 1993, 7, 1086.
12. R. Ritsema and O. F. X. Donard, in *Sample Handling and Trace Analysis of Pollutants: Techniques, Applications and Quality Assurance*, ed. D. Barcelo, Elsevier Science, Amsterdam, 1999, Chapter 21, p. 1003.
13. J. Zheng, M. Ohata, N. Furuta and W. Kosmus, *J. Chromatogr. A*, 2000, **874**, 55.

Section 3

Reaction Cells for ICP-MS

REACTION CHEMISTRY AND COLLISIONAL PROCESSES IN MULTIPOLE DEVICES*

Scott D. Tanner, Vladimir I. Baranov and Dmitry R. Bandura

PerkinElmer-SCIEX, 71 Four Valley Drive, Concord, Ontario Canada L4K 4V8

1 INTRODUCTION

In the past several years, collision cells and reaction cells have gained deserved attention for their potential in alleviating plasma source isobaric interferences. The fundamentals of ion-molecule collisional and reaction processes are understood in certain applications; organic tandem mass spectrometry [1,2], radiation chemistry [3], and fundamental kinetic studies [4-7]. Because the elemental analytical application is relatively new, there is some confusion as to the relative efficiencies of these processes. Hence there is some uncertainty in the selection of appropriate analytical conditions. It is the intent of this contribution to provide a cursory review of the microscopic interaction kinetics with a view to providing guidance on the type and pressure of collision/reaction gases and the selection of ion energies (lens voltages). Specifically, we consider the efficiency of ion-molecule reaction chemistry relative to that of Collision Induced Dissociation (CID), and consider the efficiency compromise involved in the instance that kinetic energy discrimination is used to suppress the appearance of ions produced in reactions within the cell. It should be noted that the discussion herein is limited to consideration of bimolecular processes; different energetic considerations apply to trimolecular (clustering/association) reactions.

2 ICP-MS APPLICATIONS

The first application of a collision cell for ICP-MS [8] proposed the collisional fragmentation of polyatomic ions at conventional collision energies (ca. 50 eV) with an inert gas (N_2, Ar) at relatively low pressure (equivalent to less than 1 mTorr). However, it was shown that ion loss rates are comparable to fragmentation rates, allowing the conclusion that "large gains in metal ion to molecular ion ratios will not be possible …using…a collision cell for dissociation". On the other hand, provision of a reactive gas (O_2) gave high efficiency of chemical distinction (Tb^+ was oxidized while CeO^+ did not react further). It was suggested that "ion molecule chemistry may provide a way around persistent interferences" [8]. Shortly thereafter, Rowan and Houk [9] showed that appropriate selection of the collision gas allowed specific reactive removal of isobaric interference ions. Several years later, Koppenaal et al. [10-13] showed high efficiency of chemical resolution in reactions with adventitious water in an ion trap and with H_2 added either to the trap or to a 2D octopole collision cell. Where most of these early works

*Adapted in part from Bandura et al, 2001, Fresenius' J. Anal. Chem (in press)

targeted reactive removal of argide interferences, Eiden *et al.* [14] took advantage of the specificity of oxidation to resolve atomic isobars through removal of either the interference or analyte (the definition of which depends on the application) to a different mass.

As of the Durham Conference (September 2000), three manufacturers offered collision and/or reaction cells in combination with quadrupole mass spectrometers: Platform (Micromass), ELAN DRC (PerkinElmer-SCIEX) and ExCell (TJA Solutions). Agilent has since described their 7500c. As noted in the contribution by Furuta *et al.* in this volume [15], Hitachi has combined the function of a collision/reaction cell with a mass analyzer in a 3DQMS in a manner similar to that described by Koppenaal [10-14]. The 3D cell differs in fundamental operation from the 2D cells, as confinement in the third dimension allows energy accumulation through consecutive "pumping" collisions, which facilitates high energy Collision Induced Dissociation (CID). Further, the 3D trap is commonly operated at lower pressure (and longer storage time), so that there are fewer collisions per rf cycle and hence the collisions are, on average, of higher energy. Finally, auxiliary "tickle" frequencies are often used to further heat the ion that is intended to be fragmented. While the microscopic collision dynamics discussed in this contribution apply to all of these devices, the discussion will focus on the 2D instance.

Under conditions that provide for high efficiency (and specificity) of reaction, the analyte/isobar ratio can be improved by many orders of magnitude. Figure 1 shows two spectra: one obtained under conventional ICP-MS conditions (hot plasma, no reaction gas) and one using NH_3 as the reaction gas in a Dynamic Reaction Cell. The argide ion signals (Ar^+ and ArH^+) are suppressed by up to 9 orders of magnitude, due to the efficient reaction of ArX^+ ions with ammonia. Importantly, many analyte ions, including Ca^+ and K^+, are not reactive with ammonia. Thus, the isobaric interference is essentially eliminated with minimal effect on the sensitivity to the non-reactive analyte ions. This is shown in Figure 2 for which it is seen that the elimination of the argide ions allows the determination of Ca and K at the low and sub-ppt level. This is the conventional and simple mode of operation of a reaction cell: removal of the isobaric interference through a reaction which is specific for the isobar. Alternatively, chemistry may be selected to allow modification of the analyte ion to a different (and not interfered) mass, as is discussed in the contribution by Bandura *et al.* to this volume [16].

3 COLLISIONAL FRAGMENTATION AND CHEMICAL REACTION

Strictly speaking, a collision cell is readily distinguished from a reaction cell. In a generic sense, the former is intended to promote collisional fragmentation, usually accomplished at relatively high energy and relatively low pressure (single, or just a few, collision conditions). A reaction cell is operated at higher pressure (many collisions) and hence lower energy (preferably near-thermal conditions). It can be said that a collision cell is a relatively low efficiency device designed to promote endothermic fragmentation, whereas a reaction cell is intended to promote exothermic chemical reaction. Fragmentation is non-specific. Chemical reaction, at least for thermal ion-molecule processes, is specific through the thermochemistry and kinetics.

Of course, chemical reactions can also be supported in a collision cell, and the recent trend towards operation of collision cells at higher pressures (to take advantage of collisional focusing [1,2,17]) blurs the distinction between collision cells and reaction cells. We propose a distinction based on ion energy distributions: if the energy distributions of analyte ions and ions produced in reactions in the cell are indistinguishable at the exit of the cell, the cell is a reaction cell. The corollary of this is that, in a collision cell, analyte and product ions may be distinguished on the basis of their kinetic energies. This is an important distinction.

Kinetic energy discrimination is usually enacted through adjustment of the collision/reaction cell rod offset potential to a value more negative than that of the mass analyzer rod offset, or by establishment of a potential barrier at an intermediate lens. Because the initial kinetic energy distribution of analyte ions (ions derived from the plasma) is damped by collisions, as will be discussed below, the distinct energy distributions of analyte ions and cell-produced ions becomes blurred under multiple collision conditions. Therefore, kinetic energy discrimination may be applied only under conditions where sufficiently few collisions occur that the energy distributions do not overlap. Since the efficiency of reaction is exponentially proportional to the number of collisions, kinetic energy discrimination incurs a limitation on efficiency of plasma-based isobar resolution.

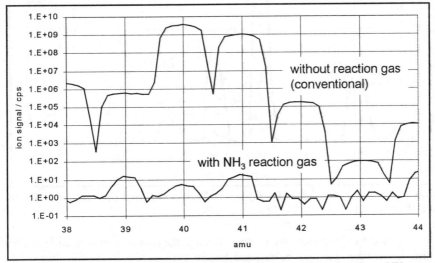

Figure 1 *Spectra for normal analytical "hot" plasma with and without NH₃ reaction gas. ELAN DRC, 1150 W, 0.92 L/min nebulizer flow, De-ionized Water (DIW) sample*

4 ION-MOLECULE REACTION EFFICIENCY AND SPECIFICITY

The efficiency of an ion-molecule reaction, at the microscopic level, is the fraction of collisions that yield reaction; it is normally defined as the ratio of the reaction rate constant to the collision rate constant. For the instance where an isobaric interference is reacted in order to remove its interference on an analyte ion, we define the extent of reaction as the fraction of the isobar signal that is removed. If 1% of the isobar signal remains under analytical conditions, the extent of reaction is 99%, assuming that the ion signal recorded is proportional to the density of the ions in the cell. We define the efficiency of interference rejection (or chemical resolution) as the ratio of improvement of the ratio of the analyte signal to the background signal for conditions without reaction to conditions with reaction. If the isobar reacts and the analyte signal remains constant as a function of reaction gas, and provided that no new isobaric interferences are produced in the cell, the efficiency of chemical resolution is the same as the extent of isobar reaction. In many instances, the efficiency required of the reaction/collision cell is many orders of magnitude. For example, the ion signal at m/z=40, dominated by Ar^+ under non-reactive collisions, is greater than 10^9 cps. When it is of interest to determine Ca at sub-ppt levels, an efficiency of some 9 orders of magnitude (99.9999999%) is required. Such efficiency requires four concomitant conditions: the isobar must react rapidly with the reaction gas, the analyte must be non-reactive with the reaction gas, new isobaric interferences produced within the cell must be avoided, and multiple collisions of the ions with the reaction gas must be provided.

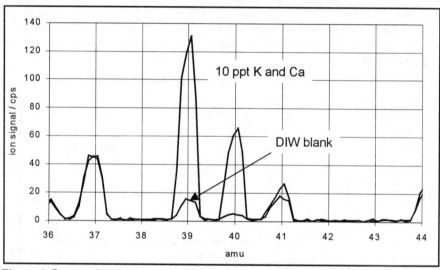

Figure 2 *Spectra (conditions as for Figure 1) for 10 ppt K and Ca, and for Deionized Water (DIW) blank*

The first two conditions are obtained through the provision of thermal conditions so that the reaction gas may be selected on the basis of the thermochemistry. The enthalpy of reaction, ΔH_r, is defined as the difference between the sum of the heats of formation, ΔH_f, of the products of the reaction and those of the reactants:

$$\Delta H_r = \sum \Delta H_f \text{(products)} - \sum \Delta H_f \text{(reactants)} \qquad (1)$$

These parameters are defined at the measurement temperature, typically room temperature. Properly speaking, the reaction energy is the free energy of reaction, ΔG_r, which includes an entropy term, ΔS_r:

$$\Delta G_r = \Delta H_r - T \Delta S_r \qquad (2)$$

However, in most small particle transfer reactions, the change in entropy is negligible and it is common to consider only the reaction enthalpy. (The entropy term is substantial for clustering/association reactions.) If the reaction is exothermic, $\Delta H_r<0$, the reaction releases energy to the environment and is termed "thermodynamically allowed". If the reaction is endothermic, $\Delta H_r > 0$, the reaction absorbs energy from the environment, and is "not allowed". When a thermal ion-molecule reaction is thermodynamically allowed, it may proceed and is usually efficient (occurs on 10% or more of collisions) if it does so because of the electrostatic interaction of the ion with the neutral. When a thermal reaction is not allowed, it generally does not take place or is very slow. The thermodynamics thus provide specificity of thermal reactions. If the reaction gas is chosen such that it has an exothermic channel with either the analyte ion or the isobaric interference ion, and thermal conditions prevail, the allowed reaction is likely to take place and the disallowed reaction is not, and chemical resolution is achieved. An example is the chemical resolution of Ca^+ from Ar^+ using NH_3 as the reaction gas. The ionization potential of Ar (15.76 eV) is greater than that of NH_3 (10.16 eV), which in turn is greater than that of Ca (6.11 eV). Thus, NH_3 is sandwiched between Ar and Ca, and charge transfer is allowed for Ar^+ but disallowed for Ca^+. The Ar^+ reaction is fast and the Ca^+ reaction is exceptionally slow (if it proceeds at all). Passing the ion beam through a cell containing NH_3 as the reaction gas therefore promotes reactive loss of Ar^+ while Ca^+ is essentially unaffected, thus achieving chemical resolution. It is important to recognize that the provision of (near) thermal conditions is essential: if the collision energy is sufficiently large to overcome the endothermicity of the disallowed (under thermal conditions) reaction, the specificity is forfeit.

The third condition is obtained either by preventing the formation of new isobaric ions within the cell (through the selection of the reaction gas and/or application of a bandpass within the cell to interrupt the secondary chemistry) or by distinction of the isobar and analyte after the cell (for example, by application of a kinetic energy barrier).

The fourth condition, provision of multiple collisions, is a result of the statistical nature of the collision events. When the average ion suffers X collisions, a fraction will suffer significantly more collisions and a fraction will suffer significantly fewer (even zero) collisions. Even for a fast reaction that occurs on every collision, a large average number of collisions is required in order to ensure that most (for example, all but 1 per 10^9) ions have at least one collision. For the bimolecular reaction of an ion A^+ with a neutral B:

$$A^+ + B \rightarrow C^+ + D \tag{3}$$

and the reaction rate is given by:

$$\frac{d[A^+]}{dt} = -k[A^+][B] \tag{4}$$

where t is the reaction time and k is the reaction rate constant (typically in units of cm^3 $molecule^{-1}$ $second^{-1}$). For a reaction cell, t is not independent of [B], but it is also not a strong function if approximately thermal conditions apply (i.e., the reaction time is a strong function of [B] only at low pressure, where the ion energies have not yet been damped). Therefore, under thermal conditions, a plot of the residual ion signal for A^+ is semilogarithmic linear with respect to the concentration of B, according to:

$$\ln[A^+] = \ln[A^+]_0 - kt[B] \tag{5}$$

where $[A^+]_0$ is the ion signal for $[B]=0$. The slope of the semilogarithmic linear decay of $[A^+]$ is proportional to $k\,t$. Of course, the function

$$k\,t\,[B] = (\text{rate constant}) \times (\text{reaction time}) \times (\text{neutral density}) \qquad (6)$$

is simply the average number of reactive collisions.

Since the extent of reaction (i.e., $[A^+]/[A^+]_0$) is exponentially dependent on the number of reactive collisions (which is linearly dependent on the total number of collisions and hence is a function of the cell pressure), the efficiency of reaction is also exponentially dependent on the rate constant. This is shown in Figure 3, where the ion signal, $[A^+]$, is plotted against cell pressure for three different reaction rate constants. These are chosen for a fast reaction that proceeds on every collision ($k/k_{ADO}=1$, where k_{ADO} is the collision rate constant and will be described later), for a moderate rate that occurs with 50% efficiency ($k/k_{ADO}=0.5$) and for a "relatively slow" reaction ($k/k_{ADO}=0.1$). Where the unit efficiency reaction yields a signal suppression of 5 orders of magnitude at ca. 10 mTorr, the 50% efficient reaction yields only 2.5 orders of magnitude, and the 10% efficient reaction provides only a half order of magnitude. The extent of reaction is thus proportional to 10^k. Thus, a difference of a factor of 10 in rate constant for the isobar and analyte ions is sufficient to provide a high degree of chemical resolution between the analyte and the isobar (an improvement of 5 orders of magnitude in the signal/background at 10 mTorr, or 10 orders of magnitude at 20 mTorr). The difference in rate constants is often much greater than this; for example, the rate constant for the reaction of Ar^+ with NH_3 is 1.6×10^{-9} cm^3 s^{-1} [18] while that for Ca^+ is $< 10^{-13}$ cm^3 s^{-1}; a difference of more than 4 orders of magnitude. Thus, Ca^+ is efficiently resolved from Ar^+ with ammonia as a reaction gas, as shown in the experimental data in Figure 4, where the Ar^+ signal is reduced by more than 8 orders of magnitude, limited by the Ca contamination of the blank (ca. 3 ppt) while the Ca^+ signal is hardly affected. It is this efficiency and specificity of ion-molecule chemistry that allows the determination of trace levels of Ca^+, as was shown in Figures 1 and 2.

5   ENERGY TRANSFER IN COLLISIONAL PROCESSES

As shown in Figures 3 and 4, the ion signal is not a linear function of pressure (flow) at low reaction gas flow. This is a result of two phenomena. One is collisional focusing: as the pressure is increased, the ions migrate to the axis of the multipole, resulting in an increase in ion transmission efficiency through the on-axis exit aperture. The detected ion signal thus increases until scattering losses dominate. The subject has been discussed by Douglas and French [17], Krutchinsky *et al.* [2] and in a recent review by Douglas [1]. The second phenomenon is related: collisions with the gas cause sequential damping of the ion energies, and the collision cross-section increases as the energy approaches thermal. The number of collisions that an ion experiences increases because of the increase of collision cross-section and because the velocity of the ion decreases (increasing the reaction time). In addition, the lower energy means that the collision energy does not contribute to the reaction energy, and the specificity provided by the thermochemistry is obtained.

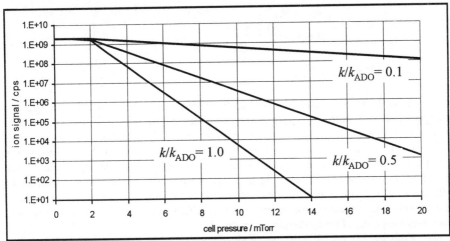

Figure 3 *Calculated ion signal dependence on cell pressure for three assumed reaction efficiencies. k is the rate constant. k_{ADO} is the collision rate constant (defined in Equation (14)). The model is described in Section 8.*

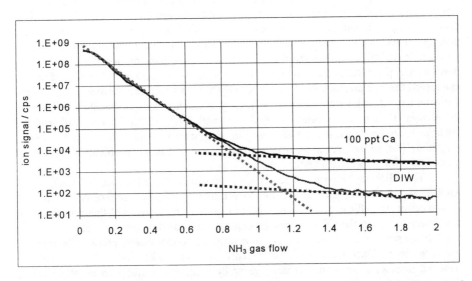

Figure 4 *Experimental reaction profiles for m/z=40 for 100 ppt Ca and for Deionized Water blank. The dashed lines approximate the linear portion of the signal slope, and are proportional to the rate constants. ELAN DRC, 1150 W, 0.92 L/min nebulizer flow*

The reader is referred to the review of collisional processes by Douglas [1]. In an elastic (no internal excitation) non-reactive collision of an ion of mass m_1 and kinetic energy E_1 with a stagnant ($E_2=0$) neutral of mass m_2, the energies after collision are given by:

$$E_1' = E_1 \left[\frac{m_1^2 + m_2^2}{(m_1 + m_2)^2} \right] \tag{7}$$

$$E_2' = E_1 - E_1' \tag{8}$$

Thus the ion loses energy according to the reduced mass of the collision partners: a larger mass ratio neutral/ion increases the rate of energy damping of the ion. Subsequent collisions reduce the ion energy to the thermal limit, at which point the ion simply executes an essentially 'random walk' through the cell. A large ion energy at the entrance to the cell (source potential plus expansion energy minus cell offset potential) requires more collisions for energy damping. Higher initial energy also results in a reduction of efficiency because the number of collisions is reduced and the probability of reaction during collision decreases, compromises the specificity of the thermal chemistry, and increases the potential for sputtering cell materials.

Equation (7) describes the loss of kinetic energy of the ion in a non-reactive elastic collision. On the other hand, if reaction takes place with transfer of a small particle (electron, proton or H-atom), the resultant product ion has a kinetic energy given approximately by Equation (8). Because the reactant neutral may be considered stagnant, the product ion generally has substantially lower kinetic energy than the reactant ion. Hence, prior to thermalization, product ions and incident (source-derived) ions have distinct energy distributions. As the number of collisions that the source-derived ions undergo increases, their energy distribution narrows and moves to lower energy until these ions are energetically indistinguishable from ions produced by reaction in the cell. Accordingly, if kinetic energy discrimination of source-derived and cell-produced ions is employed, the cell pressure must be sufficiently low as to ensure the persistence of the distinct energy distributions. Since the efficiency of reaction is exponentially dependent on the number of collisions, kinetic energy discrimination necessarily limit the efficiency of the cell. It can be shown that the efficiency in this instance is limited to about 4 orders of magnitude (for an ion/neutral mass ratio of ca. 2 and unit reaction efficiency, or a mass ratio of 20 with 50% reaction efficiency).

The efficiency of chemical resolution is limited by the efficiencies of reaction, the number of collisions (pressure) and by the production of new interferences in the cell (which offsets the efficiency of isobar reaction). It is to be recognized that production of new interferences in the cell requires the concomitant confinement of reactive precursor ions. If the precursor ion and the reactant neutral are concomitantly present, the reaction will take place. Once the isobaric interference product ion has formed, it may only be removed through subsequent reaction (generally a process unlikely to be effective) or discrimination post-cell. In the instance of kinetic energy discrimination, the efficiency of the primary reaction is necessarily low in order to retain differentiation of the kinetic energy distributions. Then, the efficiency of production of new interferences is correspondingly low, and the overall efficiency is determined by the conditions that maintain distinct energy distributions in the cell. Alternatively, it is to be recognized that, in most instances, new interferences are produced in a sequence of reactions involving at least one precursor ion having a substantially different mass than the mass of interest. Rapid removal of such precursor ions, for instance through application of appropriate bandpass conditions using the multipole stability characteristics, aborts the secondary chemistry and prevents the formation of the new interference ions [19]. With *in situ* suppression of the secondary chemistry, there is no need for kinetic energy discrimination: the cell may be operated under higher pressure (more collisions) conditions that provide higher efficiency of reaction and allow approach to thermal conditions which validate application of the ion kinetic and thermochemical databases for the selection of the appropriate reaction gas.

6 COLLISIONAL FRAGMENTATION (CID)

Collisional fragmentation of a polyatomic ion is obtained through the transfer of a portion of the collision energy to internal degrees of freedom of the polyatomic. Generally, the transfer may be to vibrational modes, and accumulation of sufficient energy in a single vibrational mode to exceed the bond strength of the mode may result in rupture of the bond. We noted in Section 2 that CID in a three dimensional ion trap may be efficient because of the confinement of the ions in three dimensions, which allows sequential pumping to the dissociation limit. It was also noted that auxiliary excitation may be used in these devices to pump the collision energy, further improving the efficiency of CID. Further, CID is promoted under conditions of fewer than one collision per rf cycle (so that the ion gains the full strength of the rf energy between collisions). In a two dimensional multipole (i.e., a collision cell or reaction cell), the CID process is relatively inefficient. Sequential pumping of the vibrational modes is ineffective because the ions are unconstrained in the third dimension, and loss rates are comparable to fragmentation rates [8]. Though CID is of paramount importance for organic tandem MS applications, the efficiency required of the process is modest: seldom are more than 90% of the polyatomic ions fragmented, since it is only necessary to produce sufficient fragment ions in order to identify the parent polyatomic.

Equations (7) and (8) describe the energetics of collisions in the LAB frame (the energies that are apparent to an outside observer). The transfer of energy to internal modes of excitation is best understood in the center-of-mass frame of reference. For an ion of mass m_1 which is greater than the neutral mass, m_2, and assuming a stationary neutral molecule, the energy in the center-of-mass, E_{COM}, is proportional to the LAB energy of the ion, E_1, according to:

$$E_{COM} = E_1 \left[\frac{m_1}{(m_1 + m_2)} \right] \tag{9}$$

For the same collision pair, the maximum amount of energy that may be converted into internal excitation, $E_{int,max}$, is given by:

$$E_{int,max} = E_1 - E_{COM} = E_1 \left[\frac{m_2}{(m_1 + m_2)} \right] \tag{10}$$

For single collision fragmentation, it is thus desirable to use a neutral having as large a mass as possible. Unfortunately, this condition also maximizes scattering losses. Let us look at the energetics of fragmentation of Ar_2^+ (a particularly favourable case, since the bond strength is only 1.2 eV [20]). Consider the case where the argon dimer ion (m/z=80) gains ca. 4 eV from the supersonic expansion through the interface, obtains a kinetic energy derived from the plasma potential offset of ca. 3 eV, the plasma interface is at ground potential and the collision/reaction cell offset potential is –1 V. Then the Ar_2^+ ion has an initial energy (assuming that it has penetrated sufficiently far into the cell that the cell offset potential defines the potential near the ion) of ca. 8 eV. We tabulate in Table 1 the maximum amount of internal excitation energy that the argon dimer ion gains in sequential collisions with various collision gases, by recursive application of Equations (7), (9) and (10), assuming no transfer of energy to the internal excitation of the neutral. Two limiting cases are considered: (1) prior collisions are elastic (no internal excitation is obtained in prior collisions), so that the accumulated internal energy is simply the maximum internal energy on each collision, and (2) the case where the maximum internal excitation is achieved on each collision so that the maximum case of pumped excitation is considered. Clearly, single collision fragmentation of Ar_2^+ is achievable (though not necessarily obtained) using Ar as the collision gas. He and H_2 are incapable of facilitating single collision fragmentation under the conditions described. Pumped sequential excitation to fragmentation using He as the collision gas requires a minimum of 4 collisions. If pumped fragmentation is an unlikely process in a two dimensional multipole, the results suggest that fragmentation using He as the collision gas is in fact achieved through incursion of the plasma gas (which includes a significant quantity of Ar, O and H) into the cell (so that the Ar plasma gas is the effective collision partner).

For the instance under consideration, the alternative process for the removal of the argon dimer ion is chemical reaction. There are no exothermic reaction channels for Ar_2^+ with either Ar or He. However, reaction with H_2 is relatively rapid (yielding ArH^+), having a thermal rate constant of 4.9×10^{-10} cm^3 s^{-1} [18]. Accordingly, it is far more likely that the removal of the argon dimer ion is achieved through chemical reaction than through collisional fragmentation. (An alternative reaction gas is CH_4, for which Ar_2^+ reacts by charge transfer with a rate constant of 5.7×10^{-10} cm^3 s^{-1}, and with which Se^+ has no bimolecular reaction [21]).

Table 1 *Maximum internal excitation energies of Ar_2^+ for sequential collisions**

gas	mass	first collision	second collision	third collision	fourth collision
limiting condition: no accumulation of internal excitation					
H_2	2	0.20	0.19	0.18	0.17
He	4	0.38	0.35	0.31	0.29
Ar	40	**2.67**	**1.48**	0.82	0.46
limiting condition: maximum accumulation of internal excitation					
H_2	2	0.20	0.38	0.54	0.70
He	4	0.38	0.71	0.99	**1.24**
Ar	40	**2.67**	**3.26**	**3.39**	**3.42**

* a bold entry indicates that the maximum energy exceeds the bond dissociation energy of Ar_2^+ (D_0=1.2 eV)

7 ION-MOLECULE REACTION KINETICS

When viewed along the vector of the ion motion, a collision will take place if the ion and neutral are within an area known as the collision cross-section. The macroscopic analogue would describe the collision cross-section for billiard balls as $\pi(r_1 + r_2)^2$, where r_1 and r_2 are the radii of the billiard balls. The rate constant for a bimolecular reaction is given by:

$$k = \int P(\upsilon) \, \upsilon \, \sigma(\upsilon) \, d\upsilon \sim \sigma \, \upsilon \qquad (11)$$

where υ is the velocity, $P(\upsilon)$ is the probability (efficiency) of reaction at the given velocity, and $\sigma(\upsilon)$ is the collision cross-section which is a function of the velocity. The approximation $k \sim \sigma \upsilon$ is valid for the case where the velocity distribution is narrow (i.e., thermal). As given in Equation (5) in logarithmic form, the ion density (assumed to be proportional to the ion signal in ICP-MS) for A^+ reacting with neutral B is given by:

$$[A^+] = [A^+]_0 \, e^{-k\,t\,[B]} \qquad (12)$$

At high interaction energy ($E_1 > 1\,eV$), the collision can be regarded as a hard sphere process. That is, the ion and molecule behave as though they are billiard balls. The collision cross-section is approximated as π times the square of the sums of the atomic (or molecular) radii. Typical hard-sphere collision cross-sections are on the order of 20 to 70 $Å^2$ (2 to 7 x 10^{-15} cm^2). Treatment in this fashion is appropriate because at sufficient energy the ion passes by the neutral fast enough that it cannot affect the charge distribution in the neutral, and the charge-neutral interaction is weak. The Hard-Sphere rate constant, k_{HS}, approximates as:

$$k_{HS} = \upsilon \, \sigma_{HS} = (2 \text{ to } 7) \times 10^{-15}\,cm^2 \, \sqrt{2kT / m_{ion}} \qquad (13)$$

where $(2kT/m_{ion})^{1/2}$ is the ion velocity.

At lower energies ($E_1 < 1$ eV), the ion affects the charge distribution within the neutral (assumed at this point to have no permanent dipole moment), inducing a dipole. The strength of the induced dipole is proportional to the polarizability, α, of the neutral. The charge-induced dipole interaction increases the collision cross-section above the hard-sphere limit. The collision rate constant, k_L, derived using this approximation is known as the Langevin cross-section [4], and approximates the rate constant appropriate for the interaction of an ion with a neutral having no permanent dipole moment.

$$k_L = \upsilon \, \sigma_L = 2 \, \pi \, q \, \sqrt{\alpha / \mu} \qquad (14)$$

where q is the charge of the electron, α is the polarizability of the neutral and $\mu = (m_1 \times m_2)/(m_1 + m_2)$ is the reduced mass of the collision partners.

If the neutral has a permanent dipole moment, it is obvious that the ion can interact with both the permanent and induced dipoles. The ion-dipole interaction yet further increases the collision cross-section. If the ion "locks onto" the dipole vector, such that the dipole remains aligned with the ion at all times during close collision, the resulting rate constant is known as the Locked Dipole limit, and this overestimates the true collision rate. A refinement, described originally by Bass *et al.* [6] and refined (and corrected) in reference [7], recognizes that thermal rotation of the neutral reduces the effect of the dipole interaction. The collision rate constant determined according to the Average Dipole Orientation theory, k_{ADO}, well approximates measured values (i.e., it provides a good estimate of the true reaction rate for fast (unit efficiency) reactions).

$$k_{ADO} = \upsilon \, \sigma_{ADO} = \frac{2 \pi q}{\sqrt{\mu}} \left(\sqrt{\alpha} + C \mu_D \sqrt{2/\pi \, k \, T} \right) \tag{15}$$

where μ_D is the permanent dipole moment, k is Boltzmann's constant, T is the temperature and C is an empirical parameter having a tabulated value that is dependent on $\mu_D/\alpha^{1/2}$ (see references [6,7]).

The first term is simply the Langevin rate constant, the second introduces an inverse square root dependence of the rate constant on the temperature (note that the rate of an ion-molecule reaction is expected to decrease with an increase in temperature). A yet further refinement, the AADO theory, includes the conservation of angular momentum [7], but yields collision rate constants that are similar to those of the ADO theory.

An impression of the importance of the kinetic energy (or rather, the importance of reducing the ion kinetic energy, preferably to near thermal ~ 0.03 eV) might be gleaned from Figure 5. Here we plot the collision cross-section as a function of kinetic energy for the reaction of Ar^+ with three reaction gases. At high energy (> 1eV), the cross-section approaches the hard-sphere cross-section (28.3 Å^2, 49.1 Å^2 and 63.9 Å^2 for H_2, CH_4 and NH_3, respectively). At lower energy, the ion-induced dipole interaction (the Langevin

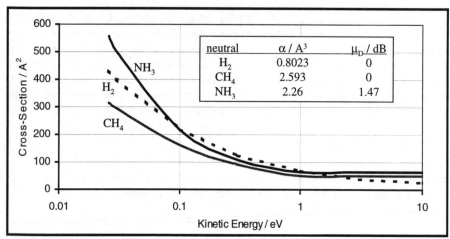

Figure 5 *Calculated collision cross-sections as a function of ion kinetic energy*

term) becomes significant, increasing the collision cross-section. Of the three gases shown, ammonia is the only neutral having a permanent dipole moment, with the result that the collision cross-section for Ar^+/NH_3 exceeds even that for the lightest gas (H_2, for which the reduced mass is the smallest and hence the cross-section should be expected to be greatest) at energies below 0.1 eV.

8 FACTORS AFFECTING REACTION EFFICIENCY

Arguably, the factors that most affect the efficiency of reaction in a multipole reaction cell include the reaction rate constants (cross-sections) and the number and energy of the collisions. These factors include by implication a number of other factors that may be selected by the user or are imposed by the cell design.

Of paramount importance is the reaction thermochemistry, which imposes a binary "go/no-go" on the reaction process. Under thermal conditions, if the reaction is exothermic, it probably proceeds and usually at a rate that approaches the collision rate. If a reaction is endothermic, the reaction usually does not take place or is very slow. Accordingly, the user should attempt to select a chemistry for which reaction with the analyte is endothermic (slow) and reaction with the isobar is exothermic (fast). Failing this, it is necessary to choose a chemistry for which the products of the analyte reaction are distinct and are not themselves interfered. Where possible, reference to thermal reaction rates reported in the literature [18] are the beacon for the analyst. Rate constants for many appropriate chemistries are known, and a concerted effort to extend the database are underway [22,23]. It should be recognized that the accuracy and relevance of reported rate constants is affected by the measurement technique. In the opinion of the authors, rate constants obtained using SIFT (Selected Ion Flow Tubes) or FA (Flowing Afterglow) techniques are preferable. In the absence of reliable kinetic data, resort may be made to calculated reaction energies. The thermochemistry related to analyte ions and feasible reaction gases are relatively well-known and tabulated [24,25].

The rate constant is, of course, a characteristic of the ion and the neutral. It is a function of the masses of the collision partners and of the electrical characteristics of the neutral (polarizability and permanent dipole moment). While the rate of reaction of an ion with a heavier neutral may seem to offer promise, it should be remembered that in a multipole reaction cell, scattering losses are also important, and these are more severe as the neutral mass approaches or exceeds that of the ion. Another factor that is less apparent is the number of degrees of freedom of the neutral, which increases with the number of atoms in the molecule. A more complex neutral may more effectively absorb the collision energy, and hence damp the ion energy more effectively. This is desirable to a point, but if the ion energy becomes truly thermal, the transmission efficiency may suffer as the ion "wanders" in the cell.

The collision rate constant plays a major role in determining the rate of energy damping (a larger collision rate means more collisions and more efficient energy damping). As the energy is damped to near-thermal, the collision rate increases with a concomitant increase in reaction efficiency.

The number and energy of collisions is, of course, strongly affected by the pressure of the reaction gas. Again, the effect is not linear at low pressure because the first step involves energy damping; after the energies are damped below ca. 1 eV, the number of collisions is nearly linearly dependent on the pressure of the gas (and so the extent of reaction is exponentially dependent on the pressure).

Other perhaps obvious factors that impact the reaction efficiency are the length of the reaction cell (nearly linear effect on the number of collisions, though there is a compromise related to transmission efficiency) and the amplitude and frequency of the rf applied to the multipole. The latter affects the rf contribution to the collision energy (the approach to thermal conditions), and also affects the number of collisions per rf cycle (a larger number leads to more effective thermalization) [26]. Of course, the amplitude and frequency of the rf also determine the Mathieu stability parameter q, which defines the low mass cutoff of the bandpass of the cell, and this affects also the extent of secondary reactions in the cell that otherwise may obfuscate the primary reaction efficiency.

We have developed a simple model to demonstrate these effects. The model makes some important simplifying assumptions that compromise the quantitative accuracy, but the qualitative aspects appear to be appropriate. The model determines the ion energy and the number of collisions that an ion experiences during transit through a multipole cell as a function of the pressure of the cell. The collisions are considered to be elastic (no internal excitation). It is assumed that only those ions having an axial energy greater than 0.03 eV succeed in transiting the cell (the model places this as a minimum energy, to avoid calculating Brownian motion). It assumes that the radial kinetic energy is equal to the axial kinetic energy; that is, it assumes that the rf contribution to the collision energy is thermalized with the axial energy. The results are calculated for an ion having m/z=40 (i.e., Ar^+), for a cell length of 12.5 cm operated at 1 MHz rf.

It can be shown that a higher initial kinetic energy requires more collisions before thermalization, and reduces the number of collisions that the ion experiences. A higher neutral mass damps the ion energy more rapidly. Interestingly, this results in an increase in the number of collisions at low pressure, but the higher thermal collision rate of a lower mass neutral eventually yields the higher number of collisions (at quite high pressure). In this model, the neutral polarizability has little effect on the rate of energy damping, but a large (positive) effect on the number of collisions. The permanent dipole moment also has little impact on the rate of energy damping, but also a significant (positive) effect on the number of collisions (perhaps slightly less significant than the polarizability in this regard). We attempt to demonstrate all of these effects simultaneously in Figure 6, which shows the calculated average kinetic energy at the exit of the cell, and the total number collisions during transit of the cell, as a function of the cell pressure for four reaction/collision gases. The kinetic energy is damped most rapidly by either CH_4 or NH_3. A cell pressure greater than 5 mTorr is required to thermalize the ions with He, and nearly 9 mTorr for H_2. In each instance, a point of inflection in the number of collisions *vs.* cell pressure is observed. At low pressure, the total number of collisions increases with pressure only slowly; above the point of inflection the number of collisions increases rapidly. The inflection points are indicated by arrows on the graphs. Clearly, the critical parameter is the reduction of the ion energy below ca. 0.2 eV (i.e., approaching thermalization). Once the ions are thermalized, the increase in collision cross-section due to the charge-induced dipole (or charge-dipole) interaction dominates. Since the reaction efficiency is exponentially dependent on the number of collisions, it is clear that it is desirable to approach thermalization, and a reaction gas having a larger polarizability and/or a permanent dipole provides significant advantage.

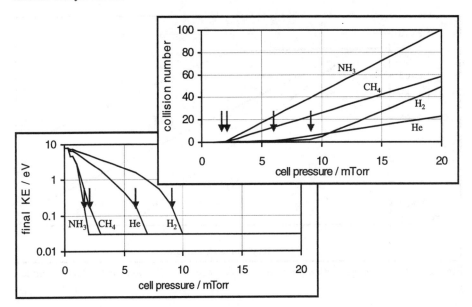

Figure 6 *Calculated final (at the exit of cell) kinetic energies and number of collisions as a function of cell pressure*

We recall that Figure 3 showed a relative plateau in ion signal at low cell pressure, followed by a semilogarithmic decay of ion signal that was proportional to the rate constant of reaction. The plateau in that instance is a direct result of the same phenomenon: the ions must first be thermalized before the reaction rate constant can be expressed in high efficiency of reaction. We now adapt the data of Figure 6 to predict "reaction profiles" for reaction of Ar^+ with H_2, CH_4 and NH_3, given in Figure 7. The exponential decay of the signals is determined for the total number of collisions multiplied by the reaction probability per collision (i.e., k/k_{ADO}). The reaction rate constants are taken from the compilation of Anicich [18]: 8.9×10^{-10} cm^3 s^{-1}, 9.8×10^{-10} cm^3 s^{-1} and 1.6×10^{-9} cm^3 s^{-1}, respectively for H_2, CH_4 and NH_3. The corresponding reaction probabilities (thermal conditions) are: 0.59, 0.88 and 0.81. The difference in reaction efficiency are enormous. For example, at 10 mTorr, NH_3 provides 5 orders of magnitude reaction, CH_4 only 3 orders and H_2 less than half an order. In particular, the efficiency of H_2 is much less than would be predicted from the comparative thermal rate constants because of the higher pressure required to thermalize the ions. It is probable that impurities in H_2 would have a disproportionate impact on the reaction efficiency. Since the reaction probability for H_2 is relatively high (0.59) under thermal conditions, the impact of the impurities is more likely to affect the rate of thermalization than the subsequent thermal reaction efficiency. (In other instances where the reaction probability with H_2 is low, the reaction efficiency of the impurities may be more important). If the impurity fraction is relatively high (for example, incursion of plasma gas into the cell), the higher mass impurities may be effective in accelerating the rate of thermalization, which would subsequently improve the efficiency of reaction with H_2.

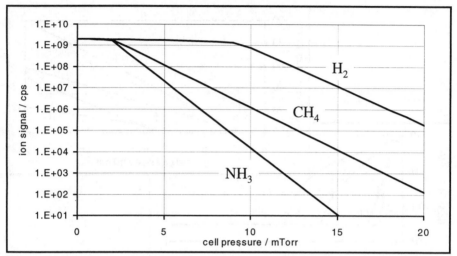

Figure 7 *Calculated reaction profiles as a function of cell pressure for three reaction gases*

Of course, the cell pressure is an adjustable parameter (and depends on the flow of gas into the cell and the conductance of the entrance and exit apertures of the cell), and operation at higher pressure improves the efficiency of the H_2 reactions. However, in the mass spectrometry application, the gas load is normally a limitation due to the constraints of the vacuum system (and consequent collisional effects in the downstream mass analyzer and in the ion optics chamber). Unless very large (and expensive) vacuum pumps are used, or the cell apertures are very small (affecting sensitivity), a reasonable upper limit to the cell pressure is of the order of 20 mTorr, for which the reaction efficiency of H_2 is predicted to be approximately 4 orders of magnitude. An alternative is to introduce the H_2 concomitantly with a higher mass non-reactive buffer gas. In this instance, the buffer gas provides conditions for thermalization at lower cell pressure, allowing the promotion of thermal reactions with H_2 (providing that kinetic energy discrimination is not employed). He is not particularly suited as a buffer gas, since it's thermalization properties are not significantly different than those of H_2 (see Figure 6). Preferred buffer gases include the inert gases, with the proviso that higher mass buffer gases increase scattering losses of lower mass analyte ions.

9 CONCLUSIONS

Ion-molecule chemistry, enacted in a pressurized multipole reaction cell, has become an important tool for the elemental analyst, and it will continue to grow in importance as new chemical methods are developed and as challenging applications are seen to be resolved on-line (rather than through sample preparation). We consider several forms of efficiency in the application. (1) The efficiency (or probability) of reaction in a single collision, which is defined for thermal processes by the ratio of the measured rate constant to the collision rate constant. This efficiency may, to first order, be considered binary, dependent on the thermochemistry. If the reaction is endothermic, it does not take place (or is very slow) under thermal conditions. If the reaction is exothermic, it may take place, and if it does it generally proceeds at close to the collision rate (with probabilities on the order of 0.1 to 1.0). (2) The extent of reaction (which can be considered an efficiency), which is determined by the ratio of the residual signal for the reacting ion to its magnitude in the absence of reaction gas. This is dependent on the operating conditions, and is strongly affected by the kinetic energy of the ions (and thus on the masses of the ions and gas, and on the cell pressure). If kinetic energy discrimination against isobaric interferences produced within the cell is employed, the extent of reaction is limited by the requirement to retain distinct energy distributions, and is thus of the order of 10^4 in a favourable instance. If the secondary chemistry is suppressed through other means (such as bandpassing the multipole), the extent of reaction can be demonstrated up to 9 orders of magnitude. (3) The efficiency of chemical resolution describes the improvement in the ratio of the signal to the background. This depends on the relative reactive efficiencies of the analyte ion and the isobaric interference, and also includes the formation of new isobaric ions within the reaction cell. When new interferences are suppressed, and an appropriate chemistry is employed (providing reaction of the interference ion and minimal effect on the analyte ion), the efficiency of chemical resolution is comparable to the efficiency of reaction of the isobaric ion.

In general, the efficiency of chemical resolution required is many orders of magnitude. Fragmentation (CID) is an inefficient process relative to ion-molecule chemistry. Ion-molecule chemistry can be very efficient, depending on the choice of chemistry, the energetics of the ions and the methods used to suppress the appearance of secondary interference ions. Higher efficiency is required when the chemistry is chosen to remove the plasma-based interference ion so that analysis of the analyte is performed at the nominal mass of the analyte. In this instance, the high efficiency is required because it is necessary to effectively remove the interference to background levels without affecting the sensitivity for the analyte. This is the common, and probably preferred, approach. An alternative is to react the analyte ion to a different mass (such as oxidation, as described in the paper by Bandura *et al.* in this volume[15]). In this instance, a lower efficiency of reaction is satisfactory provided that the analytical mass is not itself interfered.

Ion-molecule chemistry is the dominant process in achieving chemical resolution when the ion energies are near thermal (< 1eV). Thermalization increases the collision cross-section, which affects the overall efficiency exponentially. Under thermal conditions, the specificity provided by the thermochemistry may be obtained. We have attempted to demonstrate that the efficiency of the cell is also determined by the polarizability and dipole moment of the reaction gas molecules, the complexity of the reaction gas (the number of degrees of freedom), the pressure of the reaction gas, the masses of the ions and neutrals, and certain design features of the cell (including the length and the operating rf amplitude and frequency).

10 REFERENCES

1.   D.J. Douglas, *J. Am. Soc. Mass Spectrom.,* 1998, **9**, 101.
2.   A.N. Krutchinsky, I.V. Chernushevich, V.L. Spicer, W. Ens and K.G. Standing, *J. Am. Soc. Mass Spectrom.,* 1998, **9**, 569.

3.   S.G. Lias and P. Ausloos, "Ion-Molecule Reactions: Their Role in Radiation Chemistry", American Chemical Society, Washington, 1975.

4.   P. Gioumousis and D.P. Stevenson, *J. Chem. Phys.*, 1958, **29**, 294.

5.   R.A. Barker and D.P. Ridge, *J. Chem. Phys.*, 1976, **64**, 4411.

6.   L. Bass, T. Su, W.J. Chesnavich and M.T. Bowers, *Chem. Phys. Letters*, 1975, **34**, 119.

7.   T. Su and E.C.F. Su, *J. Chem. Phys.*, 1978, **69**, 2243.

8.   D.J. Douglas, *Can. J. Spectroscopy*, 1989, **34**, 38.

9.   J.T. Rowan and R.S. Houk, *Appl. Spectroscopy*, 1989, **43**, 976.

10.   C.J. Barinaga and D.W. Koppenaal, *Rapid Commun. Mass Spectrom.*, 1994, **8**, 71.

11.   D.W.Koppenaal, D.J. Barinaga and M.R. Smith, *J. Anal. At. Spectrom.*, 1994, **9**, 1053.

12.   G.C. Eiden, C.J. Barinaga and D.W. Koppenaal, *J. Anal. At. Spectrom.*, 1996, **11**, 317.

13.   G.C. Eiden, C.J. Barinaga and D.W. Koppenaal, *J. Anal. At. Spectrom.*, 1999, **14**, 1129.

14.   G.C. Eiden, C.J. Barinaga and D.W. Koppenaal, *Rapid Commun. Mass Spectrom.*, 1997, **11**, 37.

15.   N. Furuta,A. Takeda and J. Zheng, this book.

16.   D.R. Bandura, ..., this book.

17.   D.J. Douglas and J.B. French, *J. Am. Soc. Mass Spectrom.*, 1992, **3**, 398.

18.   V.G. Anicich, *The Astrophysical Journal Supplement Series*, 1993, **84**, 215. Also see http://astrochem.jpl.nasa.gov/asch/

19.   S.D. Tanner and V.I. Baranov, *J. Am. Soc. Mass Spectrom.*, 1999, **10**, 1083.

20.   R.S. Houk and N. Praphairaksit, *Spectrochim. Acta, Part B*, 2001 (accepted).

21.   Vitali Lavrov, York University, Canada, personal communication

22.   G.K. Koyanagi, V.V. Lavrov, V. Baranov, D. Bandura, S. Tanner, J.W. McLaren and D.K. Bohme, *Int. J. Mass Spectrom.*, 2000, **194**, L1.

23.   G.K. Koyanagi, V.I. Baranov, S.D. Tanner and D.K. Bohme, *J. Anal. At. Spectrom.*, 2000, **15**, 1207.

24.   S.G. Lias, J/E/ Bartmess, J.F. Liebman, J.L. Holmes, R.D.Levin and W.G. Mallard, *J. Phys. Chem. Res. Data*, 1988, **17**, 1.

25.   http://webbook.nist.gov/chemistry/

26.   V.I. Baranov and S.D. Tanner, *J. Anal. At. Spectrom.*, 1999, **14**, 1133.

ION-MOLECULE REACTIONS IN PLASMA SOURCE MASS SPECTROMETRY

Gregory K. Koyanagi and Diethard K. Bohme

Department of Chemistry and Center for Research in Mass Spectrometry, York University, Toronto, Ontario M3J 1P3, Canada

ABSTRACT

A novel inductively coupled plasma – selected ion flow tube (ICP-SIFT) mass spectrometer has been constructed to permit the study of the kinetics and product distributions of reactions of plasma source ions with neutral reagents. This has been achieved by modifying the SIFT mass spectrometer in the Ion-Chemistry Laboratory at York University to accept ions from a commercial plasma torch assembly and atmosphere-vacuum interface (ELAN Series, PerkinElmer SCIEX). The results of these measurements permit the formulation of strategies for the differentiation, by chemical reactivity, of isobaric ions (so-called "chemical resolution"). An extensive database of ion-molecule reaction rates will be useful in the selection of reagent gases used in ICP-MS reaction cells implemented to improve quantitative detection limits.

1 INTRODUCTION

During trace element analysis using plasma mass spectrometry perhaps the most serious complications are due to the existence of diatomic and tri-atomic ions generated in the plasma[1,2]. These polyatomic ions are often isobaric with one or more elements of interest, leading to increased background count rates and degraded detection limits. Several methods exist for the resolution or reduction of some of these isobaric interferences such as: cold/cool plasma techniques[3-5], plasma modification techniques[6,7], collisional dissociation[8], and high resolution mass spectrometry but none is universally applicable and all have drawbacks.

Recently, PerkinElmer SCIEX has developed and introduced the Dynamic Reaction Cell (DRC) for implementation on the Elan series of ICP-MS instruments. The DRC is a multipole band-pass reaction cell placed between the plasma interface and the principal mass analyser. The DRC permits the use of selective ion-molecule chemistry for the 'chemical resolution' of isobaric interferences. Choice of reagent gas admitted to the DRC is influenced by the reactivity of that gas towards both the interference ion and the analyte ion. Furthermore 'chemical resolution' of an isobaric overlap can be effected either by removal of the interference ion (by chemical reaction) or by addition of a known mass to the analyte ion to move it to a 'cleaner' area in the mass spectrum.

For the DRC to gain wide and routine use a complete database of ion-molecule reaction rate constants and product branching ratios for all possible ion-molecule reactions occurring within the DRC is required. The task of accumulating the required database is sizable. For example, taking the cations from Lithium to Americium, and including their oxides, carbides, nitrides, hydrides, hydroxides and chlorides as possible reagent ions one arrives at 651 potential reagent ions. Selecting 14 neutral reagent molecules (currently being studied as chemical resolving agents) which include methane, ammonia, hydrogen, oxygen, nitrogen, water, nitric oxide, nitrous oxide, carbon monoxide, carbon dioxide, methyl chloride, methyl fluoride, benzene, and nitrogen dioxide, we arrive at a total of 9114 possible primary reactions (and a significantly larger number of secondary reactions). Currently there are several compilations detailing ion-molecule reaction rates [9,10], but coverage of metal ion-molecule reactions is limited to only approximately 300 primary and secondary reactions from the above matrix.

The Selected-Ion Flow Tube (SIFT) mass spectrometer at York University has been modified to accept ions, generated in an Inductively Coupled Plasma torch, through an atmosphere/vacuum interface [11,12]. The SIFT technique is a powerful and versatile means for making accurate determinations of ion-molecule rate constants and product branching ratios. [13,14]

Early ICP/SIFT experiments in our laboratory [11,12] focused on individual case studies using 'chemical resolution' to remove specific isobaric interferences. In one years time we have managed to study a large number of metal ion-molecule reactions and these studies have led to the observation of several general trends in ion-molecule reactivity. What follows is a brief discussion of these trends and their interpretation in terms of the basic principles of ion-molecule chemistry. The application of a selected number of reactions to specific 'chemical resolution' problems also is discussed.

It is important to review the similarities and differences between the Elan 6100DRC and the ICP/SIFT instrument. In Figure 1, schematics for the two instruments are illustrated. Both instruments utilize essentially identical plasma generators and atmosphere-vacuum interfaces. While the Elan 6100DRC simultaneously selects and reacts the primary ion in the dynamic reaction cell the SIFT instrument first selects the primary reagent ion in one region and then reacts the ion in the flow tube in the presence of helium buffer gas. In the SIFT the reagent gas is a minor component of the helium buffer gas so tube pressure and flow conditions do not change during the course of an experiment. However, in the DRC the reaction gas is the sole source of pressure so cell conditions change as reaction gas flow increases. The primary effect is that of collisional scattering which slowly increases at high cell pressures. Furthermore, since the SIFT flow tube is pressurized to 0.35 Torr with helium the mass selected ion experiences approximately 10^5 collisions with He before being permitted to react with the neutral of interest. Figure 2 contains plots of reaction profile data for the ICP/SIFT and Elan 6100DRC. The similarity is encouraging and suggests that rate constant data is transportable between the two instruments.

2 FACTORS AFFECTING ION-MOLECULE REACTIONS

2.1 Ion-molecule interaction energy

Ion-molecule reactions are driven strongly by charge-induced dipole (polarisability) interactions and charge–dipole interactions. Stronger attractive forces bring reacting entities closer and provide a source of energy to overcome any activation barrier that might be present. Ion-molecule collision theories have been developed at several levels of theory and can be used to predict ion-molecule collision rate coefficients [15,16]. Pure polarization theory makes the assumption that the ion can be treated as a point charge and that the molecule is a point polarisable entity. The total effective interaction potential, V_{eff}, is then:

$$V_{eff}(r) = (L^2 / 2\,\mu\,r^2) - (\alpha\,q^2 / 2\,r^4) \tag{1}$$

where L is the orbital angular momentum between the two particles, μ is the reduced mass of the system, q is the charge on the ion, α is the spherically averaged polarisability of the neutral molecule and r is the separation between the center of mass of the ion and molecule. The predicted collision rate coefficient, k_C, often referred to as the Langevin collision rate coefficient, is then [15,16]:

$$k_C = 2\,\pi\,q\,(\alpha\,/\,\mu)^{1/2} \tag{2}$$

When the molecule also has a permanent dipole moment the total effective interaction potential of the ion and molecule becomes [16]

$$V_{eff}(r,phi) = (L^2 / 2\,\mu\,r^2) - (\alpha\,q^2 / 2\,r^4) - (q\,\mu_D\,cos\,\varphi\,/\,r^2) \tag{3}$$

where μ_D is the permanent dipole moment of the neutral and φ is the angle between the dipole moment vector of the neutral and the distance vector to the ion. The locked dipole assumption proposes that the dipole vector remains aligned with the distance vector during ion-molecule interactions (*cos* $\varphi = 1$) and then the thermally averaged collision rate, k_{LD}, is given by:

$$k_{LD} = 2\,\pi\,q\,\mu^{-1/2}\,[\alpha^{1/2} + \mu_D\,(2/\pi\,k_B\,T)^{1/2}] \tag{4}$$

where T the absolute temperature and k_B is Boltzmann's constant. However, it is generally accepted that molecular rotation is not completely quenched by the ion field and thus complete dipole 'locking' is not generally realised. So the averaged dipole orientation (ADO) description [17-19] was developed to account for the degree of dipole orientation caused by the ion field. The ADO theory uses an empirically paramaterised locking constant, c, that is a function of $\mu_D \cdot \alpha^{-1/2}$. The ADO collision rate, k_{ADO}, is then:

$$k_{ADO} = 2\,\pi\,q\,\mu^{-1/2}\,[\alpha^{1/2} + c\,\mu_D\,(2/\pi\,k_B\,T)^{1/2}] \tag{5}$$

Of course non-polar neutral species can possess a sizeable electric quadrupole moment so ion-quadrupole interactions can be significant. Su and Bowers [20,21] introduced the average quadrupole orientation (AQO) formalism to describe these cases. Similarly, ion polarisability, dipole moment and quadrupole moment can not be ignored for some ionic species. The higher order interactions generated by multipole-multipole interactions are higher order in the reciprocal of the separation distance so these interactions effect the short range interaction potential and not the magnitude of the collision parameter (collision rate coefficient). A thorough description of classical ion-molecule collision theory can be found in Reference 22.

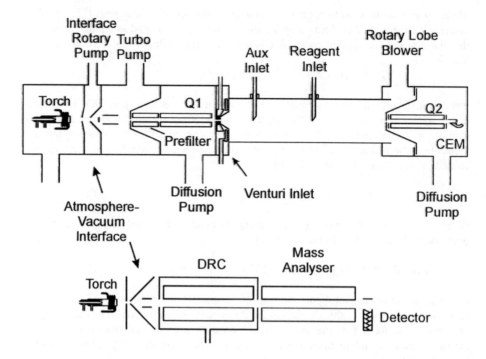

Figure 1 *Schematic diagrams for the ICP/SIFT (top) and Elan 6100DRC (bottom) instruments. In the ICP/SIFT instrument ions are produced in a plasma source and sampled by a standard atmosphere-vacuum interface (see text). The ion of interest is mass-selected in Q1, after which it is passed to the flow tube via a Venturi inlet. The flow tube is kept at a constant pressure of 0.35 Torr. Thermalised ions react with a neutral admitted at the reagent inlet as they flow downstream (to the right). Product ions are mass analysed by Q2. In the Elan 6100DRC ions are produced and sampled as above but are then reacted in a band-pass quadrupole reaction cell (DRC). Unreacted ions and product ions are detected by the mass analyser quadrupole*

2.2 Energy distribution

Once an ion-molecule collision complex has been created, the electronic structure of the two species must reform to the product electronic configuration. The collisional complex must be permitted time for this electronic structure rearrangement. Typically in ion-atom collisions an elastic or near-elastic collision occurs and the colliding entities separate before a new electronic configuration is formed, thus ion-atom reactions are generally of low probability (inefficient on a per-collision basis). However, in an ion-molecule collision, energy redistribution from the reaction coordinate to vibrational modes of the complex cause an inelastic collision which results in a longer orbital-complex lifetime. This longer orbital-complex lifetime permits reformation of the electronic wave function to that of the product channel. [23,24]

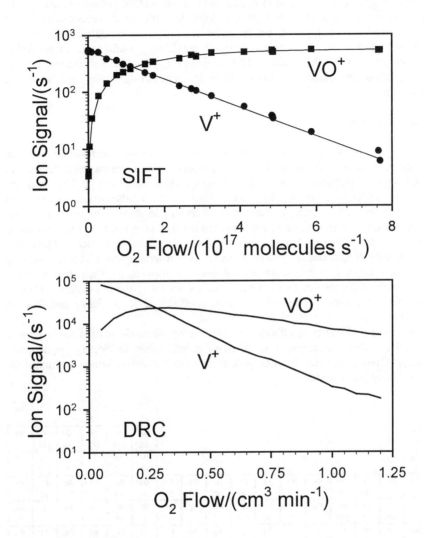

Figure 2 *Reaction profiles for V^+ reacting with O_2 as a function of O_2 flow in the respective reaction regions. ICP/SIFT data on top and Elan 6100DRC data below. Scattering loss as a function of reaction cell pressure is evident in the VO^+ signal for the DRC data*

2.3 Classical valance and electron shell considerations

Variations in outer shell electron configuration can cause significant variations in reactivity (for instance along a row in the periodic table). Atomic cations with fully filled (spherical) outer shell configurations as found in neutral noble gases, exhibit reduced reactivity. In contrast, cations with partially filled valence shells, in particular those occurring in the third, fourth, fifth and sixth period, exhibit higher reactivity. As in neutral chemistry, the concepts embodied in the octet and 18-electron rules [25] are useful tools for predicting the products of ion-molecule reactions. Observation of 'hypervalent' product ions is most often an indication that simple electrostatic clustering is occurring between the cation and neutral species.

2.4 Thermodynamics

Beyond the factors mentioned in the preceding three sections, thermodynamics is an underlying guide to whether or not a reaction proceeds. Extensive monographs exist in which enthalpies of formation for ionic and neutral species are tabulated [26,27]. Working through these tables on a reaction by reaction basis can be a tedious task, and leads to little or no insight into reactivity trends or periodicities in families of reactions. The tedium is reduced by examining specific thermodynamic properties such as ionisation energies or proton affinities d their relative values, *viz.* scales or ladders. Much like electrochemical cell reactions are broken down into half reactions, ion-molecules reactions can be broken down into half reactions. Reaction enthalpies are then derived, quite simply, from the difference of two numbers. Scales helpful to the ion-molecule chemist are the ionisation energy, proton affinity, H-atom affinity, O-atom affinity and Cl-atom affinity scales.

     In Figure 3, the O-atom affinity scale for singly-positive elements is displayed as a periodic table. This representation is extremely helpful when developing ion-molecule chemistries of reactions with O_2, N_2O, NO, or any other molecule where a terminal O-atom could be readily transferred.

H 116																	He
Li 8	Be 87											B 86	C 194	N 272	O 155	F 160	Ne
Na 13	Mg 35											Al 42	Si 115	P 191	S 125	Cl 111	Ar
K 3	Ca 77	Sc	Ti 165	V 134	Cr 77	Mn 59	Fe 75	Co 66	Ni 49	Cu 33	Zn 32	Ga 6	Ge 82	As	Se	Br 94	Kr
Rb 7	Sr 71	Y 180	Zr 208	Nb 208	Mo 114	Tc	Ru 94	Rh 56	Pd 53	Ag	Cd	In	Sn 75	Sb	Te 97	I	Xe
Cs 14	Ba 93	La 208	Hf 174	Ta 189	W 127	Re	Os	Ir 60	Pt 60	Au	Hg	Tl	Pb 53	Bi 42	Po	At	Rn
Fr	Ra	Ac	Rf	Ha													

Ce 208	Pr 192	Nd 181	Pm	Sm 140	Eu 93	Gd 181	Tb 175	Dy 144	Ho 143	Er 142	Tm 128	Yb 93	Lu 134
Th 210	Pa	U 194	Np 186	Pu	Am	Cm	Bk	Cf	Es	Fm	Md	No	Lr

Figure 3 *Periodic table of O-atom affinities for atomic cations. (kcal mol^{-1}) derived from Reference 26*

2.5 Spin electron rules

At the quantum mechanical level isolated bimolecular reactions must conserve total angular momentum. Spin conservation rules therefore apply. Consider the reaction

$$M^+ + N \rightarrow P^+ + Q \qquad (6)$$

Formally, for systems of 'light' atomic ions and neutrals the intersection (in the algebraic sense) of the sets S1 and S2 must be non-zero, where S1 is the set $\{s(M^+) + s(N), s(M^+) + s(N) - 1, \ldots |s(M^+) - s(N)|\}$ and S2 is the set $\{s(P^+) + s(Q), s(P^+) + s(Q) - 1, \ldots |s(P^+) - s(Q)|\}$ and $s(X)$ is the total electronic spin of X. The implication, for example is that a doublet species reacting with a doublet species can have the following product pairs: two triplets, two doublets, a triplet and a singlet or two singlet species (and other combinations of higher multiplicity). Perhaps more significantly, if the neutral reactant and neutral product of reaction 6 are both singlet then the reactant and product ions must have the same multiplicity. Adherence to spin conservation rules can govern the rate and branching ratio of a reaction. [28,29,30]

Table 1 *Selected properties of reactant neutrals and reaction rate coefficients for reactions with strontium and rubidium cations.*

| | Reagent Properties | | | Sr^+ Primary | k (cm^{3}molecules^{-1}s^{-1}) | |
	D^a	α^b	$(2s+1)^c$	Channel	$Sr^+(^2S_{1/2})$	$Rb^+(^1S_0)$
NO$_2$	0.32	3.02	2	O-atom x-fer	6.7×10^{-10}	2.0×10^{-13}
C$_6$H$_6$	0	10.30	1	Cluster	2.2×10^{-10}	1.3×10^{-11}
N$_2$O	0.17	3.03	1	O-atom x-fer	6.3×10^{-11}	$<1 \times 10^{-14}$
HCl	1.08	2.77	1	Cl-atom x-fer	3.4×10^{-11}	$<1 \times 10^{-14}$
CH$_3$Cl	1.87	4.72	1	Cl-atom x-fer	1.4×10^{-11}	7.7×10^{-13}
NO	0.15	1.70	2	O-atom x-fer	1.4×10^{-11}	1.6×10^{-13}
D$_2$O	1.85	1.45	1	Cluster	3.5×10^{-12}	$<1 \times 10^{-14}$
C$_2$H$_4$	0	4.53	1	Cluster	2.6×10^{-13}	4.4×10^{-14}
O$_2$	0	1.58	3		5.0×10^{-14}	$<1 \times 10^{-14}$
CH$_4$	0	2.59	1		4.6×10^{-14}	3.6×10^{-14}
NH$_3$	1.47	2.26	1		3.6×10^{-14}	$<1 \times 10^{-14}$
CO	0.11	1.95	1		$<1 \times 10^{-14}$	$<1 \times 10^{-14}$

^{a} dipole moment (Debye)
^{b} polarisability (10^{-24} cm^3)
^{c} electron spin multiplicity

3 EXAMPLE 1: REACTIONS OF STRONTIUM AND RUBIDIUM CATIONS

In Table 1 is presented reaction rate constants for Sr^+ and Rb^+ reacting with several small neutrals reagents. Some pertinent physical properties of the neutral are also listed. For the data presented in Table 1 it is clear that Rb^+ reacts only very slowly with the majority of molecules studied. The Rb^+ cation is a closed shell ion iso-electronic with Kr neutral, which is known to be relatively unreactive. The only Rb^+ reaction with any significant rate constant is that with benzene. This reaction is a simple attachment reaction and is being driven by the relatively large polarisability of the benzene molecule which gives rise to a significant charge-induced dipole interaction. Also, the benzene unit is large and possess 30 vibrational modes for reaction coordinate energy redistribution.

The strontium cation, on the other hand, has electronic configuration $[Kr]\cdot4s^1$ and has significantly different reactivity compared to Rb^+. The fastest Sr^+ reaction studied in this work is, by far, that with nitrogen dioxide. Ground state nitrogen dioxide and Sr^+ are both doublet species. Similarly NO and SrO^+, the logical products for this reaction, are doublet species. This radical-radical reaction is spin-allowed, thermoneutral and barrierless and is observed to proceed essentially at the ADO collision rate. The O-atom transfer reaction with NO is endothermic and proceeds only slowly. The O-atom transfer reaction with N_2O is spin allowed and exothermic, but proceeds one order of magnitude more slowly than that with NO_2. The reduced rate constant is due to the closed shell nature of the N_2O species which introduces an activation barrier to the bond formation between the Sr^+ and O. Cl-atom transfer from HCl and CH_3Cl are both exothermic and spin allowed but proceed at 2% and 1% of the ADO collision rate, respectively, indicating that a small barrier to the Cl-atom transfer reaction exists. Several clustering reactions were also observed. O-atom transfer from CO, O_2 and D_2O are endothermic and electrostatic clustering was observed instead. Clustering with the benzene and ethylene was also observed. Benzene has the greatest number of internal degrees of freedom and exhibits the highest clustering rate constant (though this could also be due in part to its large polarisability).

The implications of the ion-molecule chemistry discussed in this section is that O-atom (or possibly Cl-atom) transfer reactions could be used to differentiate Sr^+ and Rb^+ isobaric interferences. By using N_2O (NO_2 is corrosive and persistent) as a reagent gas in the DRC the Sr^+ ions could be reacted to SrO^+ (and therefore appear 16 mass units higher in the spectrum), while the Rb^+ ions remain unreacted.

4 SAMPLE 2: ELECTRON TRANSFER VERSUS CLUSTERING IN REACTIONS OF BENZENE

The ionisation energy scale is also an important consideration when predicting the outcome of an ion-molecule reactions. A general trend apparent in reactions with benzene is that the outcome is, with few exceptions, either simple clustering of benzene to the metal ion or e^- from the benzene to the metal. The transition from one mechanism to the other is sharply dependant on the first ionisation energy of the metal cation. In Table 2 a selection of rate constants for atomic ion-benzene reactions is presented. Of note is the sharp transition from e^- transfer to clustering that one can infer from the Au^+ reaction. In a thermal ensemble the population in the high energy tail has sufficient translational energy to overcome slightly endothermic reaction conditions thus the 0.02 eV reaction endothermicity (for the reaction $Au^+ + C_6H_6 \rightarrow Au + C_6H_6^+$) can be overcome by a small percentage of the Au^+ ions. Also of note is the slow clustering with K^+, since it has a closed shell [Ar] electron configuration.

The implication of these reaction rate coefficients and product distribution channels is that a reaction gas with high electron transfer efficiency and know ionisation energy could be used as a low-pass filter, discriminating against interference ions on the basis of first ionization potential.

Table 2 *Reactions of benzene (IE = 9.24eV) with elemental cations. Since collision rate varies significantly reaction rates are expressed as % efficiencies (=100% × $k_{observed}$ / $k_{collision}$).*

	IE(eV)	% Eff	Channel
Ar	15.8	72%	e^- transfer
K	4.34	2%	Clustering
Ca	6.11	30%	Clustering
In	5.79	37%	Clustering
Sn	7.34	70%	Clustering
Sb	8.61	62%	Clustering
Te	9.01	66%	Clustering
I	10.45	75%	e^- transfer
Pt	8.96	86%	Clustering
Au	9.22	84%	80% cluster, 20% e^- transfer
Hg	10.43	67%	e^- transfer
Tl	6.11	24%	Clustering

Table 3 *Selected physical properties of Lanthanide cations and observed reaction rate coefficients, k, for lanthanides reacting with N_2O and O_2 to produce LnO^+ at 295±2 K and 0.350±0.01 Torr.*

Ion	k (N_2O)[a]	k (O_2)[a]	O-Atom Affinity[b] (kcal mol^{-1})	M^+ Ground State Electronic Configuration
La^+	5.9×10^{-10}	4.3×10^{-10}	208 ± 3	$5d^2$
Ce^+	5.2×10^{-10}	5.1×10^{-10}	208 ± 4	$4f^1 5d^2$
Pr^+	2.8×10^{-10}	4.3×10^{-10}	192 ± 5	$4f^3 6s^1$
Nd^+	2.3×10^{-10}	3.3×10^{-10}	181 ± 4	$4f^4 6s^1$
Pm^+	—	—	—	$4f^5 6s^1$
Sm^+	1.4×10^{-10}	2.8×10^{-10}	140 ± 4	$4f^6 6s^1$
Eu^+	6.9×10^{-11}	$\leq1\times10^{-13}$	93 ± 6	$4f^7 6s^1$
Gd^+	5.3×10^{-10}	4.9×10^{-10}	181 ± 4	$4f^7 5d^1 6s^1$
Tb^+	1.2×10^{-10}	3.8×10^{-10}	175 ± 4	$4f^9 6s^1$
Dy^+	2.7×10^{-11}	2.7×10^{-10}	144 ± 10	$4f^{10} 6s^1$
Ho^+	1.7×10^{-11}	2.4×10^{-10}	143 ± 4	$4f^{11} 6s^1$

Er$^+$	1.3×10^{-11}	2.5×10^{-10}	142 ± 6	$4f^{12}6s^1$
Tm$^+$	4.4×10^{-12}	$\leq1\times10^{-13}$	128 ± 2	$4f^{13}6s^1$
Yb$^+$	6.5×10^{-13}	$\leq1\times10^{-13}$	93 ± 3	$4f^{14}6s^1$
Lu$^+$	2.6×10^{-10}	7.7×10^{-11}	134 ± 2	$4f^{14}6s^2$

[a] in cm^3 molecule^{-1} s^{-1}
[b] derived from reference 26

5 EXAMPLE 3: LANTHANIDES WITH N$_2$O AND O$_2$

Several elements in the lanthanide series possess an array of naturally occurring isotopes. This combined with their propensity for creating the monoxide in the plasma source makes the mass spectrum quite cluttered between 135 Daltons and 191 Daltons. For instance, peaks arising from as many as four lanthanide elements can occur at several masses. Measurement of lanthanide isotope ratios and elemental distributions is important to the geological sciences, and so it would be beneficial to simplify the mass spectrum in this region. The O-atom affinity of several lanthanides is quite high (see Figure 3) and so right away removing the oxygen via bond dissociation was ruled out. Instead, controlled oxidation of the atomic metal ion to the monoxide was attempted. Since N$_2$ has a low O-atom affinity (40 kcal mol^{-1}), N$_2$O was chosen as an oxidizing agent.

Reaction rate constants for N$_2$O reacting with the lanthanide cations are shown in Table 3. Despite the fact that all reactions are exothermic by more than 52 kcal mol^{-1} some reactions were observed to proceed only very slowly, indicating an activation barrier exists between reactants and products (see Figure 4). Work by Schilling and Beauchamp [31] and by Cornehl et al.[32] suggests that this barrier arises from the necessity to promote the lanthanide to a bonding $f^{n}d^1s^1$ electronic configuration to permit bonding with the closed shell N$_2$O. Oxidation, when it occurred, led exclusively to the monoxide. At very high flows of N$_2$O electrostatically bound clusters of N$_2$O, viz. LnO$^+\cdot$(N$_2$O)$_n$, were observed.

Reaction rate coefficients for O$_2$ reacting with the lanthanide were also determined (see Table 3). For the production of the monoxide cation these reactions can be broken into two categories: exothermic and thermo-neutral/endothermic. It was observed that when a reaction was exothermic it proceeded at the collision rate, regardless of the quantity of energy released in the reaction. There was one example of a thermo-neutral reaction (Tm$^+$) which proceeded via sequential O$_2$ clustering at approximately 1% of the collision rate. Reactions of O$_2$ with Eu$^+$ and Yb$^+$ to produce the monoxide are endothermic and O-atom transfer was not observed. Again sequential O$_2$ clustering was observed. Molecular oxygen is a di-radical (two unpaired electrons) and it would appear that in the case of O$_2$ reacting with Ln$^+$ the bonding electron can be supplied by the O$_2$. Again, primary reaction chemistry led to formation of the monoxide, while in the case of Eu$^+$, Tm$^+$ and Yb$^+$ very slow formation of the dioxide was observed at high O$_2$ flow rates. The dioxide formation is attributed to the formation of a simple electrostatic complex between the lanthanide cation and the O$_2$ molecule.

Exclusive covalent formation of the monoxide is logical because the preferred oxidation state for a lanthanide is +3. Thus, in the LnO$^+$ species Ln is formally (+3) and O is formally (−2). The covalently bound LnO$_2^+$ species would require the Ln to be formally (+5), which is unlikely.

The ion-molecule chemistry determined for lanthanides reacting with O_2 and N_2O suggests that controlled oxidation to the monoxide is possible, for most species. This reduces the 'clutter' in the lanthanide region significantly. Further experiments, perhaps with ozone or NO_2 or a combination of gases, are indicated.

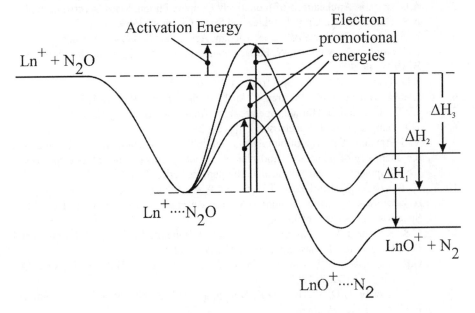

Figure 4 *Potential energy surface for the reaction of Ln⁺ + N₂O illustrating the barrier created by the orbital promotion energy required to populate the (proposed) reactive $f^x d^1 s^1$ electron configuration.*

6 CONCLUSIONS

In the above series of examples it is shown that a few simple rules can be used to rationalise the majority of outcomes in metal ion-molecule chemistry. Similarly, where reaction rate constant data is lacking, these same rules can be used to predict, *a priori*, the outcome of an ion-molecule reaction.

Although metal ions exhibit a diverse range of ion-molecule chemistries, it is precisely this diversity which will permit judicious selection of reagent gases capable of resolving multiple isobaric interferences.

Work to date has found several reagents that can be targeted for further study as viable 'chemical resolution' agents for use in the ion-molecule reaction cells. Groups such as those at Ohio State University[33], USA, the University of Reading[34], UK, and at PerkinElmer SCIEX[35] continue to develop the reaction cell technique into a routine analytical procedure.

7 REFERENCES

1. Inductively Coupled Plasma Mass Spectrometry, ed. A. Montaser, J. Wiley, New York, 1998.
2. A.L. Gray, in Applications of Inductively Coupled Plasma Mass Spectrometry, Ed. A.R. Date and A.L. Gray, Blackies and Sons, Glasgow, 1989, ch 1.
3. K. Sakata and K. Kawabata, Spectrochim. Acta., Part B, 1994, **49**, 1027
4. N.S. Nonose, N. Matsuda, N. Fudagawa and M. Kubota, Spectrochim. Acta., Part B, 1994, **49**, 995.
5. S.D. Tanner, J. Anal. At. Spectrom., 1995, 10, 905.
6. J.W.H. Lam and J.W. McLaren, J. Anal. At. Spectrom., 1990, **5**, 419.
7. J.W.H. Lam and G. Horlick, Spectrochim. Acta., Part B, 1990, **45**, 1313.
8. D.J. Douglas, Can. J. Spectrosc., 1989, **34**, 38.
9. V.G. Anicich, A survey of bimolecular ion-molecule reactions for use in modeling the chemistry of planetary atmospheres, cometary comae, and interstellar clouds: 1993 Supplement, Astrophysical J. Supplement Series, **84** (1993) 215.
10. Y. Ikezoe, S. Matsuoka, M. Takebe and A.A. Viggiano, *Gas-phase ion-molecule reaction rate constants through 1986*, Ion Reaction Research Group of the Mass Spectroscopy Society of Japan, 1987.
11. G.K. Koyanagi, V.V. Lavrov, V. Baranov, D. Bandura, S. Tanner, J. McLaren and D.K. Bohme, Int. J. Mass Spectrom., **194** (2000) L1.
12. G.K. Koyanagi, V.I. Baranov, S.D. Tanner and D.K. Bohme, J. Anal. At. Spectrom., **15** (2000) 1207.
13. G.I. Mackay, G.D. Vlachos, D.K. Bohme and H.I. Schiff, Int. J. Mass Spectrom. Ion Phys., 1980, **36**, 259.
14. A.B. Raksit and D.K. Bohme, Int. J. Mass Spectrom. Ion Processes, 1983/84, **55**, 69.
15. P.M. Langevin, Ann. Chim. Phys., **5** (1905) 245.
16. G. Gioumousis and D.P. Stevenson, J. Chem. Phys., 1958, **29**, 294.
17. T. Su and M.T. Bowers, J. Chem. Phys., 1973, **58**, 3027.
18. T. Su and M.T. Bowers, J. Amer. Chem. Soc., 1973, **95**, 1370.
19. T. Su and M.T. Bowers, Int. J. Mass Spectrom. Ion Phys., 1973, **12**, 347.
20. T. Su and M.T. Bowers, Int. J. Mass Spectrom. Ion Phys., 1975, **17**, 309.
21. T. Su and M.T. Bowers, Int. J. Mass Spectrom. Ion Phys., 1976, **21**, 424. *N.B.* This is an erratum for the preceding reference.
22. T. Su and M.T. Bowers, chapter 3, *Gas Phase Ion Chemistry*, M.T. Bowers, ed., Academic Press, New York, 1979.
23. W.J. Chesnavich and M.T. Bowers, chaper 4 , *Gas Phase Ion Chemistry*, M.T. Bowers, ed., Academic Press, New York, 1979.
24. W.L. Hase, Acc. Chem. Res., 1998, **31**, 659
25. N.V. Sidgwick, *The Electronic Theory of Valency*, Cornell University Press, Ithica, N.Y., 1927.
26. JANAF Thermochemical Tables, Second Edition, Nat. Stand. Ref. Data Ser., Nat. Bur. Stand. (U.S.), **37**, 1971.
27. S.G. Lias, J.E. Bartmess, J.F. Liebman, J.L. Holmes, R.D. Levin, and W.G. Mallard, Gas-phase ion and neutral thermochemistry, J. Phys. Chem. Ref. Data, Volume 17, Supplement 1, 1988.

28.   P.B. Armentrout, Annu. Rev. Phys. Chem., 1990, **41**, 313.
29.   D.E. Clemmer, N. Aristov and P.B. Armentrout, J. Phys. Chem., 1993, **97**, 544.
30.   D. Schroder, S. Shaik and H. Schwarz, Acc. Chem. Res., 2000, **33**, 139.
31.   J.B. Schilling and J.L Beauchamp, J. Am. Chem. Soc., 1988, **110**, 15.
32.   H.H. Cornehl, C. Heinemann, D. Schroder and H. Schwarz, Organometallics, 1995, **14**, 992.
33.   See J. Olesik and co-workers, this volume.
34.   See L. Simpson and co-workers, this volume.
35.   See D. Bandura and co-workers, this volume.

8 ACKNOWLEDGEMENTS

The financial support of MDS SCIEX, the National Research Council of Canada and the Natural Sciences and Engineering Research Council of Canada is gratefully acknowledged. The authors would like to thank D.R. Bandura, V.I. Baranov and S.D. Tanner for their invaluable knowledge and assistance during the development phase of this project.

ION-MOLECULE CHEMISTRY SOLUTIONS TO THE ICP-MS ANALYTICAL CHALLENGES*

D. R. Bandura[1], S. D. Tanner[1], V. I. Baranov[1], G. K. Koyanagi[2], V. V. Lavrov[2] and D. K. Bohme[2]

[1] PerkinElmer-SCIEX, 71 Four Valley Drive, Concord, Ontario L4K 4V8, Canada
[2] Department of Chemistry, York University, Toronto, Ontario K1A 4V8, Canada

1 INTRODUCTION

First observations of gas-phase ion-molecule reactions in mass spectrometry date as early as 1912, when J. J. Thomson noted the presence of an ion of mass 3 in his first "positive rays" spectrometer with H_2 in the ion source [1]. Since then, ion-molecule reactions have been extensively studied with use of many types of mass spectrometers. Yet the idea of using ion-molecule reactions to reduce isobaric interferences in ICP MS by modifying the composition of the ion beam extracted from plasma, first suggested in 1989 by Douglas [2] and Houk [3], was utilized in commercial instruments only in the late 90's. Since most of the high intensity interfering species for argon ICP MS contain Ar, simple suppression of argon-containing interferences has been the main application of reaction/collision cells in ICP MS. Ar^+, Ar_2^+, ArO^+, ArN^+, $ArCl^+$, ArC^+, all having relatively high first ionization potentials, can be efficiently removed by electron-transfer reactions with a neutral of lower ionization potential. A choice of gases having ionization potentials between that of the argides and the analyte ions is available; the commonly used gases are H_2, NH_3, CH_4. Other reactions like proton-transfer and H-atom transfer [4] can also be efficient provided that newly formed species are adequately resolved from the analyte one mass below either spectroscopically (which requires high abundance sensitivity) or chemically by further reactions.

The majority of reported results have been obtained by removal of the interfering ion from the analyte mass of interest. Yet the first demonstration of possible analytical benefits of ion-molecule reactions in ICP MS was done by a shift of Tb^+ to m/z=175 by oxidation with O_2, while CeO^+ was oxidized at a much slower rate [2]. Bollinger and Schleisman observed formation of AsO^+ when detecting As in HCl matrix by ICP-DRC-MS (presumably from O-atom transfer reaction with water impurities in N_2 gas used), and raised the possibility of using AsO^+ for the determination of As [5].

In this contribution we present the results of a study of oxidation, hydroxylation, chlorination and O-atom abstraction reactions that can be used for analyte or interference shift to a different m/z. Correlation between reaction data available from the literature, thermochemical predictions and measured DRC ICP-MS reaction profiles are considered for the range of ions and neutrals.

*Adapted in part from Bandura *et al*, 2001, Fresenius' J. Anal. Chem. (in press)

2 THERMOCHEMICAL AND KINETIC CONSIDERATIONS

In the absence of kinetic information, guidance in the selection of a reaction gas for resolving a particular interference can be had by calculating the enthalpy change of candidate reactions, using gas-phase ion and neutral thermochemistry data (available from, for example, Lias et.al [6]). For O-atom transfer reactions of the analyte M^+ and interference $Int.^+$ with nitrous oxide, for example,

$$M^+ + N_2O \longrightarrow MO^+ + N_2 \qquad (1)$$
$$Int^+ + N_2O \longrightarrow IntO^+ + N_2, \qquad (2)$$

the enthalpy change of the reactions is calculated as

$$\Delta H_r(1) = \Delta H_f(MO^+) + \Delta H_f(N_2) - \Delta H_f(M^+) - \Delta H_f(N_2O), \qquad (3)$$

$$\Delta H_r(2) = \Delta H_f(IntO^+) + \Delta H_f(N_2) - \Delta H_f(Int^+) - \Delta H_f(N_2O), \qquad (4)$$

where $\Delta H_f(X)$ is the heat of formation of species X. From the available thermochemical data, one could compile a table of enthalpy changes of the reactions with a particular gas, as, for example, Table 1 for N_2O. If $\Delta H_r < 0$, the reaction is allowed and may proceed, unless other factors (for example, spin multiplicity conservation rule [7]), forbid it.

Table 1 *Enthalpy of reactions with nitrous oxide*

Ion (M^+)	$\Delta H_r(MO^+)$	Ion (M^+)	$\Delta H_r(MO^+)$	Ion (M^+)	$\Delta H_r(MO^+)$	Ion (M^+)	$\Delta H_r(MO^+)$
Al+	-1.71	F+	-120.31	NO+	-21.94	Sm+	-99.61
AlO+	-53.81	FO+	-45.61	Na+	27.19	Sn+	-35.11
B+	-45.51	Fe+	-35.41	Nb+	-167.61	Sr+	-31.41
BO+	-79.61	Ga+	34.39	Nd+	-141.41	Ta+	-149.01
Ba+	-52.81	Gd+	-140.61	Ni+	-8.51	Tb+	-134.61
Be+	-46.61	Ge+	-41.81	Np+	-145.61	Te+	-56.61
Bi+	-1.61	H+	-76.21	O+	-114.71	Th+	-169.61
Br+	-53.81	HO+	-64.51	O2+	22.89	ThO+	-70.61
C+	-153.87	H2+	-142.31	P+	-150.61	Ti+	-125.31
CO+	-82.75	H2O+	-42.61	PO+	-30.61	TiO+	-34.91
Ca+	-37.21	Hf+	-133.61	Pb+	-13.21	Tm+	-87.61
Ce+	-167.61	Ho+	-102.61	Pd+	-9.61	U+	-153.61
Cl+	-70.61	Ir+	-19.61	Pr+	-151.61	UO+	-142.61
ClO+	-34.61	K+	36.99	Pt+	-19.61	V+	-94.21
Co+	-25.61	La+	-167.61	PtO+	-54.61	W+	-87.11
Cr+	-37.41	Li+	31.99	Rb+	32.99	WO+	-91.11
CrO+	-29.61	Lu+	-93.61	Rh+	-15.61	Y+	-139.61
Cs+	25.99	Mg+	5.59	RhO+	-53.61	Zn+	7.59
Cu+	8.19	Mn+	-18.61	Ru+	-53.61	Zr+	-167.91
Dy+	-103.61	Mo+	-73.61	S+	-85.41	ZrO+	-23.31
Er+	-101.61	MoO+	-73.61	SO+	-45.81		
Eu+	-52.61	N+	-232.48	Si+	-75.01		

Ideally, ΔH_r for reactions (1) and (2) should have opposite sign, so that one of the reactions is thermodynamically forbidden, while the other is allowed. For ions highlighted in bold in the Table 1, oxide ion formation is thermodynamically forbidden. Thus, oxidation of interference would be required to resolve those ions. For all other ions, reaction rate constants of the analyte ion and the interference have to be compared, and, depending on which reaction is faster, bare ion or its oxide is to be selected as analyte. Rate constants for many useful ion-molecule reactions are available from the database developed at JPL [8]. A recently initiated research program on the ICP-SIFT apparatus at York University, Toronto, is extending the database to the atomic and molecular ions that are of particular interest to ICP-MS analysts, with more than 400 reactions studied in the last 7 months [9,10].

Most rate constants reported in the literature are valid for reactions under thermal conditions at the temperature of the measurement (normally 300 K). Hence, the conditions at which the reaction cell is operated are extremely important. Reaction specificity, expected from thermochemical and reaction kinetics data, can be maintained only if the contribution of the ion kinetic energy, obtained either during plasma expansion, post-skimmer acceleration, or from rf-heating in the reaction cell, into reaction collision complex energy, is small. More detailed consideration of the effects of collisional energy damping on the operation of reaction cells can be found in the contribution by Tanner *et al* to this volume [11], while the effect of rf-heating was considered in detail by Baranov and Tanner in [12]. Due to the low kinetic energy with which ions enter the cell, high number of collisions and low rf and dc fields, the DRC is a near-thermal ion-molecule reactor. Table 2 shows data comparing the reaction rate constants obtained by G.K.K. and V.V.L. on the ICP-SIFT apparatus, with the rate of total ion loss (i.e. sum of scattering and reactive losses) obtained with DRC. By the rate of ion loss we consider coefficient b in the ion signal intensity I decay approximated by $I = 10^{\,a - bx}$, where x is reaction gas flow in arbitrary units (see for example Hf^+ reaction profiles in Figure1). DRC quadrupole stability parameters used while obtaining data for Table 2 were $q = 0.5$, $a = 0$. These values were used for all DRC experiments reported in this paper unless stated otherwise. Note that the conversion of the requested DRC gas flow (arb.un.) to sccm depends on the mass flow controller calibration and the gas used. For N_2O introduced through the channel B controller 1 arb.un.= 0.493 sccm.

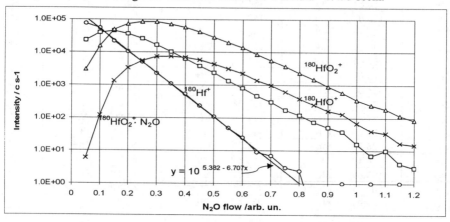

Figure 1 *ICP-DRC-MS profiles for reaction of Hf^+ (10 µg/L sample) with N_2O*

Table 2 *ICP-DRC-MS ion loss rate and ICP-SIFT rate constants for reactions with N_2O*

Ion	SIFT rate constant/cm^3 $molec^{-1}$ s^{-1}	SIFT Products	DRC Ion Loss Rate b, arb.un.$^{-1}$ (see text)	DRC Products
Zr^+	7.7E-10	$ZrO_m^+ \cdot (N_2O)_n$, m=1,2 n=0-3	-6.0	ZrO^+, ZrO_2^+
As^+	6.7E-10	$AsO^+ \cdot (N_2O)_n$, n=0,1	-4.5	AsO^+
Nb^+	6.5E-10	NbO^+, $NbO_2^+ \cdot (N_2O)_n$, n=0-3	-5.3	NbO^+
Hf^+	6.3E-10	HfO^+, $HfO_2^+ \cdot (N_2O)_n$, n=0-3	-6.7	HfO^+, HfO_2^+
Ta^+	6.1E-10	TaO^+, $TaO_2^+ \cdot (N_2O)_n$, n=0-3	-6.4	TaO^+, TaO_2^+
Ge^+	3.6E-10	$GeO^+ \cdot (N_2O)_n$, n=0,1	-4.0	GeO^+
Ir^+	2.9E-10	IrO_n^+, $IrO_2^+ \cdot N_2O$, n=1-3	-4.4	IrO^+
V^+	2.4E-10	$VO_m^+ \cdot (N_2O)_n$, m=1,2 n=0-3	-2.9	VO^+, $VO2^+$
Co^+	2.1E-12	$Co^+ \cdot (N_2O)_n$, n= 1- 3, $CoO^+ \cdot (N_2O)_n$, n = 0 - 3	-2.0	CoO^+
Cr^+	1.5E-13	$Cr^+ \cdot (N_2O)_n$, n=1, 2 $CrO^+ \cdot (N_2O)_n$, n= 0-3	-0.9	Not detected
Se^+	1.8E-12	SeO^+	-2.1	SeO^+
Zn^+	2.4E-12	$Zn^+ \cdot N_2O$	-1.2	$Zn \cdot N_2O^+$, ZnO^+/ZnN^+

As can be seen from Table 2, reactions that are slow under thermal SIFT conditions, are generally slow under DRC conditions. Faster reactions proceed fast in the DRC, too. We observed this trend for all ion-neutral combinations so far. In our opinion, the databases of the thermal reaction rate constants can be reliably used for selecting appropriate ion-molecule chemistry for ICP-DRC-MS.

3 RESOLVING ISOBARIC INTERFERENCES BY ATOM-TRANSFER REACTIONS

Electron-transfer reactions for isobaric interference removal can be efficient only when the ionization potential of the analyte ion (IP_A) is lower than that of the interfering ion (IP_{Int}). Relatively easy cases are those involving argon-containing interferences that usually have high IP. In other cases, although $IP_A < IP_{Int}$, the difference between the IPs is too small or the IPs are too low, so the choice of the neutral is challenging. One example of such interference is that of $^{90}Zr^+$ (IP=6.84 eV) and $^{90}Y^+$ (IP=6.22 eV) on $^{90}Sr^+$ (IP=5.695 eV). Although Sr has lower IP than Zr and Y, a relatively low mass reaction gas with IP of 5.8 – 6 eV cannot be found. However, removal of interferences by reactions other than charge transfer can still be possible. It was shown by Eiden et al that interfering Y^+ and Zr^+ can be suppressed by oxidation with O_2, while Sr^+ loss is low [13], allowing an opportunity for detection of ^{90}Sr in the presence of a 100-fold excess of either ^{90}Zr or ^{90}Y. Calculations of $+\Delta H_f$ according to (3) for the O-atom transfer reactions give 47.8 kcal/mol, -88.7 kcal/mol and –60.4 kcal/mol, respectively, confirming that the thermo-chemistry is beneficial in this case. The reaction rate constant reported for Zr^+ (5.5 E-10 cm^3 $molecule^{-1}$ s^{-1}) [8], shows that the reaction should be fast, and, at least for

$^{90}Zr^+$, more than 100 fold analytical advantage should be achievable if sufficient number of collisions at near-thermal velocity are provided. Figure 2 shows profiles for Sr^+, Zr^+ and Y^+ reactions with O_2 measured on ICP-DRC-MS. It shows >5 orders of magnitude suppression for Zr^+ and >4 orders suppression for Y^+, while collision/reaction losses for Sr^+ are only about factor of 2 at O_2 flow of 1.2 arb.un (corresponding to 0.83 sccm).

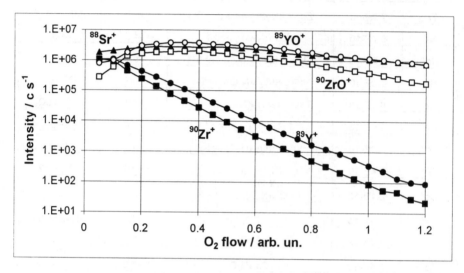

Figure 2 *Reaction profiles for Sr^+, Zr^+, Y^+ (50 μg/L) with O_2*

Mass spectra of the mixture of Sr, Zr and Y at 50 μg/L (Figure 3) show that determination of $^{90}Sr^+$ with ICP-DRC-MS should be possible in a presence of >10000 times higher abundance of $^{90}Y^+$ and $^{90}Zr^+$.

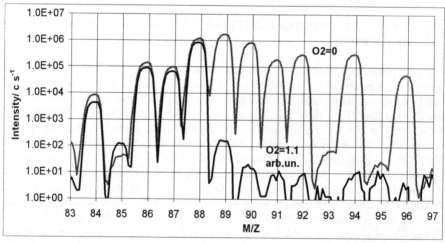

Figure 3 *Mass spectra of a mixture of Sr,Zr,Y (50 μg/L) at O_2 flow 0 and 1.1 arb.un.*

An example of isobaric interference where a reaction gas for electron-transfer cannot be found is the isobaric overlap of $^{87}Rb^+$ (IP = 4.18 eV) and $^{87}Sr^+$ (IP = 5.70 eV), the isotope pair that is used in Rb-Sr geochronology. To resolve ^{87}Rb and ^{87}Sr spectroscopically, a resolution of 286,800 is required. Usually ion-exchange techniques are used to chemically separate Rb from Sr in the sample before the analysis. Gas-phase thermochemical predictions for the reactions of Rb^+ and Sr^+ with N_2O show $\Delta H_r (Sr^+) < 0 < \Delta H_r(Rb^+)$ (see Table 1). DRC reaction profiles (see Figure 4), show that Sr^+ reacts fast (DRC ion loss rate *ca.* −5), while Rb^+ does not react.

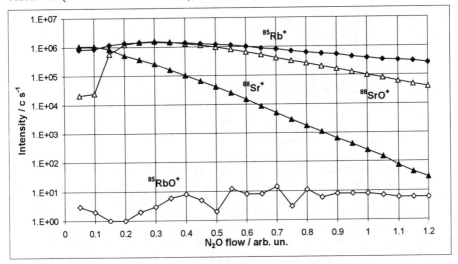

Figure 4 *ICP-DRC-MS profiles for reactions of Sr^+ and Rb^+ with N_2O*

Naturally, the choice of ion to consider as interference is arbitrary in this case. To measure $^{87}Rb^+$ in the presence of Sr, interfering $^{87}Sr^+$ may have to be suppressed by many orders of magnitude, and the efficiency of the reaction must be high. The number of collisions required to achieve high efficiency must be high (see Tanner *et al* contribution to this issue, [12]), which for the relatively heavy reaction gas (N_2O) can cause significant collisional losses of the analyte. In our case, at N_2O flow of 1.2 arb.un. (0.59 sccm), Rb^+ signal is suppressed about four times while Sr^+ is removed by *ca.* 4.5 orders of magnitude. However, to measure $^{87}Sr^+$ in the presence of Rb, high reaction efficiency is not required. Even 50 % efficient reaction converts ½ of the analyte ions into a new analyte ion at a different m/z. Hence, relatively low flow and low number of collisions can be used. In fact, as can be seen from Figure 4, collisional focusing can cause "net efficiency" to be high even for an incompletely oxidized analyte: the response for SrO^+ at a flow of 0.15-0.45 arb.un. is higher than for Sr^+ in non-pressurized mode. Typical spectra measured for a sample containing Rb and Sr at 50 µg/L (Figure 5) show that at N_2O flow of 0.35 arb.un. (0.17 sccm) collisional focusing provides enhancement of both Rb^+ and SrO^+ signals compared to Rb^+ and Sr^+ signals measured at zero N_2O flow. The accuracy of Sr isotope ratios measured as ratios of SrO is not significantly altered by minor isotopes ^{17}O (0.038%) and ^{18}O(0.2%), especially for the $^{87}SrO/^{86}SrO$ ratio. Impurities of H_2O in the reaction gas, though, promote formation of *ca.* 3 % $SrOH^+$ (mass peak at m/z=105 in Figure 5).

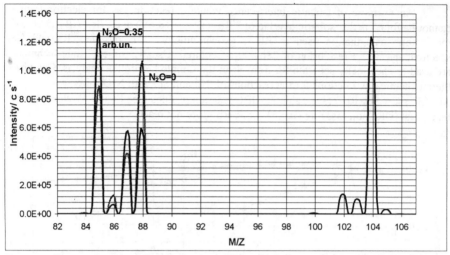

Figure 5 *Spectra for Rb + Sr sample (50 µg/L) at N₂O flow 0 and 0.35 arb.un.*

Non-interfered measurements of Sr isotope ratios in the presence of Rb by partially oxidizing Sr is possible if there is no interferences at the new m/z (102, 103, 104). Thus, the reactions of Ru^+, Rh^+ and Pd^+ must also be considered. All three reactions are thermodynamically allowed. Ru reacts relatively slowly, with a DRC ion loss rate of -1.6. Rh, however, does not react, as the electron spin multiplicity change during O-atom transfer reaction for Rh^+ forbids it [14]. For samples with significant Rh content, other chemistries have to be found. On such candidate reaction is Cl-atom transfer from methyl chloride, which is endothermic for Rb^+ ($\Delta H_r = +79$ kcal/mole) and exothermic for Sr^+($\Delta H_r = -27$ kcal/mole). The Sr^+ loss rate is almost as high as with N₂O (-4.3 vs –4.5 with N₂O), however a higher fraction of ions is lost due to scattering by the heavier gas (50.5 Da vs 44 Da) (see Figure 6).

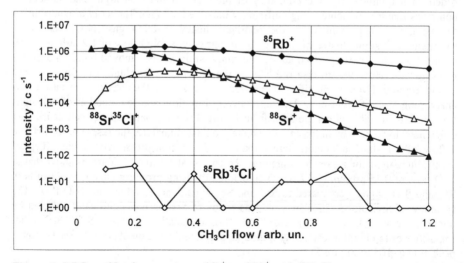

Figure 6 *DRC profiles for reactions of Sr^+ and Rb^+ with CH_3Cl*

Although the efficiency of the reaction of chlorination seems to be close to that of oxidation with N_2O, the maximum of the abundance-corrected response for $SrCl^+$ is about $1/9^{th}$ of the maximum signal for SrO^+. The cause of reduced sensitivity of $SrCl^+$ measured at $q = 0.5$ is losses of the precursor ion Sr^+ for which the corresponding effective q is 0.7, which is already sufficiently close to a stability boundary. Operating the DRC at a lower q is beneficial for higher sensitivity towards $SrCl^+$, but is undesirable for the reasons discussed below.

4 SEQUENTIAL ION-MOLECULE CHEMISTRY IN DRC

4.1. Intercepting unwanted sequential reactions

When the mass of an atom that is added to a precursor ion by an atom-transfer reaction is comparable to the mass of the precursor, the DRC quadrupole should be operated at a relatively low q to simultaneously confine the precursor and transmit the product ion. However, the efficiency of suppression of secondary ion-molecule chemistry when the DRC is operated with wider band-pass is lower. Figure 7 shows mass spectra for a de-ionized water (DIW) sample blank measured at CH_3Cl flow of 0.3 arb.un. (corresponding to 0.132 sccm) at different values of q.

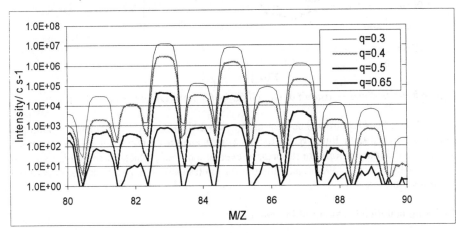

Figure 7 *ICP-DRC-MS spectra for DIW sample measured at CH_3Cl flow of 0.3 arb.un.*

When mass 83 is measured at $q = 0.3$, the low-mass boundary of stability is at about m/z =27, and most dominant precursor and product ions (Cl^+, Ar^+, CH_3Cl^+) are stable in the cell. These ions can further react with CH_3Cl forming new interferences. Abundance ratios of the most prominent mass peaks at 83, 85 and 87 a.m.u.(Figure 7), is close to those for $CHCl_2^+$, which can be possibly formed by the sequential reactions:

$$Ar^+ + CH_3Cl \rightarrow CH_3Cl^+ + Ar \qquad \Delta H_r = -104.8 \text{ kcal/mole,} \quad (5a)$$
$$\text{--->} CH_2Cl^+ + Ar + H, \qquad \Delta H_r = -62.5 \text{ kcal/mole,} \quad (5b)$$

$$CH_2Cl^+ + CH_3Cl \rightarrow CHCl_2^+ + CH_4, \qquad \Delta H_r = -15.4 \text{ kcal/mole,} \quad (6)$$

These in-cell produced interferences are suppressed by ca. 4 orders of magnitude at q=0.65, when the low-mass cut-off boundary is at m/z=59 and the dominant ions are unstable. Thus, when measurement of Rb^+ is required in the presence of Sr, high q must be selected in order to intercept formation of new polyatomic interferences. Figure 8 shows the Rb isotope ratio measured at CH_3Cl flow of 0.3 arb.un. as a function of q for the sample containing 50 µg/L of Rb and for the blank (DIW). At low q most of the signal at m/z=85 and 87 is defined by the interferences. At a q of 0.6 and higher, the interferences are suppressed by more than 4 orders of magnitude, and the $^{87}Rb/^{85}Rb$ ratio corresponds to the tabulated value.

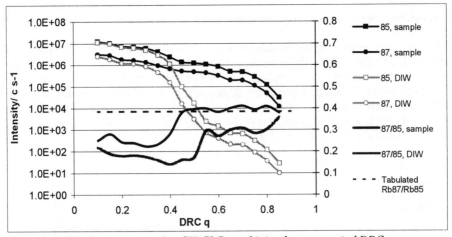

Figure 8 *Rb isotopes measured at CH_3Cl flow of 0.3 arb.un. at varied DRC q*

When Sr is required to be measured in the presence of Rb, chlorinated Sr isotopes can be measured at a lower q, as the measured analytes (m/z =119 –125) are much heavier. The dominant ions are suppressed even at a q=0.35, which corresponds to a low-mass cut-off at ca. m/z=49 and an optimum for ion transmission, q=0.5, at the mass of the precursor ions.

4.2 Using beneficial sequential ion chemistry

Sequential chemistry in the DRC can be used in order to shift the products of the primary reaction to a different m/z. In order to convert SrO^+ produced by O-atom transfer from N_2O and potentially interfered by non-reacting Rh^+, to $SrOH^+$, methane was introduced via channel A while N_2O entered the cell via channel B. The H-atom transfer from CH_4 is exothermic for $SrO^+(\Delta H_r = -28$ kcal/mole). The reaction proceeds at a rate of ca. -2.5 (1 arb.un. for CH_4 = 0.5 sccm) (Figure 9). Methane is relatively light, so additional scattering of the relatively heavy product is small and $SrOH^+$ can be measured at sensitivity similar to SrO^+ (14 MHz/ppm) even at the relatively high CH_4 flow (0.9 arb.un.) that is required to completely convert SrO^+ to $SrOH^+$.

Complete resolution of $^{87}Sr^+$ and $^{87}Rb^+$ is shown in Figure 10, where residual signal from $^{88}Sr^+$ is seen at ca. 1E-4 level of $^{85}Rb^+$ contained in a sample at equal concentration (50 µg/L). At the N_2O flow of 1 arb.un. used in this measurement, RbO^+

signal is more than 4 orders of magnitude lower than SrO$^+$. Full chemical resolution of ^{87}Sr and ^{87}Rb at the 0.01 % level of peak height corresponds to a spectral resolution of ca. 1,100,000 (f.w.h.m.) for ideal Gaussian peak shape.

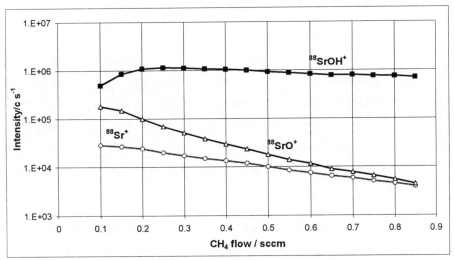

Figure 9 *Profile of reaction of SrO$^+$ and Sr$^+$ (50 µg/L)with CH$_4$; N$_2$O flow = 0.5 arb.un.*

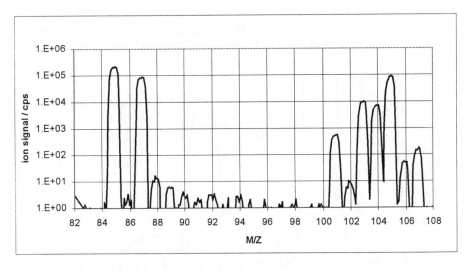

Figure 10 *Mass spectrum of Sr+Rb sample (50 ppb) measured at N$_2$O flow of 1 arb.un. and CH$_4$ flow 0.7 arb.un.*

Sequential chemistry can also be used for producing dioxides in the instance where the analyte oxide m/z is interfered. An example of such situation is the interference of ^{35}Cl^{16}O$^+$ on ^{51}V$^+$. V$^+$ is easily oxidized by O-atom transfer from N$_2$O (see Tables 1 and 2), however ClO$^+$ reacts with N$_2$O forming ClO$_2$ $^+$(see Figure11a). Also, for Zn-containing samples, ^{67}Zn$^+$ would interfere with ^{51}VO$^+$, Zn$^+$ having low reactivity towards

N_2O. Further oxidation of VO^+ to VO_2^+ provides better background equivalent concentration and detection limit at m/z=83 than at m/z=67 for the sample matrix of 3.3 % HCl. Note that the minor isotope of Kr (IP=14.0 eV) at m/z=83 is removed by charge-exchange with N_2O (IP=12.89 eV).

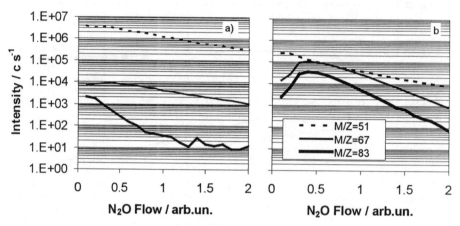

Figure11 *Reaction profiles with N_2O for the samples containing 3.3% HCl (a) and V at 10 µg/L (b) in DIW at q=0.4*

As $^{51}V^+$ is shifted 32 a.m.u. up, relatively low q = 0.4 is used in order to simultaneously confine it and transmit VO_2^+. Also, autolens was kept at an optimum voltage for $^{51}V^+$ in order to optimize its transmission into the cell.

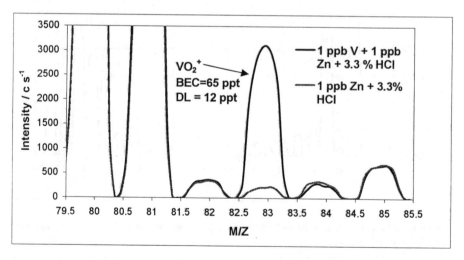

Figure 12 *Mass spectra for sample of V at 1 µg/L in 3.3 % HCl + 1 µg/L Zn matrix, measured at N2O flow of 0.5 arb.un. and q=0.4*

5 OXIDATION REACTIONS FOR As⁺ AND Se⁺

As⁺ reacts fast with N_2O (Table 1), forming AsO⁺ at m/z=91. Potential interference from $^{91}Zr^+$ is slight because Zr^+ reacts very fast (DRC ion loss rate –6.0). $Ar^{35}Cl^+$ does not react with N_2O by O-atom transfer, but is removed at a rate of –7.9, likely by charge-exchange (Figure13).

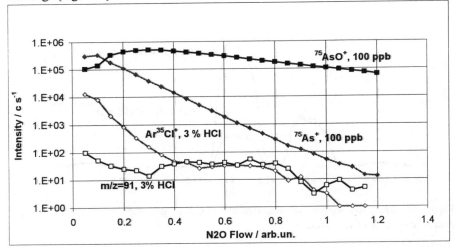

Figure 13 *As⁺ and ArCl⁺ reaction profiles with N_2O*

Similarly, As⁺ reacts with O_2 by O-atom transfer. Another isobaric interference at m/z =75, $^{40}CaCl+$, in our experience, produces lower background equivalent concentration at m/z=91 for AsO⁺ in reactions with O_2 than with N_2O. Oxygen also produces lower scattering losses. Reaction profiles for As⁺ with O_2, in a matrix containing 1% HCl + 10 mg/L Ca matrix are shown in Figure 14.

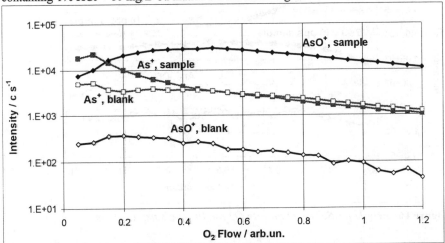

Figure 14 *Reaction profiles with O_2 for As⁺ for a sample containing 10 µg/L As + 10 mg/L Ca + 1 % HCl and a method blank*

CaCl$^+$ does not react by electron transfer with O$_2$ or N$_2$O as its IP=5.6 eV is lower than the IPs of those gases, and, in fact, of most of ready-available reaction gases. Thus, shifting the analyte appears to offer the best opportunity in this case. Oxidation of As$^+$ allows its detection at the ppt level even in a challenging matrix, such as Ca + HCl (see Figure15). Note that potential interference from ^{91}Zr+ is efficiently removed by its oxidation (at DRC ion loss rate of – 4.5).

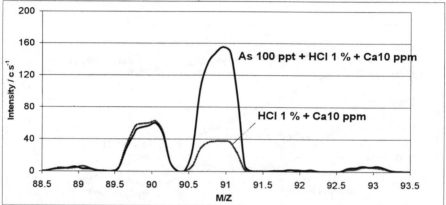

Figure 15 *Mass spectra for the sample containing 100 ng/L of As in 1 % HCl + 10 mg/L matrix at O$_2$ flow of 1.2 arb.un.*

Oxidation with N$_2$O or O$_2$ works effectively also for shifting Se$^+$, isotopes of which are interfered by Ar$_2^+$, Ar$_2$H$^+$ and Kr$^+$, to SeO$^+$. Reaction with O$_2$ is slower than with N$_2$O,

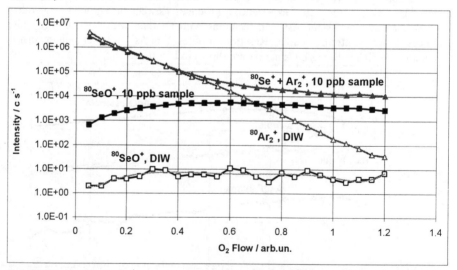

Figure 16 *Profile for Se$^+$ reaction with O2 for 10 µg/L sample*

However, the background equivalent concentration and detection limits were better, probably due to lower collisional losses. Note that Ar$_2^+$ (IP=14.5 eV) and Kr$^+$ (IP=14 eV) react with O$_2$ by electron transfer (IP=12.1 eV) (Figure16). As the reaction of Se$^+$ with O$_2$ is slow, Se isotopes can be measured either as bare ions or as oxide ions. Although reaction of Ar$_2^+$ is much faster than that of Se$^+$ (DRC ion loss rate – 4.4 vs –0.6, respectively), oxidation provides the better BEC and detection limits. Only at a very high flow is Ar$_2^+$ removed to an extent that the Se BEC at m/z=80 becomes comparable to that at m/z=96. This is a clear demonstration that even low efficiency reactions can be beneficial in the DRC when the analyte is shifted to a different m/z. Conversion of only *ca.* 10 % of the Se$^+$ to SeO$^+$ provides single digit ppt BECs, because the m/z to which the analyte is shifted is not interfered (Figure 17).

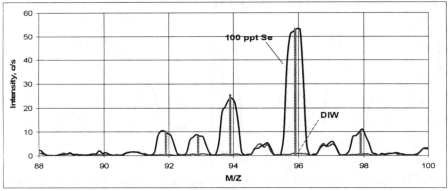

Figure 17 *Mass spectra for 100 ng/L Se sample measured at O$_2$ flow of 0.6 arb.un.*

6 OPPORTUNITIES FOR CaO$^+$/Fe$^+$ INTERFERENCE CHEMICAL RESOLUTION

CaO$^+$ and CaOH$^+$ interference on Fe$^+$, Ni$^+$, Cu$^+$ and Zn$^+$ is one of the most challenging in current ICP-MS. The low ionization potentials of these polyatomics (6.9 eV and 5.7 eV) in comparison to the IPs of the analytes (7.9 eV, 7.6 eV, 7.7 eV and 9.4 eV, respectively), precludes using cold plasma conditions, as well as electron-transfer reactions. Oxidation of the analyte ions with N$_2$O is thermodynamically allowed only for Fe$^+$ and Ni$^+$.

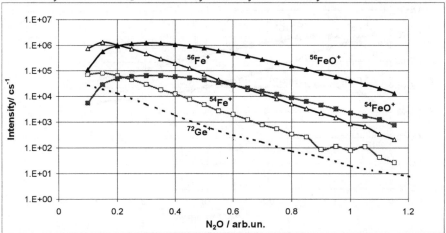

Figure 18 *Fe$^+$ and Ge$^+$ oxidation reaction profiles with N$_2$O, q=0.45*

Reaction with Fe proceeds at a DRC ion loss rate of –3.9, FeO^+ being a major product (Figure18). Potential interference of $^{72}Ge^+$ is removed by oxidation at a similar rate.

Thus, Fe can be detected as FeO^+ at low ppt BEC at m/z =72(Figure19).

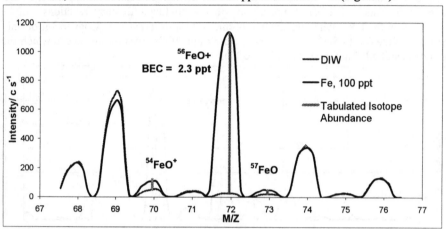

Figure 19 *Mass spectra for Fe at 100 µg/L measured at N2O flow =0.3 arb.un , q=0.45*

We could not locate any data on thermodynamics nor reaction rate constants for the possible reactions of CaO^+ with N_2O. Our data shows that Ca^+ oxidizes in the DRC, forming both CaO^+ and CaO_2^+, with the DRC analyte loss rate for Ca^+ of *ca.* - 4.0 and for CaO^+ of *ca.* –2.6 (Figure 20). Thus, there is some indication that the oxidation rate for CaO^+ is lower than that for Fe^+, although the measurement involves CaO^+ ions formed predominantly in the DRC. Measurement of oxidation rate of plasma formed CaO^+ ions is troublesome –under any plasma conditions we could not produce more than 0.1% of CaO^+.

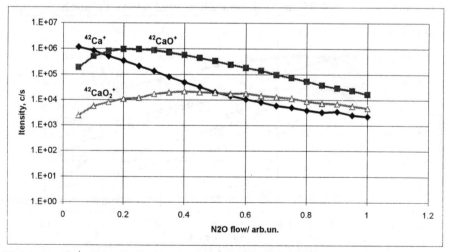

Figure 20 *Ca^+ (10 µg/L) oxidation reaction profile with N_2O, q=0.45*

Another possible way to resolve the CaO^+ interference on Fe^+ is by O-atom abstraction reaction with CO. The reactions are thermodynamically favorable:

$$CaO^+ + CO \longrightarrow Ca^+ + CO_2, \qquad \Delta H_r = -50 \text{ kcal/mole}, \qquad (7)$$

$$Fe^+ + CO \rightarrow FeO^+ + C, \; \Delta H_r = 182 \text{ kcal/mole}, \qquad (8)$$

$$Fe^+ + CO \rightarrow FeC^+ + O, \; \Delta H_r = 163 \text{ kcal/mole}, \qquad (9)$$

Preliminary data show *ca.* 10-fold improvement in BEC for Fe^+ measured in the presence of 50 mg/L Ca in the sample (Figure 21). Reactions of ArO^+ and Fe^+ with CO have been studied using the ICP-SIFT and showed that Fe^+ is unreactive towards CO ($k<2E-13$ cm^3 molecule^{-1} s^{-1}), while ArO^+ reacts by O^+ transfer ($k=5.0E-10$ cm^3 molecule^{-1} s^{-1}) [9]. Thus, CO appears to be an attractive gas to remove both ArO^+ and CaO^+ isobaric interference on Fe^+. The purity of CO becomes very important, as any traces of water or other oxidizing gases in CO can cause CaO^+ formation. The data shown here are preliminary; further study of O-atom transfer from CaO^+ and $CaOH^+$ is required.

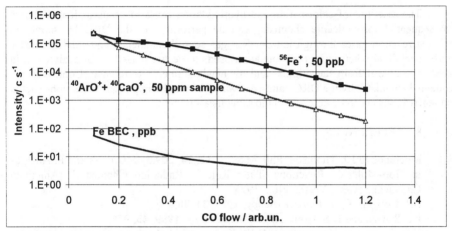

Figure 21 *Reaction profiles for Fe^+ and CaO^+ with CO. The signals presented are blank (DIW) subtracted*

7 CONCLUSIONS

Ion-molecule chemistry, recently implemented in commercial ICP-MS, is an important tool in resolving isobaric interferences. To date, most of the reactions used involve reactive removal of the interfering ion from the m/z of the analyte by (mainly) charge transfer and H-atom transfer reactions. We have shown here that O-atom transfer reactions can be controllably used with the ICP-DRC-MS. It has been shown that for the reactions studied, thermochemical and kinetic data can successfully predict the ion-molecule chemistry in the DRC. Comparison of the reaction rate constants measured on ICP-SIFT with those of the DRC show that in general, reactions that are slow under thermal SIFT conditions are slow under DRC conditions, and that faster reactions proceed fast in the DRC, too.

Many challenging interferences that cannot be removed from the analyte m/z, can be resolved by shifting the analyte to a non-interfered m/z by atom transfer reactions. In comparison to interference removal reactions, the analyte shift reactions can be useful even if they are relatively slow – provided the rate of the reaction is higher than the scattering loss rate and that the new m/z is non-interfered. In many instances, possible interference at a new m/z is removed by reaction with the same reaction gas. In some instances, the same gas can be used for removal of the interference from the m/z of the analyte and shifting some of analyte ions to the m/z+16 mass channel. This provides a choice of mass channels for the measurement; the choice can be made depending on the application.

For isobaric atomic ion resolution, shift of one of the ions to a different m/z can deliver chemical resolution that is equivalent to a spectral resolution of > 1,000,000 (f.w.h.m.) at high abundance sensitivity (ideal Gaussian peak shape).

Oxygen atom transfer reactions may offer a potential means to resolve CaO^+ interference on Fe^+ - either by removing O-atom from CaO^+ or by converting Fe^+ to FeO^+.

The reactions of chlorination and hydroxylation can also be used. It was shown that sequential ion-molecule chemistry can be performed in the DRC by supplying different reaction gases through independent gas control channels. Sequential chemistry can be beneficial when more than one atom needs to be transferred to an analyte ion in order to resolve it from the original and /or newly formed interference. Control of the sequential chemistry by the DRC band-pass was shown to be important when non-reacted and reacted ions are to be detected in the same measurement.

8   REFERENCES

1.   Reference to J.J. Thomson, Phil. Mag., 1912, **24**, 209 by S.G. Lias and P. Ausloos in "Ion-Molecule Reactions: Their Role in Radiation Chemistry", American Chemical Society, Washington, 1975.
2.   D.J. Douglas, *Can. J. Spectroscopy*, 1989, **34**, 38.
3.   J.T. Rowan and R.S. Houk, *Appl. Spectroscopy*, 1989, **43**, 976.
4.   G.C. Eiden, C.J. Barinaga and D.W. Koppenaal, *J. Anal. At. Spectrom.*, 1996, **11**, 317.
5.   D.S.Bollinger and A.J. Schleisman, in *Plasma Source Mass Spectrometry: New Developments and Applications*, eds G. Holland and S.D. Tanner , Royal Society of Chemistry, Cambridge, UK, 1999, 80.
6   S.G. Lias, J.E. Bartmess, J.F. Liebman, J.L. Holmes, R.D.Levin and W.G. Mallard, *J. Phys. Chem. Res. Data*, 1988, **17**, 1.
6.   G.K. Koyanagi, ..., this book.
7.   V.G. Anicich, *The Astrophysical Journal Supplement Series*, 1993, **84**, 215. Also see http://astrochem.jpl.nasa.gov/asch/
8.   G.K. Koyanagi, V.V. Lavrov, V. Baranov, D. Bandura, S. Tanner, J.W. McLaren and D.K. Bohme, *Int. J. Mass Spectrom.*, 2000, **194**, L1.
9.   G.K.Koyanagi, V.I. Baranov, S.D. Tanner and D.K. Bohme, *J. Anal. At. Spectrom.*, 2000, **15**, 1207.
10.   S.D.Tanner, ..., this book.
11.   S.D. Tanner and V.I. Baranov, *J. Am. Soc. Mass Spectrom.*, 1999, **10**, 1083.
12.   G.C. Eiden, C.J. Barinaga and D.W. Koppenaal, *Rapid Commun. Mass Spectrom.*, 1997, **11**, 37.

13. G.K. Koyanagi, V.V. Lavrov, D.K. Bohme, V. Baranov, D. Bandura, S. Tanner, J.W. McLaren, Paper FP 36, 2000 Winter Conference on Plasma Spectrochemistry, Ft. Lauderdale, Florida, January 10-15, 2000.

A REACTION MECHANISM FOR SOLVING THE OXIDE PROBLEM IN ICP-MS ANALYSIS OF THE NOBLE METALS

Lorna A. Simpson[1], Maryanne Thomsen[2] and Brian J.Alloway[1].

[1]Postgraduate Research Institute for Sedimentology, University of Reading, Whiteknights, PO Box 227, Reading, RG6 6AB. UK

[2]Perkin Elmer Instruments, Post Office Lane, Beaconsfield, Bucks. HP9 1QA. UK

1 INTRODUCTION

Concerns over increasing Pt levels in the environment have generated interest in quantifying noble metal (Ru, Rh, Pd, Ag, Os, Ir, Pt, Au) concentrations in a range of environmental matrices. The low natural abundance of this group necessitates low detection limits, which are not achievable by many common analytical methods. Inductively Coupled Plasma Mass Spectrometry (ICP-MS) is one of the few spectroscopic techniques with the required sensitivity. ICP-MS offers improved detection limits over conventional methods, often by several orders of magnitude for the noble metals. Additionally, the technique offers rapid, multi-element analysis, allowing a range of other elements to be analysed simultaneously.

Removal of any of the interferences for the noble metals would represent a significant improvement in analysis protocols. The analysis of the noble metals is subject to a number of oxide-based interferences, which can be removed by utilisation of Dynamic Reaction Cell (DRC) technology. Quantification of the oxide-generation effect has been largely ignored, despite evidence that suggests that it may, in some analyses, constitute a major source of interference. This effect potentially causes a significant source of error for the noble metals.

Initial work has outlined instrumental parameters that affect oxide generation in ICP-MS analysis, and correction strategies are discussed. Data are presented to show the percentage oxides generated for several environmentally important isotopes in relation to the standard CeO/Ce ratio. In contrast to conventional ICP-MS analysis, where the formation of oxides is undesirable and operating conditions are optimised to limit their production to a worst-case scenario of $< 3\%$ of the total ion population, this work outlines a novel approach, which focuses upon promoting the formation of oxides through the introduction of an oxidizing gas into a DRC-ICP-MS instrument. The success of the methodology is explored through analysis of results from Pt, Au and Ag isotopes, which are severely interfered with by HfO, TaO, ZrO and NbO.

2 OXIDE BASED INTERFERENCES

Oxide interferences are due to the presence of entrained air, or matrix components[1], and are caused by incomplete dissociation or recombination in the tail flame of the plasma. Oxide species occur 16, 17 and 18 atomic mass units (amu) above the parent ion, corresponding to the three isotopes of oxygen (^{16}O, ^{17}O and ^{18}O). Figure 1 illustrates the oxide peak for Ce, which is the most strongly forming oxide element.

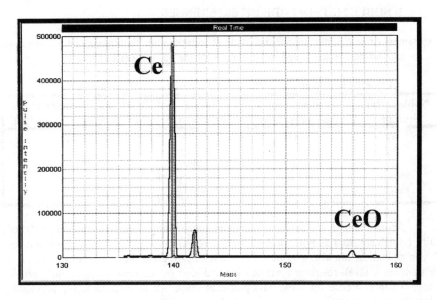

Figure 1 Scan of the *Ce and CeO peak*

The ELAN 6000 is routinely optimised to limit oxide production to <3% of the total ion population for Ce, since it is in effect the worst-case scenario. This is achieved in tandem with limiting doubly charged species by selecting a compromise value for nebulizer gas flow rates. The source of error from oxide interferences is often considered to be minimal, and few authors have made further attempts to quantify the error, choosing merely to limit it to a known percentage of the CeO/Ce total ion population through manipulation of instrument operating parameters, despite evidence that suggests that there may, in some analyses, be a major source of interference.

There are two notable exceptions:

1. Falkner and Edmund[2] published a method for preconcentration of Au utilising a cyanide complexing reaction [Au(Cn)$_2^-$]. The authors approached the Pt hydride and Ta oxide interferences on 197 Au by making the assumption that the species were not present at levels above 1% and 5% respectively, and calculating a correction factor on that basis.

2.   Sylvester and Eggins[3], in their study of the PGE levels in basalts, employed a correction strategy based upon analysis of solutions prior to analysis, which contained the potentially interfering species but not the analytes themselves. They found corrections did not exceed 3%, except in the case of HfO on Pt and TaO on Au. (^{194}Pt 24%, ^{195} Pt 6.5%, and ^{197}Au 3.8%).

Apart from these examples, (which both consider noble metals), the oxide effect has been assumed to be negligible.

3 NOBLE METAL OXIDE INTERFERENCES.

Assessment of oxide interference and oxide abundance data highlighted the following as the most significant oxide-based interferences for the noble metals (Table 1).

Table 1 *Significant oxide-based interferences in noble metal analysis*

Noble metal isotope	Relative abundance (%)[4]	Oxide interference	Relative abundance of the oxide forming species (%)[4]
^{107}Ag	51.839	^{91}Zr^{16}O	11.22
^{109}Ag	48.161	^{93}Nb^{16}O	100
^{197}Au	100	^{181}Ta^{16}O	99.988
^{194}Pt	32.9	^{178}Hf^{16}O	27.297
^{195}Pt	33.8	^{179}Hf^{16}O	13.629

Applications in the precious metals industry highlighted the problem of Zr ball mills, (which are used as a practical alternative to Pt) causing contamination of samples. Zr contamination can also originate from borosilicate glass digestion vessels, (such as carius tubes[5]), alkali-resistant glass vessels and alumina ceramic mortars and pestles[6]. The ^{181}Ta interference on ^{197}Au is significant since Au is monoisotopic, preventing selection of an alternative isotope for analysis.

Preliminary experiments were undertaken to quantify the generation of oxides for each of these interfering elements. This was achieved through measurement of 10 ppb solutions of each interferent, in a data-only method, analysing each solution at the mass, and also at its mass + 16amu, where the oxide would occur. This experiment was repeated for varying nebulizer gas flow rates (to simulate 1%, 2%, 3%, 4% CeO/Ce ratios), to confirm that the relationship remained linear. Results confirmed the predicted linearity, and Figure 2 illustrates the results for the percentage oxides generated normalized to CeO/Ce.

Although the percentage oxides generated may initially seem minimal, the significance becomes apparent upon consideration of crustal abundance data. The average abundance of both the analyte and interference in crustal rocks illustrates the problem (Table 2).

The interfering analytes are consistently >2 orders of magnitude more abundant, which makes the contribution from their oxides a significant interference, especially in geological samples.

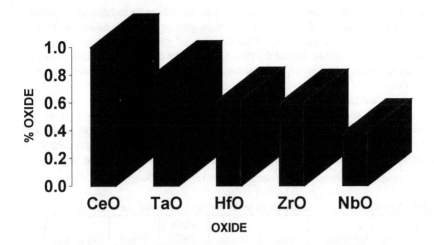

Figure 2 *Histograms to illustrate the percentage oxides generated for the interfering elements, normalized to 1% CeO.*

Table 2 *Average abundance in crustal rocks*

Element	Average crustal abundance (ppm)[7]
Ta	2
Au	0.003
Hf	3
Pt	0.005
Zr	150
Nb	20
Ag	0.05

4 SOLUTIONS TO OXIDE INTERFERENCES

A range of methods have been employed to correct for oxide-based interferences.
1.   Matrix removal
2.   Plasma modifications
3.   High resolution mass spectrometry
4.   Mathematical corrections
5.   Post plasma techniques[8]

   Matrix-removal techniques include utilisation of on-line chelating exchange resins or techniques such as ultrasonic nebulization, which removes the water, which is responsible for the oxide generation from the plasma. Plasma modifications include cool plasma techniques, which tend to suffer from greater baseline noise and matrix intolerance, manifesting in instability. Optimisation of plasma operating conditions could include reducing the sample uptake rate, reducing the nebulizer gas flow rates and even chilling the spray chamber. Mathematical corrections are the traditional approach to the problem.

Analysis using high-resolution instruments is another possible method of analysis. However, such instruments require increased resolving powers to separate oxide-based interferences at the heavy end of the spectrum[9], which encompasses the range of the noble metals. Each increase in resolution is inherently connected to a concomitant loss in sensitivity. Table 3 illustrates the calculated resolution required to separate the oxide interferences being considered here; Resolution = $m/\Delta m$, [10]

Table 3 *Resolution required to separate the noble metal oxide interferences*

Isotope	amu	Interferent	amu	Oxide	Mass Diff.	Mass	Resolution
^{107}Ag	106.91	^{91}Zr	90.91	106.90	0.0045	107	23604. 68
^{109}Ag	108.90	^{93}Nb	92.91	108.90	0.0035	109	31457.43
^{194}Pt	193.96	^{178}Hf	177.94	193.94	0.0240	194	8068.54
^{195}Pt	194.96	^{179}Hf	178.95	194.95	0.0240	195	8111.82
^{197}Au	196.97	^{181}Ta	180.95	196.94	0.0236	197	8334.74

The final group of solutions to the oxide problem are post plasma techniques. These have the advantage over many other methods in that they do not interfere with plasma chemistry or the instrument operating conditions. The Dynamic Reaction Cell is one such example of a post-plasma technique.

5 DYNAMIC REACTION CELL

The DRC provides a form of chemical resolution and was described in its simplest form as consisting of an "enclosed cell pressurised with a reactive gas through which the ion beam is passed"[11]. The reaction gas enters the cell as a neutral species and is mixed with the ion beam colliding with the ions. The ultimate choice of reaction gas is critical and should combine a very high efficiency for the so-called 'purifying reaction', i.e. the complex reactions between the ion beam and reaction gas, in the cell for the interfering species the analyst is seeking to remove, and conversely a low reaction efficiency for the analyte(s) of interest. The efficiency of the reaction is related empirically to the rate constant for that reaction; therefore, the reaction gas is chosen so that the rate constant for the reaction of the analyte ion is significantly smaller than that for the interference ion[12].

The DRC consists of a standard quadrupole ICP-MS with the DRC inserted between the ion lens/shadow stop assembly and the main quadrupole (Figure 3). The cell is an enclosed vessel and the DRC chamber can be vented and the gas evacuated to allow the ELAN 6100 to operate as a conventional ICP-MS. The DRC then operates merely as an elongated form of RF-only prefilter[14]. The principal improvement between the ELAN 6100 DRC and other systems employed by similar ICP-MS instrumentation is that, owing to the inclusion of devices that scan a bandpass range, which varies in synchronisation with the mass analyser, sequential chemical reactions are largely suppressed. This maintains low background counts and thereby improves the signal-to-noise ratios achievable[14], and allowing more complex reaction gases to be utilised than is possible with conventional collision cells[15]

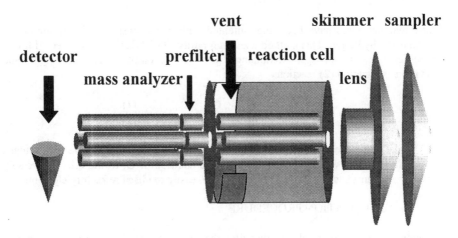

Figure 3 *Schematic of a Dynamic Reaction Cell[13]*

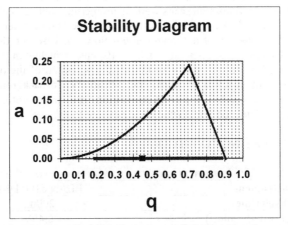

Figure 4 *Stability Diagram to Calculate Bandpass.*
q=0.45,a=0

The bandpass is controlled by manipulation of RPa for a high-mass cutoff by control of the DC voltage applied to the rods, and for a low-mass cutoff control of RPq, through the application of RF. The bandpass or low- and high-mass cut-offs are calculated using a simple EXCEL programme, which incorporates a stability diagram (Figure 4.) The EXCEL programme calculated that for the bandpass parameters used in this work RPa = 0, RPq =0.45 (the default DRC parameters), the low bandpass cut-off for 109Ag would be at 54amu, and the high mass cut-off was left open (infinity).

6 OXYGEN REACTION GAS

Many reaction gases have been experimented with in the past, principally ammonia, methane and hydrogen. This method uses high purity (99.999%) oxygen, selected for its properties as a strongly oxidising gas, to promote oxidation (1) and oxidation/association (2) reactions

$$\mathbf{M + O_2 \Rightarrow \quad MO + O} \qquad (1)$$
$$\mathbf{M^+ + O_2 \Rightarrow \quad MO_2^+} \qquad (2)$$

The noble metals are chemically inert elements and initial simple experiments confirmed that they would not oxidise with the addition of O_2 to the cell. Conversely however elements such as Zr, Hf, Nb and Ta are, easily oxidised refractory elements.

7 OPTIMISATION PROCEDURE

Composite scans were produced that plot the matrix blank, the spiked solution and Estimated Detection Limits (EDL), allowing the user to select an optimum cell gas flow rate. The widest parameter range was entered for cell gas flow optimisation (0.5 – 3.0 mL/min with 0.05 steps), and the matrix blank solution was aspirated, then optimised. This procedure was repeated for a solution containing the matrix blank +10 ppb of an analyte spike. The two optimisation scans were then superimposed using the "composite view" feature in the ELAN NT operating software, and the optimum cell gas flow rate was determined. The selection of optimum cell gas flow was based on the point where counts for the interfering species are below 10 cps, which can be considered background, and EDL's are the best, with the least suppression of the original signal. The EDL are based upon ratios of signal (net sensitivity) to background noise (the square root of the blank), with a 1 second integration time, and therefore give a good estimate of the actual detection limits which it is possible to achieve.

8 DRC OPERATING CONDITIONS

Table 4 *DRC Instrument operating parameters/specifications*

Instrument	ELAN 6100 DRC
Dwell time	200ms
RPq (Low mass cutoff)	0.45V
Rpa (high mass cutoff)	0V

Cell path voltage (CPV)	-18V
Cell rod offset (CRO)	0V
Quadrupole rod offset (QRO)	-7.5V
Nebulizer gas flow	0.97 L/min
Spray chamber / nebulizer	Quartz Cyclonic/ Meinhard A3
Lens voltage	Optimised individually per element.
RF power	1300
Gas channel: B-channel with no getter device fitted.	Oxygen 99..999% pure

9 RESULTS

The new mechanism was first tested on ^{109}Ag, the main oxide of Nb being the interfering oxide. Figure 5 shows clearly that no AgO$^+$ is created, but the interference is initially reduced, then removed by the introduction of relatively small volumes of O_2 reaction gas. There was some small loss in the Ag signal, which is caused by collisional scattering in the cell environment, but the total removal of the interference improved the EDL.

The optimisation procedure was repeated for ^{107}Ag, which is subject to a ZrO interference (Figure 6), and the optimum cell gas flow rate was identical. The arrow on the optimisation profiles indicates the optimum cell gas flow rates and is selected by user interpretation of the scans. It can be seen that the new reaction mechanism fulfils all the parameters for a successful cell gas reaction. The interference reacts fully with the O_2 reaction gas, but as predicted not with the noble metal analyte. A scan of the spectrum (Figure 7) confirms the success of the reaction, superimposing the isotopic 'fingerprint' of Ag, and demonstrates that the match with the isotopic ratios is correct. No peaks are evident from the interfering refractory oxide; the hydroxide species are produced by trace impurities in the oxygen reaction gas. This reaction mechanism cannot be recommended in this instance, since the amount of H_2 present depends solely on the purity of the reaction gas used and will consequently vary. The ELAN 6100 DRC instrument includes a 'getter' device that is designed to dry gases prior to analysis to remove such impurities. However, passing oxygen over the getter would only serve to saturate the device. Isotope-ratio experiments were carried out to ensure that a complete removal of the interference was occurring and that the isotopic ratios of the analyte were not affected by the reaction. Table 5 demonstrates excellent agreement for replicate analysis between the natural isotopic abundance and that measured utilising the DRC. Generally we see a factor-of-two improvement in isotope-ratio precision when employing cell gas, owing to collisional damping effects causing the removal of small-scale signal fluctuations [16,17]

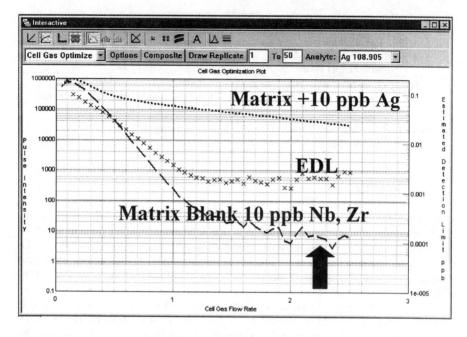

Figure 5 *Optimisation Profile for ^{109}Ag, with NbO interference*

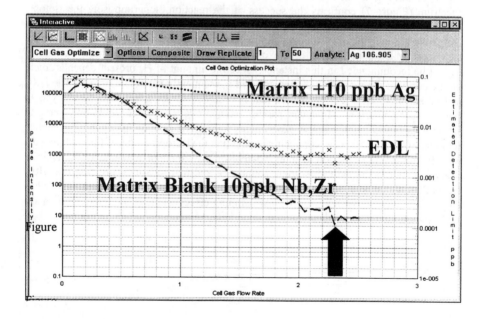

Figure 6 *Optimisation profile for ^{107}Ag, with ZrO interference*

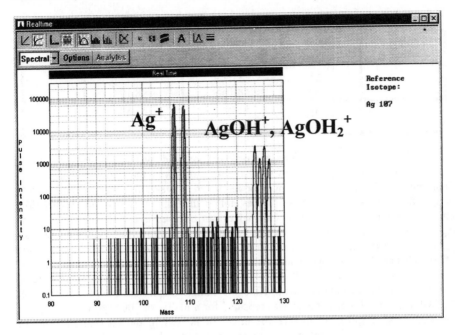

Figure 7 *Scan of the Spectrum with an Optimised Cell for Ag*

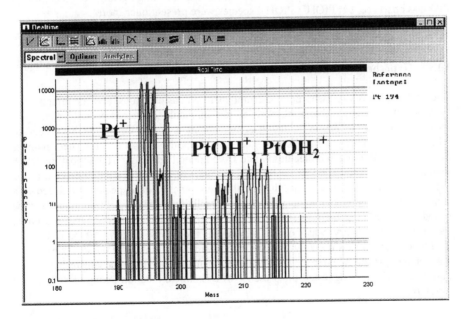

Figure 8 *Scan of the Spectrum with an optimised Cell for Pt*

Table 5 *Isotope ratio results for replicate analysis of Ag*

Replicate	Isotope	Natural abundance (%)	DRC measured abundance (%)	% RSD
1	^{107}Ag	51.839	51.81	0.06
	^{109}Ag	48.161	48.19	
2	^{107}Ag	51.839	51.82	0.1
	^{109}Ag	48.161	48.18	

The optimisation procedure was repeated for gold, which is interfered with by TaO. The optimisation profile was similar to that of silver, the intensity at a cell gas flow rate of 2.25Ml/min was slightly above 10 cps, however this was almost certainly due to the memory effects that Au causes, rather than the DRC requiring higher flow rates. The Au and TaO interferent and analyte pair behaved in exactly the same way as Ag. The scan was less clear since there is only one isotope and therefore no isotopic pattern to confirm the results. Total removal of the TaO interference, and the generation of $AuOH^+$ and $AuOH_2^+$, hydroxide species was observed, but no AuO^+ was produced. The optimisation scans for ^{194}Pt, the most abundant isotope of Pt, which is interfered with by the most abundant isotope of Hf, again exhibited optimum cell gas flow rates of 2.25Ml/min. The Pt optimisation was repeated for a matrix of 100 ppb, with 10 ppb of the analyte as a spike, to observe the effect of increasing the concentration of the interferent. The optimum cell gas flow rate was found to be comparable to that for 10 ppb of matrix, and the Pt scan (Figure 8.) for that reaction illustrates the removal of the HfO interference. There is a perfect fit with the isotopic fingerprint. No generation of PtO was observed, but $PtOH^+$, $PtOH_2^+$ species were present once more.

10 REACTION MECHANISMS

10.1 Hypothesis 1- Thermodynamic and Kinetic Considerations

The first hypothesis is based upon thermodynamic considerations, and the effects on a species kinetic energy that a number of ion molecule collisions would cause. The introduction of an effective oxidizing agent, in this example O_2, would have been to facilitate a vast number of collisions. The initial hypothesis was that the sheer number of reactions with oxygen molecules would have meant that the interferent lost too much kinetic energy to remain stable in the cell environment, and ultimately to make it to the detector. Tanner[18] suggested that the initial kinetic energy of a species is reduced to a small fraction of its initial value after only 5-8 collisions. This hypothesis was easily be tested by considering the scattering rates of the analyte and interferent, they would be identical if the theory was correct. However, the scans clearly demonstrate very different rates, so it is unlikely that the interferent could be scattered so dramatically, but not the analyte of interest. This observation made the first mechanism hypothesis highly unlikely and suggested the presence of a chemical reaction taking place with the interferent.

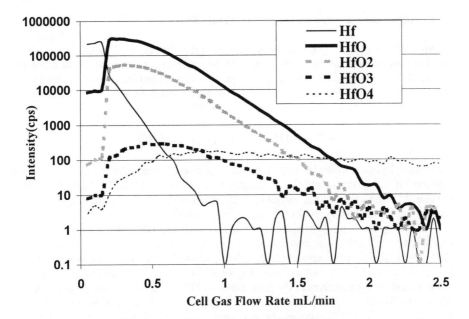

Figure 9 *Graph to show the generation of multiple Hf oxide species, throughout an optimisation procedure*

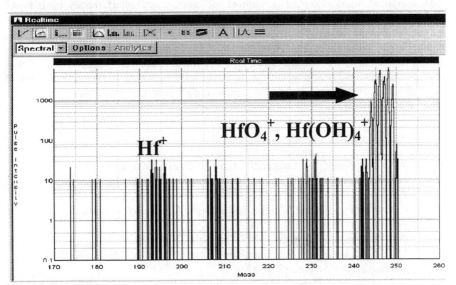

Figure 10 *Scan showing the removal of HfO⁺, and the formation of Hf multiple oxide and hydroxide species*

10.2   Hypothesis 2 – Formation of New Oxide Species

Detailed analysis of the optimisation profiles indicated that at very low flow rates, i.e. at cell gas flow rates where the mass flow analyser had just begun to control flow we see an initial increase in the interfering species profile. The second hypothesis is based upon the concept that due to the efficiency of the oxidizing reactions we were witnessing the generation of new multiple oxide species. For example:

$$Hf^+ \Rightarrow HfO_2^+ \Rightarrow HfO_3^+ \Rightarrow HfO_4^+$$

New multiple-oxide species would be difficult to detect because the species occur at the high end of the spectrum and the parent ion + 16, 32, 48 or 64 amu would often occur outside the mass range of the instrument. Complete scans of the entire mass range for Nb, Ta, and Zr showed no peaks for such multiple oxides.

The optimisation procedure was repeated with the 100 ppb Hf solution, but on this occasion ^{178}Hf was monitored alongside Pt^{194}, which in the matrix blank represents HfO^+, and additionally masses 210, 226, 242 and 258 which represent HfO_2^+, HfO_3^+, HfO_4^+, and HfO_5^+ respectively. Figure 9 shows the formation of the multiple-oxide species and demonstrates the persistence of HfO_4 in a fully-optimised cell. A full mass-scan for Hf with an optimised cell revealed a series of peaks at masses 244 - 250 amu, at the limit of the instruments analytical range, which can be attributed to the production of a mixture of HfO_4^+ and $Hf(OH)_4^+$ (Figure 10).

11 CONCLUSIONS

The analytical significance of a number of oxide-based interferences in ICP-MS analysis has been highlighted and discussed, especially for the noble metals. The success of oxygen as a reaction gas in promoting oxidation reactions, which remove oxide based interferences has been reproducibly demonstrated. Consideration of the mechanism behind the reaction leads to a hypothesis based upon generation of new multiple oxide(s) and ozone species. A new reaction mechanism has been demonstrated that improves detection limits for noble metal analysis by removal of oxide interferences, which can represent a significant source of interference.

12 REFERENCES

1   W.G. Diegor and H.P.Longerich. *Atom.Spectros.*, 2000, **21(3)**, 11.
2   K.K. Falkner and J.M. Edmond, *Anal.Chem.*, 1990, **62,** 147.
3   P.J. Sylvester and S.M. Eggins, *Geostand Newsl.*, 1997. **21(2),** 215
4   Commision on Atomic Weights and Isotopic Abundances. Pure and
        Appl.Chem., 1991, **63(7),** 991.
5   D.G.Pearson and S.J.Woodland, Chem.Geol.,2000, **165**, 87
6   M.Zeif and J. Mitchell, Contamination Control in Trace Element Analysis
        Chem.Analysis Series., 1976, **47**
7   A.W. Rose, H.E. Hawkes and J.S. Webb, Geochemistry in Mineral Exploration
        2nd Ed. Academic Press Inc.London. Chapter 2, p30

8   D.S. Bollinger .and A.J. Schleisman, in Plasma source Mass Spectrometry: New Developments and Applications. ed. G. Holland and S.D Tanner, Royal Society of Chemistry, 1999 Chapter 1, p.80.

9   N. Jakobowski, ISAS.Pers.Comm. 2000

10   J Mendham, J.D.Barnes, R.C.Denny and M.J.K.Thomas, 2000, Revised Ed.- Vogels Textbook of Quantitative Analysis, 6th Ed, Prentice Hall, p.76

11   K. Neubauer and U. Vollkopf At. Spectrosc., 1999. **20(2)**, 64.

12   S.D.Tanner , and V.I. Baranov, At. Spectrosc.,1999, **20(2)**, 45.

13   Perkin Elmer DRC Training course notes 2000.

14   S.D.Tanner and V.I. Baranov, in Plasma source Mass Spectrometry: New Developments and Applications.ed. G. Holland and S.D.Tanner, Royal Society of Chemistry.1999, Chapter 1, p.46.

15   U.Vollkopf, V.I. Baranov , and S.D.Tanner, in Plasma source Mass Spectrometry: New Developments and Applications. ed G.Holland and S.D.Tanner, Royal Society of Chemistry.1999, p. 63.

16   D.R. Bandura and S.D.Tanner, At. Spectrosc.,1999, **20(2)**, 69.

17   D.R.Bandura, V.I.Baranov and S.D.Tanner, J.Anal.At.Spectrom., 2000, **15**, 921.

18   S.D.Tanner, D.R.Bandura and V.I. Baranov, Presentation at the 7th International Conference on Plasma Source Mass Spectrometry, University of Durham 10-15th September 2000

ACKNOWLEDGEMENTS

A University of Reading Studentship and Perkin Elmer Instruments have funded this research jointly. The authors thank Scott Tanner, Dmitry Bandura and Vladimir Baranov at Perkin Elmer Sciex for many valuable discussions.

Section 4

Applications

APPLICATION AND QUALITY OF ICP-MS ANALYSIS

Thomas Prohaska[1,2], Gerhard Stingeder[1], Simon M. Nelms[2], Christophe Quétel[2], Christopher Latkoczy[3], Stephan Hann[1], Gunda Köllensperger[1] and P. Taylor[2]

[1]University of Agricultural Sciences, Institute of Chemistry, Muthgasse 18, A-1190 Vienna, Austria
[2]European Commission Joint Research Center, Institute for Reference Materials and Measurements, Retieseweg, B-2440 Geel, Belgium

[3]Laboratory for Isotope and Trace Element Research – LITER, Old Dominion University - Department of Chemistry and Biochemistry, 4541 Hampton Blvd., Norfolk, VA 23529-0126 USA

1 INTRODUCTION

The history of mass spectrometry can be considered to have begun at the Friday evening meeting of the Royal Institution, April 30, 1897 when J.J. Thomson reported the m/z ratio of an electron (5.69×10^{-9} g/coulomb). He said after that event that

"At first there were very few who believed in the existence of these bodies smaller than atoms. I was even told long afterwards by a distinguished physicist who had been present at my lecture at the Royal Institution that he thought I had been 'pulling their legs'."[1]

At the beginning of the 20th century, he went on to construct the first mass spectrometer (then called a parabola spectrograph) for the determination of the mass to charge ratios of ions. Thomson received the 1906 Nobel Prize in Physics *"in recognition of the great merits of his theoretical and experimental investigations on the conduction of electricity by gases"*. Mass spectrometry started to play a more and more important role and mass spectrometers were even used for the separation of Pu for the construction of the first atomic bomb within the Manhattan Project.

The use of the ICP as ion source for mass spectrometry was a conception to a market for the next generation. Inductively coupled plasma atomic emission spectrometry was at that time developing rapidly in research laboratories around the world, some years before the launch of commercial ICP-AES systems in 1974.[2] In 1970 Knewstupp and Hayhurst , who had been involved extensively with Sugden at Cambridge in the mass spectrometry of atmospheric flames [3, 4] considered a high temperature plasma to be suitable as ion source for subsequent analysis by a quadrupole mass analyser. In 1971 a feasibility study was started at the University of Liverpool. The first report was made in April 1974 at the Society for Analytical Chemistry.[5] The capillary arc was seen soon as a limited ion source and it was clear that it had to be replaced by a more effective source.[2] The Institute of Geological Sciences in London (now: the British Geological Survey) was highly interested in such a system and its director S.H.U. Bowie submitted an EU funded project but due to policy changes the project was dropped temporarily. Gray's publication in 1974 [6] aroused interest in this topic and in early 1976 the Ames Laboratory started a project with R.S. Houk as the involved researcher. Due to the finally agreed funding the project in the UK restarted and finally collaboration with the Ames Laboratory was agreed. The first mass spectrum was produced by Houk in 1978[7]. Research in the UK was at this time based at the University of Surrey and the project started in early 1979. The main problem was interfacing the ICP operated at ambient pressure to a mass separator operated at high vacuum. Both laboratories experienced serious limitations but quite soon good spectra with high sensitivity could be demonstrated by using the ICP as an ideal ion source.[8, 9] After that, the development was rapid. The Sciex Elan was the first commercial instrument shown to the public at the 1983 Pittsburgh Conference.[2] The VG Plasma Quad, based on the second instrument built at Surrey had been announced but not exhibited as a commercial instrument at the International Mass Spectrometry Conference in Vienna in summer 1982 and was first exhibited in May 1983 at the ASMS meeting in Boston.[2]

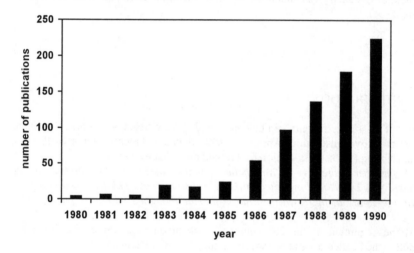

Figure 1 *Number of publications 1980 – 1990*

After that, the analytical market faced a boom of ICP-MS instruments, seen as a panacea in analytical elemental trace analysis and widely regarded as an interference free technique. Even though, it was quite soon realised that most publications focused on one major problem: spectral interferences.[10] Flexibility, convincing figures of merit and steady improvement of instruments combined with persuasive manufacturers (and decreasing instrumental costs) resulted in a leading role of ICP-MS. This is underlined by the recognition of isotope dilution ICP-MS as primary method of measurement by CCQM (Comité Consultatif pour la quantité de matière)[11]. The technique found its way into many fields of application, such as environmental sciences, geochemistry and biology but also gained importance for medical, forensic and historical applications, in food and agricultural chemistry and in nuclear research in both academic research and industry. Even if ICP-MS seems to be a settled and mature technique, more novel instrumentation has been launched to the market recently (Table 1 shows the highlights of development) than ever before in its 20 years history. The number of papers has grown almost exponentially within the first ten years (Fig. 1), with currently more than 500 publications per year.

Table 1 *Some main instrumental developments in ICP-MS*

1978	1st publication on ICP-MS [7]
1983	First commercial instruments
1989	High resolution (HR) ICP- sectorfield MS [12, 13]
1992	Multicollector (MC) ICP - sectorfield MS [14]
1994	ICP- time of flight (TOF) MS [15]
1999	HR-MC-ICP-SMS

2 ICP-MS ANALYSIS 1994-2000

The following section gives an overview about the development in ICP-MS during the last 7 years and highlights the main fields of applications. The data for the year 2000 are estimated values, calculated from the development during the period 08/1999 – 08/2000

2.1 Instrumental development 1994 – 2000

2.1.1 ICP-MS instruments Figure 2 shows an overview of the current instrumentation and the development of the ICP-MS market over the last 7 years.

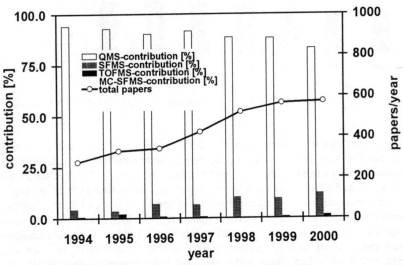

Figure 2 *ICP-MS Instrumentation 1994 – 2000. The line reflects the total number of papers/year*

It is evident that still more than 80 % of the ICP-MS instruments, which are used in scientific papers, are quadrupole based instruments, even if it is obvious that other instrumental techniques are gaining more and more importance. Especially the development in the field of high mass resolution (i.e. double focussing magnetic sector field instruments) is significant and in about 10 % of all papers HR-ICP-SFMS was used. Time of flight (TOF) ICP-MS showed a significant number of papers in 1995, one year after the introduction by Hieftje et al. [15]. After that, the number of papers decreased. This was in the years when new instruments were sold and installed in the laboratories. In the succeeding years, an increase of papers is visible and a significantly increased number of papers can be expected for 2000/2001. A boom is as well visible for multi collector ICP-MS. The urgent need for highly precise and accurate isotope ratio analysis in combination with significant instrumental developments led to a contribution of about 4% to the number of papers in 2000. The recent boom in collision cell and dynamic reaction cell techniques will increase the number of papers based on quadrupole ICP-MS within the next years.

    2.1.2 Hyphenated techniques   Fig. 3 shows the development of hyphenated techniques in the last 10 years, which include both online speciation and matrix separation techniques.

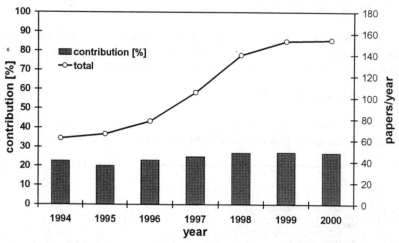

Figure 3 *Number of papers in the field of hyphenated techniques published in 1994-2000*

    The contribution of hyphenated techniques is more than 20% to the total number of papers. The first publications concentrated mainly on the theoretical background of the hyphenation and focused on As and Se speciation. In 1994 still 70 % of the papers dealt with the As and Se problem whereas in 1996 already more than 50% of the papers showed results on other systems (mainly Pb, Sn, Hg and Cr speciation). In the mid 90's new chromatographic techniques were coupled to ICP-MS, like capillary zone electrophoresis, first presented by Olesik and co-workers in 1995. [16] The new techniques were again used for As and Se speciation (2/3 of the total number of papers on hyphenated techniques). Both new techniques and new elemental systems led to a significant increase of the numbers of papers in the field of hyphenated techniques in 1996/1997.

2.1.3 *Laser Ablation ICP-MS* Fig. 4 shows the development of LASER ablation ICP-MS, which gained more and more importance as direct solid sampling device.

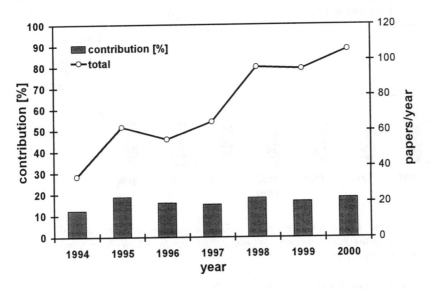

Figure 4 *Number of papers in the field of LA-ICP-MS published in 1994-2000*

LA-ICP-MS shows a steady increase of numbers of contributions within the last 7 years. The contribution to the total numbers of papers increased to about 20 %. This shows significantly the importance of laser ablation in current research. Electro-thermal vaporisation e.g. coupled to ICP-MS (ETV-ICP-MS) as another solid sampling device only contribute 3 % to the total numbers of papers.

2.1.3. *Isotope ratio measurements using ICP-MS.* Fig. 5 shows the numbers of papers that present results from isotope ratio measurements. Isotope ratio measurements were used for both tracer analysis and for isotope dilution analysis. The numbers of papers show a significant increase within the last 7 years indicating the importance of isotope ratio measurements in analytical chemistry. First papers investigated the basic theories and optimisation of isotope ratio measurements focussing on the well-defined Pb system. In 1994, almost ¾ of the total number of papers, which presented results on isotope ratio measurements, were about Pb and the number decreased to ¼ in 1997. The significant development of MC-ICP-MS for highly precise isotope ratio measurements [14] and the use of both magnetic sector field single collector ICP-MS [17] and ICP-TOF-MS [18] increased the number of papers again from 1996 on and research groups focused again on the "old" Pb system using these new techniques. This is reflected by the fact that almost ²/₃ of the total number of isotope ratio papers present again results on Pb since 1998.

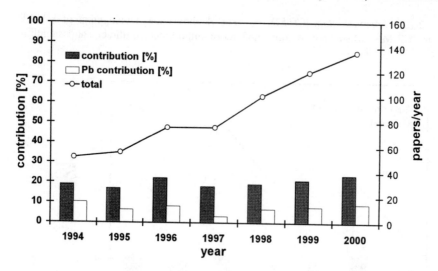

Figure 5 *Number of papers in the field of isotope ratio measurements published in 1994-2000. The white bars represent the number of papers, which show results on Pb isotope ratio measurements*

2.2 Elements measured by ICP-MS 1994-2000

The pattern of elements, which were analysed by ICP-MS show more or less a similar contribution without dramatic change throughout the last few years. As shown in chapter 2.1., hyphenated techniques focussed mainly on As and Se whereas isotope ratio measurements were mainly performed on Pb isotopes. Pb is still the main element of interest in ICP-MS analysis, even though the percentage of papers decreased from more than 12 % to about 7 % within the last 7 years. Other elements of interest are still the "common" elements like Fe, Cr, Ni, Zn, Cd, Mn and Cu, which make up about 50 % of all papers published in ICP-MS analysis, even though their individual contribution to the total numbers of papers decreased with exception of Cd, which showed growing interest due to its toxicity, the ease of measurement, the number of isotopes and the excellent limits of detection. Another toxic element of interest was Hg. Increased numbers of publications within the last years can be found also for platinum group elements and rare earth elements. The number of papers dealing with these two groups of elements almost doubled within the last 7 years. Also increased interest was found for U.

2.3 Fields of applications

Early directions in the 80's were mainly driven by the history and funding of the early steps in ICP-MS and main fields of research were environmental science (funding of the US research by the Environmental Protection Agency) and geological studies (UK funding by the British Geological Survey). [2]

Now, more than 20 years after the first publication on ICP-MS, environmental (36 %) and geological studies (14%) still represent ½ of the studies in scientific journals. About 20 % of the papers have a medical or biological background, 5 % are pure industrial applications and about 4 % of the publications show results on historical investigations. The number of reviews almost doubled (2 % of all papers) within the last 7 years, mainly due to the fact that more and more techniques were developed, which asked for comprehensive reviews due to their novelty. It is obvious that the number of contribution dealing purely with theoretical background decreased from 30 % to 20 %. This is a typical indication that on the one hand new instrumentation is still being developed but more and more fields of applications are opened up by using the excellent features of ICP-MS.

3 QUALITY OF ICP-MS ANALYSIS

3.1 Total combined uncertainty

The first published data proved that elemental analysis was in many cases limited by the techniques used due to their limits of detection and the instrumental precision. One question is still to develop an analytical method with the highest quality possible but more important is the use of a method with the adequate quality for the specific application. Still, values are reported in most of the 500 papers in ICP-MS analysis per year without focusing on the "quality" of the data. The "quality" was in many cases difficult to quantify in a comparable way. Decreasing limits and thresholds as set by governmental legislation or transnational regulations require accuracy, comparability and traceability of analytical measurements in environmental analytical chemistry. The meaning of traceability and issues of how traceability to the SI can be achieved, have lately been the subjects of considerable international debate. According to the definition of the International Vocabulary of Basic and General Terms in Metrology (VIM) document, traceability is "the property of a result of a measurement or the value of a standard whereby it can be related to stated references, usually national or international standards through an unbroken chain of comparisons all having stated uncertainties". [19] When this standard is the "Système international d'Unités" (SI) via SI quantities and units, this can be formulated as SI traceable. Thus, traceability to SI can only be ensured via the use of reference materials certified for element concentration (and also possibly for isotopic composition) - for the development and the validation of analytical methods - and via the assessment of the real uncertainty of the measurement following the EURACHEM guidelines for combined uncertainty. [20] Total combined uncertainty represents a parameter associated with the result of a measurement, that characterizes the dispersion of the values, that could be reasonably be attributed to the measurand and where all contributors to the uncertainty are taken into account. [21] This is of importance, since many investigations show that instrumental precision is not the limiting factor of the "quality" of a measurement anymore. Assessment of the "quality" is more and more established in laboratories by determining total uncertainties, which are no longer just academic exercises anymore but asked for by e.g. accreditation bodies. Actually, still only 5 % of the papers published per year show total uncertainty budgets.

Error propagation is performed using the method of partial differentiation, even though a numerical approach as presented by Kragten et al. [22] is more practical. This method is applied in a dedicated software program specially prepared for easy assessment of the total uncertainty calculation (GUM Workbench®). [23] The main advantage of a total combined uncertainty is the establishment of a realistic uncertainty of analytical results, which are comparable for different laboratories. Since all sources of errors are taken into account and their contribution to the total uncertainty can be determined, the main contributors to the uncertainty can be sourced and reduced accordingly. As an example, the following chapter shows a total uncertainty budget for an ID-MS analysis.

3.2 Total combined uncertainty of an ID-MS analysis

The quantification of trace elements by isotope dilution ICP-MS provides an example of a total uncertainty budget. Isotope dilution inductively coupled plasma mass spectrometry (ID-ICP-MS) is a powerful strategy capable of highly accurate and precise determination of element concentration.[24,25] IDMS was recognized by the CCQM (Comité Consultatif pour la Quantité de Matière) as a primary method of measurement, if carried out correctly.[11] The CCQM defines a primary method of measurement as a method, which is completely described and understood, having the highest metrological qualities and for which the results with a complete uncertainty statement can be given. [11] Primary methods and hence IDMS are unique tools in chemical measurements which can lead to results with very small combined uncertainty from an a-priori point of view when applied properly.[26,27]

The example given shows the quantification of Pb in a sediment sample using ID-MS by means of multi-collector ICP-MS. The details of sample preparation and performance of the ID-MS investigation is presented in detail elsewhere [28] and within this paper we will only focus on the contributors of the sources of uncertainty.

Main contributors to uncertainty in ICP-MS analysis are the instrumental precision, the correction for mass bias, the correction for instrumental background, interference correction (if necessary) and the correction for deadtime (in the case of ion counting systems). Table 2 presents the total combined uncertainties (relative standard uncertainties = RSU) of the blend ratio and the contribution of instrumental precision, mass bias correction and correction for background to the RSU as calculated for USN-MC-ICP-SFMS for a Pb isotope ratio $(^{208}Pb/^{206}Pb = 1:1)$ of a digested and spiked sediment sample with a total Pb concentration of 5 ng g^{-1} in solution.

Table 2 *Main contributors to the total uncertainty of the measurement of a spiked Pb blend ratio $(^{208}Pb/^{206}Pb = 1:1)$. Total Pb concentration is 5 ng g^{-1}*

Instrument	Instrumental precision	Mass bias correction	Background correction	RSU (k=1)
USN-MC-ICP-SFMS	34.7 %	45.6 %	19.7 %	0.14 %

This uncertainty on the measured blend ratio was propagated accordingly for calculating the Pb concentration in the sample using the ID-MS equation which was developed from the basic ID-MS equation for this specific measurement elsewhere.[28] Figure 6 shows the contributors to the total combined uncertainty on the final result. The figure demonstrates clearly that the measurement of the blend ratio (R_B) is not the main limiting factor and only contributes about 0.5 % to the total uncertainty. In this specific example, the digestion method is the main factor and requires therefore further investigation and improvement. This picture demonstrates as well unequivocally the usefulness of the total uncertainty budget, since it helps immediately to identify main sources of errors. Other analytical problems might have other sources of error (e.g. sampling, stability) and the contribution of the analytical measurement to the total uncertainty shows if the applied analytical method has adequate quality or needs further improvement.

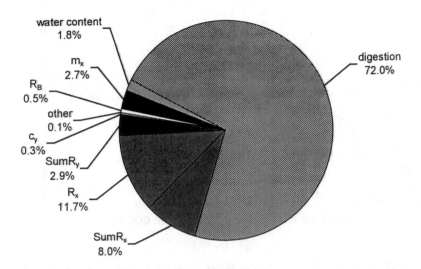

Figure 6 *Contributors to the total combined uncertainty of an ID-ICP-MS analysis of Pb in a sediment sample using MC-ICP-SFMS.*
Water content = moisture content of the sample; m_x = amount of the sample; R_B = blend ratio; c_y = spike concentration; $SumR_y$ = sum of the ratios in the spike; R_x = ratio in the sample; $SumR_x$ = sum of the ratios in the sample; digestion = uncertainty caused by the digestion of the sample.

3.3 International Measurement Evaluation Programme (IMEP)

IMEP is a large-scale interlaboratory comparison, coordinated by IRMM since 1988. The aim of the program is to picture objectively measurement performance of laboratories on "real life samples" and is open to all laboratories. Table 3 shows planned IMEP rounds. The results of participants are displayed (without identity) against reference values traceable to the SI (obtained independently from participants' results). The program demonstrates the degree of equivalence of results of international laboratories and is complementary to collaborative studies and proficiency testing schemes, which have a weaker metrological basis, but also different objectives (e.g. monitoring of a laboratories proficiency, improvement of measurement methods). Recently IMEP results have started to be used by accreditation bodies to carry out result oriented evaluation of laboratories. [29]

Figure 7 shows a typical IMEP result of the participating laboratories.

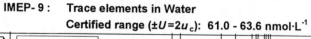

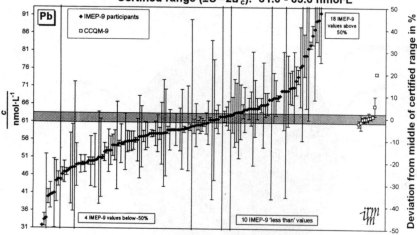

Results from all IMEP-9 participants and CCQM-9.

Figure 7 *Results from the IMEP 9 round - Trace Elements in Water and the CCQM Key Comparison K2* [30]

The figure represents the typical scattering of laboratory results. It is important to point out, that even if the value might be far off the certified value, the result might still overlap within its uncertainty, indicating that for the measurement result of this particular laboratory the quality of that result would still be adequate for the problem. Moreover, the result of the CCQM Key Comparison, indicated on the right side of the graph, represents the results of metrological laboratories using IDMS as analytical tool. The results show, that very accurate values with very small uncertainties can be obtained serving as reference.

Table 3 *Planned IMEP rounds* [29]

IMEP round	Description	time schedule
IMEP 12	Trace elements in water	2000 - 2001
IMEP 15	$^{10}B/^{11}B$ isotope ratios in water	2000
IMEP 17	Trace and minor constituents in human serum	2000 - 2001
IMEP 18	S in fuel	2001
IMEP 19	Cd in rice	??
IMEP xx	B in biological materials	??
IMEP xx	PCB's in pig fat	2001

4 CONCLUSION

Since its first introduction, ICP-MS boomed significantly and the boom is still visible after almost 20 years. It is obvious that ICP-MS is a very fascinating analytical method, fast and reliable, which has opened up many fields of applications. Even though it is already a very mature technique, much effort is being put into further development making ICP-MS a more and more universal tool. Still, new visions and ingenious strategies are strongly required to face the challenging developments of the 21st century. And we should not be dazzled by the shining chromium high tech appearance of our warships and watch every single step towards the grail of scientific truth more carefully the closer we get and realise that our distance grows with our approach.

5 REFERENCES

1. J.J. Thomson, *The London, Edinburgh, and Dublin Philosophical Magazine and J. of Sci.*, 1907 **XLVII**.

2   A. Gray, *J. Anal. Atom. Spectrom.*, 1986, **1**, 403.

3   P.F. Knewstubb and T.M. Sugden, *Proc. R. Soc. Chem.*, 1960, **255**, 520.

4   T.M. Sugden, in R.I. Reed, ed, *Mass Spectrometry*, Academic Press, London, 1965.

5   A.L. Gray, *Proc. Soc. Anal. Chem.*, 1974, **11**, 182.

6   A.L. Gray, *Analyst*, 1975, **100**, 289.

7   R.S. Houk, V.A. Fassel, G.D. Flesch, A.L. Gray and E. Taylor, *Anal. Chem.*, 1980, **52**, 2283.

8   R.S. Houk, V.A. Fassel, G.D. Flesch, H.J. Svec, A.L. Gray and C.E. Taylor, *Anal. Chem.*, 1980, **52**, 2283.

9   A.R. Date and A.L. Gray, *Analyst*, 1981, **106**, 1255.

10 E.H. Evans and J.J. Giglio, *J. Anal. Atom. Spectrom.*, 1993, **8**, 1.

11 Comité Consultatif pour la Quantité de Matière, *Rapport de la 1er session*, 1995, Edité par le BIPM, Pavillon de Breteuil, F-92312 Sèvres Cedex, France.

12 N. Bredshaw, E.F. Hall and N.E. Sanderson, *J. Anal. At. Spectrom.*, 1989, *4*, 801.

13 M. Morita, H. Ito, T. Uehiro and K. Otauka, *Anal. Sci.*, 1989, **5**, 609.

14 A. Walder and P. Friedman, *J. Anal. At. Spectrom.*, 1992, **7**, 571.

15 D. P. Myers, G. Li, P. Yang and G.M. Hieftje, *J. Am. Soc. Mass Spectrom.*, 1994, **5**, 1008.

16 J.W. Olesik, J.A. Kinzer and S.V. Olesik, *Anal. Chem.*, 1995, **67(1)**, 1.

17 F. Vanhaecke, L. Moens and R. Dams, *Anal. Chem.*, 1996, **68**, 567.

18 P.P. Mahoney, S.J. Ray and G.M. Hieftje, *Appl. Spectrosc.*, 1997, **51**, 16A.

19 ISO/VIM, *International Vocabulary of basic and general terms in metrology*, International Organisation for Standardisation, Geneva, Switzerland, 1993.

20 *Quantifying Uncertainty in Analytical Measurement*, EURACHEM, London, 1995.

21 ISO/GUM, *Guide to the expression of uncertainty in measurement*, International Organisation for Standardisation, Geneva, CH, 1995.

22 J. Kragten, *Analyst*, 119, **1994**, 2161

23 R. Kessel, *GUM Workbench*®, Metrodata GmbH, D-79639 Grenzach- Wyhlen, Germany.

24 K.G. Heumann, in *Inorganic Mass Spectrometry*, ed. F. Adams, R. Gijbels and R.J. Grieken, Wiley, New York, 1988.

25   P. De Bièvre, *Fres. J. Anal. Chem.*, 1994, **350**, 277.
26   J.D. Fassett and P.J. Paulsen, *Anal. Chem.*, 1989, **61**, 643A.
27   J.R. Moody and M.S. Epstein, *Spectrochim. Acta, Part B*, 1991, **46**, 1571.
28   T. Prohaska*, C.R. Quétel, C. Hennessy, D. Liesegang, I. Papadakis and P.D.P. Taylor, *J. Env. Monitoring*, in press
29   http://www.irmm.jrc.be
30   I. Papadakis, *IMEP 9 Participants report*, IRMM, Geel, 1999

6 ACKNOWLEDGEMENTS

I am grateful to Grenville Holland and Scott Tanner for the opportunity of the invited lecture at the Durham Conference 2000. My colleagues at the different institutions are thanked for their friendship and input in my work throughout the years by having them as co-authors of this paper.

TRACE ELEMENTS IN HONEY SAMPLES BY MEANS OF ETV-ICP-MS

M. Bettinelli, S. Spezia, N. Pastorelli, C. Terni

ENEL Produzione Spa – Laboratory of Piacenza, Via Nino Bixio 39, 29100 Piacenza, Italy.

1 INTRODUCTION

In an effort to find more indicators that are representative for the environmental pollution, bee honey has been proposed (1-5). Data for the analysis of bee honey are scarce. Due to the rather complex composition of honey, and the low content of trace elements, the sample preparation and analysis are quite difficult and time consuming, especially if sequential analytical techniques, such as AAS, are utilized.

In the literature several preparation techniques of honey matrices have been reported, based on acid dissolution in batch, ashing with specific reagents, or acid dissolution in PTFE vessels by means of a microwave oven. (6)

Recently, Caroli (7) and Bettinelli (8) described the possibility of analyzing 12-20 trace elements in honey by ICP-MS or ICP-AES after acid solubilization of matrix in microwave system with detection limits of ng/g for many elements . In this last paper Bettinelli reports a comparison between the results obtained using different analytical techniques (GF-AAS, ICP-AES and ICP-MS) and evaluates the precision of the method better than 10% with detection limits, for the USN-ICP-MS configuration, ranged between 0.1 ng/g to 10 ng/g.

The main problems cited by the authors are related to the heterogeneity of the matrix and to the contamination of the honey from trace elements that can occur in many different ways during the manipulation of the sample.

In the present study we verify the possibility to use the Electrothermal Vaporization coupled to the Inductively Coupled Plasma – Mass Spectrometry (ETV-ICP-MS) for the determination of trace elements in honey samples after plain dilution with water. The method has been evaluated by comparing the results with those by using USN-ICP-MS after MW dissolution of matrix.

2 EXPERIMENTAL

2.1 Instrumentation

A standard Perkin Elmer 5000 ICP-MS (Perkin Elmer Sciex Instruments, Concord Ontario, Canada), equipped with two mass-flow controllers and a HGA-600MS electrothermal vaporizer with a model AS-60 autosampler . The experimental conditions for ETV- Elan 5000 configuration are reported in Table 1.

An ultrasonic nebulizer model CETAC U5000AT (Cetac Technologies Inc.Omaha, NE, USA) was used as an alternative injection system to the pneumatic nebulizer of ICP-MS. The operating conditions (Table 2) have been optimized in order to obtain good sensitivity with acceptable precision. The calibration was done with five standard solutions containing 0.5, 1, 2, 5, 10 µg/L and the same acid concentrations as in the samples.

The microwave digester used for closed vessel high pressure sample digestion was the Milestone (FKV, Sorisole, Bergamo, Italy) MLS 2100 MEGA model with operating conditions listed in Table 3.

Table 1 *Instrumental parameters for ETV- ICP-MS*

Power RF (W)	1100
Argon Plasma Flow (L/min)	14,6
Argon Auxiliary Flow (L/min)	0,85
Argon Nebulizer Flow (L/min)	0,85
Argon ETV Flow (L/min)	0,30
Skimmer and sampling cones	Pt
Dwell Time (ms)	17
Resolution (amu)	0.8
Point per peak	1
Acquisition signal	Peak hoping
Processing mode	Integrated

2.2   Reagents

Nitric acid (65% m/V), was Suprapur reagents (E.Merck, Darmstadt, FGR) while hydrogen peroxide (30% m/V) was Baker analyzed-reagent grade (Baker). Standard solutions. Multielemental standard solutions were prepared from 10 mg/L multielemental atomic absorption standards, "ICP-MS Calibration Standard n. 2 and n. 4" (Perkin Elmer), by dilution in water containing the same amount of acids as the samples. The internal standard was prepared from 1000 mg/L Yb solution, Spectrosol (BDH, Poole, England). 10 µg/L Yb were added as internal standard to the blank , to all the solutions used for the instrumental calibration and to the honey samples.

Aliquots of 20-µL samples and 10-µL modifier solutions [15µg Pd + 10 µg $Mg(NO_3)_2$] were used for the determination of "P" Group elements after appropriate dilution of the corresponding solutions purchased from Perkin Elmer (1% of Pd and Mg solutions).   High-purity water was produced by passing distilled water through a deionizing system (Milli-Q, Millipore , Bedford, MA, USA).

2.3   Certified Reference Materials

No Certified Reference Materials exist for honey matrix, therefore the validation of the method was done by analyzing two real samples collected in a 'remote' unpolluted area in the Central part of Italy that were tested in a previous Interlaboratory comparison (9).

In consideration of the purpose of this study, two very different types of honey were chosen, namely Eucalyptus (solid and sticky) and Robinia (viscous and sticky). In order to obtain a more homogeneous and fluid material, samples of about 10 g of both honeys were diluted with 20 g of high purity deionized water. The mixture was then heated up to 50°C in an ultrasonic bath for 30 min. The blank of the procedure was prepared by treating 10 g of high purity water in the same way as the honey sample. At the final solutions were added 10 µg/L of Yb as internal standard.

Table 2 *Instrumental parameters for USN-ICP-MS analysis*

Parameters	
PF power	1150 W
Plasma argon	15.0 L/min
Nebulizer flow	0.950 L/min
Auxiliary flow	0.950L(min
Sample flow rate	1 mL/min
Nebulizer	Ultrasonic
Nebulizer Temperature	140 °C
Data	Peak hop transient
Resolution	normal
Reading time	150 ms
Dwell time	50 ms
Sweeps / replicates	3
Number of replicates	5
Sample read delay	50 s
Autosampler wash delay	60 s
Calibration mode	External calibration
Curve fit	Linear through zero

Interelemental corrections :

$$^{75}As = {}^{75}As - 3.087 \, {}^{77}Se + 2.619 \, {}^{82}Se \qquad {}^{51}V = {}^{51}V - 3.09 \, {}^{53}Cr + 0.353 \, {}^{52}Cr$$

Table 3 *Operating Program for Milestone 2100 Mega*

Power (Watts)	Time (min)
250	1
250	0
250	5
400	5
600	5

(2 g of sample; 5mL HNO_3 / H_2O_2 (4+1 V/V); final vol. 25mL)

2.4   Sample preparation

The analysis of trace elements by ETV-ICP-MS was done directly on the sample diluted with water. USN-ICP-MS was applied after digestion of honey in microwave oven. Sub-samples of about 2 g of the viscous solution were transferred into the MW teflon vessels, weighed and digested with 5 mL of $HNO_3 + H_2O_2$ (4+1 v/v) mixture according to the procedure reported by Bettinelli et al. (11). After digestion samples were cooled at room temperature and made up to 25 mL with high purity de-ionized water in volumetric glass flasks.

3 RESULTS AND DISCUSSION

3.1   Preliminary study

Two different ETV programs have been developed for the determination of Co, Cu, Mn, Ni, and V (so-called elements of " W" Group) and for As, Be, Cd, Pb, Sb, Se, Te and Tl (so-called elements of " P" Group "). In the Tables 4 an 5 are reported the thermal programs used for the two different groups of elements while, in Figures 1 and 2 are shown the ETV signals for the analytes of interest.

The ETV signals for the Eucalyptus honey, reported in Figure 1, shown high concentration of Mn and Cu (in the order of 200 – 2000 ng/g). The elements of "P" Group are , on the contrary, at the same level of the blank with the only exception of Be and Sb as shown in Figure 2.

Table 4 *ETV program for the determination of Co, Cu, Mn, Ni, and V in honey*

Step n.	Temperature (C°)	Ramp Time (s)	Hold Time (s)	Internal flow (Ar mL min^{-1})	Alternative gas (O$_2$ mL min^{-1})
1	120	3	10	300	
2	500	30	50	-	50
3	600	15	20	300	
4	2650	0	6	0 *	
5	2700	1	5	300	
6	20	1	10	300	

20 µL of sample; 10 µg L^{-1} internal standard (Yb); pyrocoated tube; atomization from the walls.

Table 5 *ETV program for the determination As, Be, Cd, Pb, Sb, Se, Te and Tl in honey*

Step n.	Temperature (C°)	Ramp Time (s)	Hold Time (s)	Internal flow (Ar mL min^{-1})	Alternative gas (O$_2$ mL min^{-1})
1(*)	120	3	10	300	
2	1000	10	20	300	
3	20	1	10	300	
4(**)	120	3	10	300	
5	500	30	50	-	50
6	600	15	20	300	
7	2500	0	6	0 *	
8	2700	1	5	300	
9	20	1	10	300	

20 µL of matrix modifier Pd/Mg(NO₃)₂ added at the step n.1(*), 20 µL of sample added at the step n.4(**); 10 µg L⁻¹ internal standard (Yb) added to the sample; pyro-coated tube with the L'vov platform ; atomization from the platform.

Figure 1   *ETV-ICP-MS signals for Cu and Mn in Eucalyptus honey*

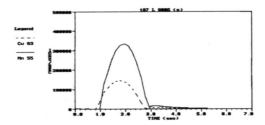

Figure 2   *ETV-ICP-MS signals for As, Be, Cd, Te, Sb and Se in Eucalyptus honey*

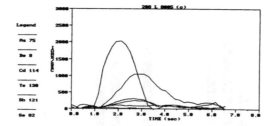

3.2   Instrumental Calibration

The instrumental calibration has been performed in the range 0-20 µg/L (at three concentration levels 5 , 10 , 20 µg/L), both in water solution that in honey matrix. In order to verify possible matrix effects due to the different nature of the honey samples, three types of honey have been considered (Acacia, Eucalyptus, Robinia). The calibration was perform following the standard addition method with and without the presence of internal standard.

The calibration curves for Co reported, as example, in Figure 3, are indicative of a slight but significant matrix effect probably due to the different viscosity of various honeys with respect to water solution. The more sensible difference was given , in effect, by Eucalyptus honey which is the most solid and sticky of the honeys here considered. The use of Yb as internal standard brings the instrumental response for the honey matrix very close to that for water solution, even if a little residual matrix effect is still observed.

For this reason, in order to obtain accurate results, the two honey samples object of this study were analyzed with the standard addition method. Figure 4 displays the detection limits (LOD), absolute, in pg, or relative to a volume of 20 µL, in ng/g. For all the elements the detection limit is of the order of 0.1 – 0.3 ng/g therefore significantly lower than that of the other analytical techniques (11).

Figure 3   *Co Calibration in water and honey matrices with and without internal standard*

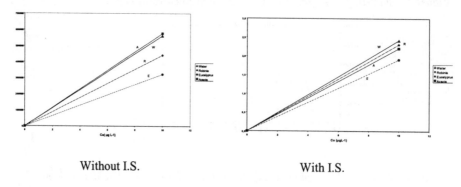

Without I.S.   With I.S.

Figure 4   *Detection Limits for the ETV-ICP-MS determination*

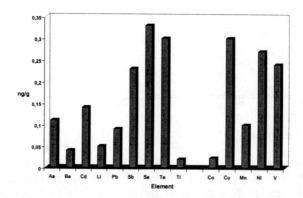

The use of internal standard, moreover, improved the precision for the determination of trace elements in honey as showed in figure 5. The relative standard deviations (RSD%) for the elements of "W" Group resulted better than 9- 10%, and for the elements of "P" Group better than 2- 5 %. The experimental precision found in honey matrix is therefore very close to that obtained in water solution.

3.3   Comparison with other techniques

The two honey samples were analyzed using two different analytical procedures consisting in : (1) the plain dilution of the sample with high purity water and the direct determination by ETV-ICP-MS; (2) the acid dissolution in a microwave oven and the analysis by USN-ICP-MS. The results given in Tables 6 and 7 show a substantial agreement for all the elements, while some discrepancies in the case of Mn and Ni in

Eucaliptus sample and of Cu in Robinia sample determinations. The number of analytes determined by ETV-ICP-MS clearly shows a better power of detection of this technique with respect to the MW digestion and USN-ICP-MS determination. The simple pretreatment of the sample reduces the possibility of environmental contamination ; the blank values are lower and , as consequence, the detection limits result better.

Figure 5 *Precision Data for the ETV-ICP-MS determination*

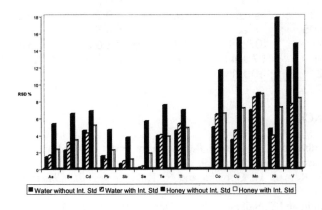

■ Water without Int. Std ▨ Water with Int. Std ■ Honey without Int. Std ☐ Honey with Int. Std

4 CONCLUSION

ETV-ICP-MS appears a very sensitive technique for the determination of trace elements in honey. The reduced sample treatment and the high instrumental sensitivity assure detection limits adequate to the very low concentrations of trace elements present.
The possibility to reduce the matrix effect by choosing an opportune thermal program and the multielemental capability of ICP-MS make the coupling between these two techniques a real and powerful tool for the determination of trace elements in honey.

Table 6 *Determination of trace elements in Eucaliptus honey using ETV-ICP-MS and USN-ICP-MS*

Element	ETV-ICP-MS ngg^{-1}	USN-ICP-MS ngg^{-1}	Element	ETV-ICP-MS ngg^{-1}	USN-ICP-MS ngg^{-1}
As	1.45 ± 0.08	< 0.5	Pb	91 ± 5	111 ± 5
Be	0.18 ± 0.09	n.a.	Sb	< 0.2	n.a
Cd	0.65 ± 0.06	0.8 ± 0.2	Se	< 0.3	n.a.
Co	2.1 ± 0.1	n.a.	Te	< 0.3	n.a.
Cu	142 ± 5	148 ± 19	Tl	0.75 ± 0.04	n.a.
Mn	1730 ± 20	2203 ± 14	V	2.5 ± 0.2	3.4 ± 0.6
Ni	7.0 ± 0.5	16 ± 2			

Table 7 *Determination of trace elements in Robinia honey using ETV-ICP-MS and USN-ICP-MS*

Element	ETV-ICP-MS ngg^{-1}	USN-ICP-MS ngg^{-1}	Element	ETV-ICP-MS ngg^{-1}	USN-ICP-MS ngg^{-1}
As	1.10 ± 0.08	< 0.5	Pb	23 ± 1	31 ± 5
Be	0.38 ± 0.02	n.a.	Sb	< 0.2	n.a
Cd	0.49 ± 0.02	< 0,1	Se	< 0.3	n.a.
Co	0.41 ± 0.05	n.a.	Te	< 0.3	n.a.
Cu	68 ± 2	103 ± 1	Tl	0.13 ± 0.08	n.a.
Mn	104 ± 4	97 ± 4	V	1.0 ± 0.3	0.72 ± 0.13
Ni	31 ± 2	30 ± 1			

5   REFERENCES

1. L. Leita, G. Muhlbachova, S. Cesco, R. Barbattini and C. Mondini, *Environm. Monitor. and Assessment*, 1996, **43**, 1–19.
2. Z. Dobrzanski, A. Roman, H. Gorecka and R. Kolacz, *Bromat. Chem. Tosksyk.*, 1994, **27**, 157–160.
3. A. Uren, A. Serifoglu and Y. Saikahya, *Food Chem.*, 1998, **61**, 185–190.
4. M. Accorti, R. Guarcini, G. Modi and L. Persano-Oddo, *Apicoltura*, 1990, **6**, 43–45.
5. M. K. Wallwork-Barber, R.W. Ferenbauch and E. S. Gladney, *Am. Bee J.*, 1982, **122**, 770–772.
6. P. Fodor and E. Molnar, *Microcim. Acta*, 1993, **112**, 113–118.
7. S. Caroli, G. Forte, A.L. Iamiceli and B. Galoppi, *Talanta*, 1999, **50**, 327–336.
8. M. Bettinelli and C. Terni, in *Applicazioni dell'ICP-MS nel laboratorio chimico e tossicologico*, C. Minoia, M. Bettinelli, A. Ronchi, S. Spezia, ed. Morgan, Milano 2000, pp. 341–349.
9. S. Caroli and G. Forte, *Esiti del Secondo Circuito per l'analisi di elementi in Traccia In Campioni di Miele*, Istituto Superiore di Sanità, Rome, 1999.

DETERMINATION OF TRACE ELEMENTS IN ICE CORE SAMPLES BY LASER ABLATION INDUCTIVELY COUPLED PLASMA MASS SPECTROMETRY

H. Reinhardt, M. Kriews, O. Schrems, C. Lüdke[1], E. Hoffmann[1] and J. Skole[1]

Alfred Wegener Institute for Polar and Marine Research, Am Handelshafen 12, D-27570 Bremerhaven, Germany.

[1]Institute for Spectrochemistry and Applied Spectroscopy, Albert-Einstein-Straße 9, D-12489 Berlin, Germany.

1 INTRODUCTION

The snow and iceshields of the polar regions serve as a climate archive and deliver a useful insight back to about 250.000 years of earth climate history[1,2]. The aim of our investigation reported here was to establish a new method for the determination of trace elements in ice cores from polar regions with Laserablation Inductively Coupled Plasma Mass Spectrometry (LA-ICP-MS)[3]. Primarily, the construction of a cryogenic laser ablation chamber and the optimization of the analysis system for the sample matrix were the main goals. This paper reports preliminary results from measurements of frozen ice samples, the achievable signal intensities, standard deviations and calibration graphs as well as the first signal progression of ^{208}Pb in an 8000 years old ice core sample from Greenland.

1.1 The Advantages Of The New LA-ICP-MS Application

LA-ICP-MS offers the possibilty of direct analysis of deep frozen ice core samples from the polar regions. The main advantages are:
- low risk of contamination in contrast to conventional solution ICP-MS
- a high spatial and time resolution from 200 to 1000 μm (recording of depth profile) dependent on the spotsize of the laser crater

With such a high spatial resolution it will be possible to detect seasonal variations of the element distribution and composition also in deep ice layers where annual layers have a thickness of only about 1 mm. Up to now, element analytical determinations of ice core samples were only possible with molten ice samples. After a special sample preparation including enrichment procedures and addition of chemicals, the samples were analysed with chemical-physical methods[4,5]. As concentrations in such samples are very low (ppt-level), the risk of contamination during sample preparation is very high. Due to the need of relative high volumina in solution analysis, only a small spatial (cm-level) and therefore a reduced temporal resolution of molten ice core samples is possible.

1.2 Tracer Elements In Ice Cores

The distribution and composition of trace elements in annual layers of ice cores can for example provide information about the changes from cold to warm periods (e.g., increase of mineral dust or seasalt), special events in earth history like volcanic eruptions (dust horizons in ice core layers) or pointing to sources for mineral dusts[6,7,8]. The detection of seasonal variations of mineral dust and seasalt concentrations can give valuable hints for the dating of the ice cores.

1.2.1. Elements Of Interest. Sodium and magnesium are tracer elements for seasalt, aluminum and iron are mineral dust tracers. The seasalt and the mineral dust concentrations in arctic snow and ice samples vary with the season. Due to the progression of the polar front in summer and winter time, airmasses bring more or less seasalt or mineral dust into the Arctic which are deposited on snow and ice. Lead, zinc and cadmium are tracers for anthropogenic sources or indicators of contamination effects. The concentrations of these elements can also show seasonal variations.

2 EXPERIMENTAL SECTION

2.1 Experimental Setup

Figure 1 illustrates the experimental setup for the direct analysis of deep frozen ice samples by Laserablation ICP-MS.

2.1.1. Laser System For many applications the use of UV wavelengths (e.g. 266 nm) for laserablation is more suitable because UV radiation can be focused to a smaller spot than longer wavelengths radiation and because it is stronger absorbed by many materials than visible and IR laser radiation[9]. Whereas the absorption coefficient of ice at UV wavelengths is about two orders of magnitude lower than at 1064 nm[10], i.e. ice is more translucent and less material would be ablated. Based on that the laser used was a modified Nd:YAG (DCR-11, SPECTRA PHYSICS) operated at 1064 nm. A power measurement unit and an aperture system are installed behind the beam exit to control laser energy and spotsize.

2.1.2. Sample Alignment For an exact alignment of the sample inside the laserablation chamber, an diod laser (red) is integrated on the optical axis of the Nd:YAG laser beam. With support of a CCD color camera the laser beam is focussed on the sample surface and the ablation process can be viewed.

2.1.3. Laser Ablation Chamber To enable a direct analysis of solid ice samples at a temperature of −30°C a special laserablation chamber was constructed[11]. −30°C is the standard temperature for the storage of ice core samples because no changes inside the ice takes place at these low temperatures. The chamber has a large volume to enable the measurement of ice core samples cut as discs (diameter: 10 cm) or segments with one or more annual layers depending on the depth of the ice origin. The inner part of the chamber consists of high purity copper and contains a cooling canal for the cooling liquid (silicon oil). The outer shell consists of teflon and enables a good insulation against the copper block. The carrier gas (argon) is cooled to avoid droplets of water at the optical window and to prevent melting processes at the sample surface during the ablation process. A computer controls the sample stage of the laserablation chamber in xyz-orientation as well as the laser system.

2.1.4. *ICP-MS System*   The measurements were carried out using the quadrupole based ICP-MS System ELAN 6000 from PERKIN ELMER/SCIEX. An argon flow rate of 1.2 l min⁻¹ carries the ablated material into the plasma (tube length = 100 cm), for ionization and subsequent analysis in the mass spectrometer.

2.2 Operating Conditions

Tables 1 and 2 give an overview over the optimized operation parameters used with the ICP-MS system and the Nd:YAG laser

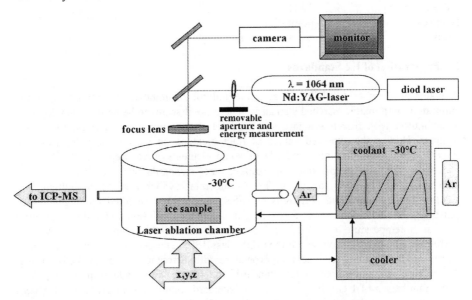

Figure 1 *Experimental setup for the laser ablation of solid ice samples*

Table 1 *Operating conditions for the ICP-MS system*

ICP mass spectrometer	Perkin Elmer Sciex Elan 6000
RF Power	1400 W
plasma gas	15 L min⁻¹
auxilary gas	0.8 L min⁻¹
carrier gas	1.2 L min⁻¹
lens setting	autolens mode on
measuring mode	peak-hopping
isotopes measured	^{17}OH, ^{19}OH, ^{23}Na, ^{24}Mg, ^{27}Al, ^{44}Ca, ^{64}Zn, ^{103}Rh, ^{114}Cd, ^{208}Pb
dwell time	20 ms
detector mode	dual (pulse and analogue)

Table 2   *Operating conditions for the Laser system*

laser sampler	Perkin-Elmer
	Laser Sampler 320
wavelength	1064 nm
mode	Q-switch
Q-switch time	220 μs
excitation lamp energy	50 J
laser energy	300 mJ
pulse frequency	10 Hz
laser scan mode	point, line
focus	on sample surface

2.3 Preparation of Ice Standards

A major problem of LA-ICP-MS as a method for quantitative determination, is the calibration with matrix matched standards. Therefore it seems to be an easy task to make a calibration with frozen standard solutions for the quantitative determination in ice samples. Due to the expected values of trace elements in ice core samples from the Arctic[1,2], low concentration standards of some elements were prepared. We found a procedure for the preparation of suitable ice standards with different element concentrations (10 ng/L to 100 μg/L). Commercial available ICP-MS multielement solutions (PE 1 and PE 2: PERKIN ELMER company) were filled up in pre-cooled purified petri dishes (diameter: 5 cm, height: 1 cm) after dilution and addition of nitric acid at a temperature of –30°C in our ice laboratory. All preparation steps were performed under cleanroom conditions (US-class 100). The petri dishes and the sample carrier run through a special cleaning procedure with different acid bassins before using. To prevent inhomogenities in ice standards the fill up was made step by step to a maximum height of 1 cm. In 5 ml steps the standard solutions were given into the petri dishes to enable a shock freezing of the solution. It seems, that, on one hand, an enlarged thickness leads to inhomogenously frozen samples, whereas a thinner layer is pervious for the laser beam, and the surface of the sample carrier will be hit. Furthermore it seems, that a higher concentrated solution (over 100 μg/L) leads to strong inhomogenously frozen ice and delivers more unstable signals. The ice standards are stored under a clean bench at –30°C.

2.4 Ice Sample Preparation

2.4.1. Different Sample Shapes   There are two possible ways to cut an ice core sample for trace element determination with LA-ICP-MS: Ice core discs and segments (Figure 2). The cutting of the cores is carried out in an ice laboratory at –30°C under clean room conditions. A ceramic working bench with ceramic knives or knives made of high purity molybdenum were used. The samples were frozen on object slides and have a maximum thickness of 1 cm. To enable focussing of the laser beam during the measurement, a plane sample surface is required. The discs have a diameter of 10 cm, the segments a length of 10 and a width of 5 cm.

2.4.2. Ablation Patterns   The ice core disc is more suitable for contamination studies. To have a look on contaminations coming from the ice core driller[2], the point scan mode was chosen (laser spots move from wall towards the center). At the point scan mode, the laser shot a certain time at the same point on the surface of the ice sample. The other possible shape is the ice core segment which is more suitable to record depth profiles. The age and therefore the depth of the ice has a horizontal direction instead of vertical direction at a disc. The point scan mode was used as well as the line scan where the laser shot along a defined line, figure 2:

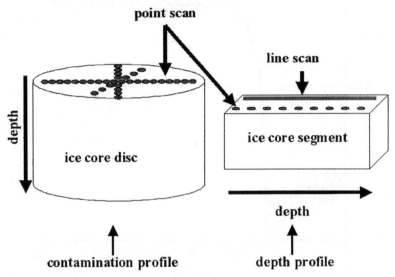

Figure 2 *Different ablation pattern along an ice core disc and an ice core segment*

3 RESULTS

3.1 Signal Progression

In figure 3 the progression of a [103]Rh and a [208]Pb signal from a line scan of a 100 µg/L ice standard are plotted. The arrow indicates where the laser started to fire. After a short forerun of about 30 seconds the signal becomes stable. A defined line on the sample surface is scanned several times. The focus of the laser beam is orientated on the sample surface. The removal of ablated material from the ice surface is approximately 30 µg per lasershot. For [103]Rh we could reach a maximum signal intensity of about 400.000 counts per second, for [208]Pb 600.000 counts per second under optimized conditions (see table 1 and 2). It is demonstrated, that the measurement of frozen ice standards leads to a linear signal dependence as known from the work with solutions. Due to the relative large dimension of the laserablation chamber, the wash out time is about 7 minutes dependent on the standard concentration.

3.2 Calibration Studies

In figure 4 the [208]Pb signal progression of ice standards at different concentrations is demonstrated. The measurement time is 150 seconds which is shown to be an adequate time with the laser signal at its most stable. A 10 ng/L (ppt) ice standard signal can be distinguished clearly from the signal of a blank.

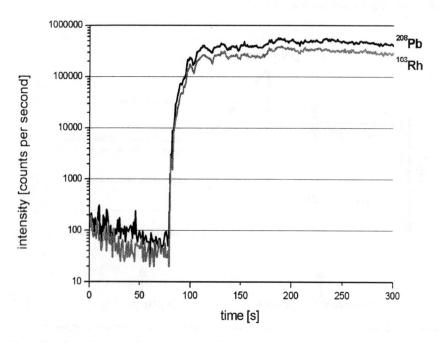

Figure 3 *Signal progression of a 100 µg/L ice standard*

3.2.1. Calibration graphs   The element lead could be calibrated very well with 5 standards from 10 ng/L up to 100 µg/L plus the blank (n = 6). The resulting regression coefficient was R = 0.99966, the measured time t = 150 s. Table 3 lists the resulting coefficients of correlation (R), where n is the number of used ice standards for the calibration curve, n<6 indicates that lower standards were not used for calibration. An example for an element which is difficult to calibrate is calcium. Here, the lower concentrated ice standards resulted in very instable signals and no good linear dependence was obtained. The relative high signals of [44]Ca in the blank and the low concentrated standards inhibit a calibration in the ng/L level (n = 3; 1,10 and 100 µg/L). Altogether, calibration graphs of 7 isotopes were performed: [23]Na, [24]Mg, [27]Al, [44]Ca, [64]Zn, [114]Cd, [208]Pb.

Table 3 *Coefficients of correlation (R) for the calibration graphs*

internal standard	^{23}Na	n	^{24}Mg	n	^{27}Al	n	^{44}Ca	n	^{64}Zn	n	^{114}Cd	n	^{208}Pb	n
none	0.9987	4	0.9995	6	0.9991	4	0.9999	3	0.9996	6	0.9993	4	0.9999	6
^{17}OH	0.9985	4	0.9993	6	0.9914	4	1	2	0.9997	4	0.9991	4	0.9997	6

3.2.2. Internal Standard   As known from solution ICP-MS, the use of an internal standard like rhodium is important to reduce the influence of signal variations coming from the drift of the ICP-MS. In real ice samples there is not a possibility to spike the sample. The main reasons for the relative high signal variations are inhomogenities of the ice standards and a different removal of material from the ice surface due to unevenness or fluctuations in laser energy. Table 4 shows the relative standard deviations in percent obtained with no internal standard as well as ^{17}OH and ^{19}OH as internal standards. The measuring time of the 10 ppb standard was 60 seconds. The ^{17}OH and ^{19}OH signals originated from the water of the ice sample and depend on the removal of material during the ablation process. With the use of ^{17}OH as internal standard an improvement of about 3 to 5 percent of RSD could be achieved. The RSDs shown in table 4 are higher than the achieved values in figure 5. The values in figure 5 were obtained with more homogeneous ice standards (preparation is shown in part 2.3.).

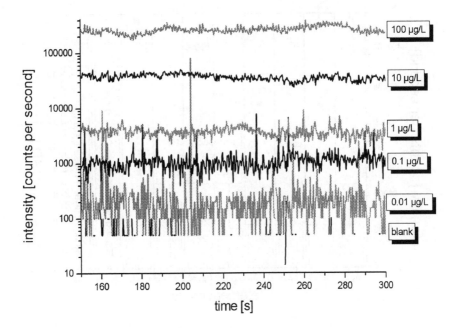

Figure 4 ^{208}Pb *signal progression of different ice standards*

Table 4 *Relative standard deviations (%) with and without internal standard*

internal standard	^{23}Na	^{24}Mg	^{27}Al	^{44}Ca	^{64}Zn	^{114}Cd	^{208}Pb
none	17	17	16	9	17	17	17
^{17}OH	14	14	14	3	15	15	14
^{19}OH	15	14	14	5	15	15	15

3.3 Daily Performance Ice Standard

It is useful to prepare a frozen daily performance standard with a selected content of elements (Be, Mg, Co, Ni, In, Ce, Pb, Bi, U) to check the performance of the measurement devices. Figure 5 illustrates the reachable intensities, relative standard deviations and the formation ratio of oxides and doupled charged ions of a 10 µg/L daily performance ice standard. The RSDs are in the range of about 3 to 6 percent (bars). The number of repetition measurements was 6. The black boxes indicate the reached intensities of the elements. Due to the fact of an relative dry laser aerosol the oxide rate was less than 0.2 per cent, the double charged ions were below 1 per cent.

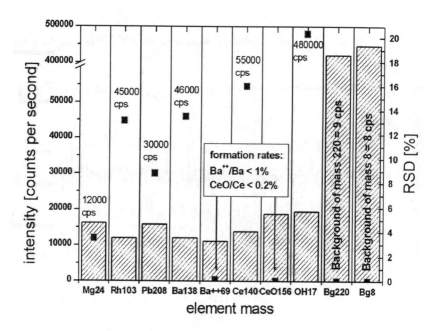

Figure 5 *Signal intensities and RSDs of a 10 µg/L daily performance ice standard*

3.4 Background Noise

Generally, aerosols which are produced by laser ablation are dry and lead to an increased background noise at the mass 220. During laser ablation of ice samples, argon which is used as carrier gas seems to take up water vapour from the probe. A comparison of different sample introduction systems shows a background of 10 – 40 cps for the laser and hence, the values are below the ones produced by a microconcentric nebulizer (60-100 cps); MCN 6000, CETAC company. On the other hand, results obtained from a cross flow nebulizer are drastically lower (1-3 cps). Therefore, the relatively low background noise caused by the laser aerosol shows a positive effect on the detection limits. According to the alignment of the microconcentric nebulizer a plasma power of 1400 W has been chosen for the laser aerosol.

3.5 ^{208}Pb Signal Progression Of A Real Sample

Figure 6 shows a signal progression of an analysed real sample (ice core segment). The sample is from Greenland from a depth of 1100 m and is about 8000 years old. After 40 seconds of forerun the laser begun to fire along a defined line. At the end of the sample (6 cm), the laser turns and goes back along the same line. The solid line shows the smoothed signal progression of lead. Similar signatures were obtained for the move fore and backwards the line scan of the laser. A calculation into concentrations with the help of the calibration graph for lead results in 10 to 20 ng/L for the sample. This value is in a good agreement with data from Boutron[1] for the GRIP core in this depth.

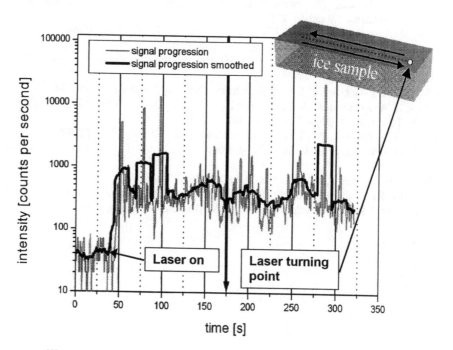

Figure 6 ^{208}Pb *signal progression of a real ice sample (Greenland)*

4 DISCUSSION AND OUTLOOK

Preliminary results for direct measurements of frozen ice samples by LA-ICP-MS have shown high counting rates and stable signals. The data gave promising indications that element signatures in polar ice cores could be obtained. In comparison to solution analysis with a Cross Flow Nebulizer the counting rates are lower by the factor of 3, however the measurement occurs directly from the solid. In spite of a special sample preparation the freezing process of the standard solution is the main cause for inhomogenities inside the ice sample and leads therefore to variations of the analyt signal. In the future we will try to prepare more homogeneous ice standards by shock-freezing at −195°C with liquid nitrogen. Examinations with a cryo-electron microscope will be carried out to check if shock-frozen ice standards are amorphous and therefore more homogeneous.

5 REFERENCES

1. C. F. Boutron, S. Hong, J.P. Candelone, *Proceedings of the International Conference: Heavy Metals in the Environment*, University of Hamburg, 1995, **1**, 28.

2. M. Legrand, R. Delmas, in *European Research Course on Atmospheres: Topics in Atmospheric and Interstellar Physics and Chemistry*, Grenoble, 1994, Chapter 19, p.387.

3. M. Kriews, H. Reinhardt, E. Hoffmann, C. Lüdke, Declaration of Invention AWI 01/0799 DE, file number DE 199 34 561.9-52, Deutsches Patent- und Markenamt, München, 1999.

4. M. Kriews, I. Stölting, J. Kipfstuhl, O. Schrems, *Proceedings of the 14. ICP-MS User Meeting*, University of Mainz, 1998, P10.

5. S. Matoba, M. Nishikawa, O. Watanabe, Y. Fujii, *J. Environ. Chem.*, 1998, **8**, 421.

6. J. M. Palais, S. Kirchner, R.J. Delmas, *Annals of Glaciology*, 1990, **14**, 216.

7. G. A. Zielinski, P. A. Mayewski, L. D. Meeker, K. Grönvold, M. S. Twickler, K. Taylor, *J. Geophys. Res.*, 1997, **102**, 26625.

8. G. A. Zielinski, M. S. Germani, *J. Archaeolog. Science*, 1998, **25**, 279.

9. D. Günther, S. E. Jackson, H. P. Longerich, *Spectrochimica Acta Part B*, 1999, **54**, 381.

10. S.G. Warren, *Applied Optics*, 1984, **23**, No. 8, 1206.

11. M. Kriews, H. Reinhardt, I. Beninga, E. Dunker, Declaration of Invention AWI V/99 KRBD, Deutsches Patent- und Markenamt, München, 1999.

RARE EARTH ELEMENT CONCENTRATIONS FOR THE MOBILE-ALABAMA RIVER SYSTEMS

E. Y. Graham[1], W. B. Lyons[2], K. A. Welch[2], T. Jones[1] and J. C. Bonzongo[3]

[1]Department of Geological Sciences, University of Alabama, Tuscaloosa, Alabama, 35487-0338, USA
[2]Byrd Polar Research Center, The Ohio State University, Columbus, Ohio 43210-1002
[3]Department of Environmental Engineering Sciences, A. P. Black Hall, University of Florida, Gainesville, Florida 32611-6450, USA

1 INTRODUCTION

Over the past decade much work has been done to determine the factors controlling the rare earth element (REE) concentration in rivers, estuaries and oceans. The primary geochemical factors that have been identified are pH, ionic strength, alkalinity, colloidal content, redox conditions, and, for rivers, the geology of the drainage basin. (1, 2, 3, 4, 5, 6) Various sites were sampled along the major tributaries of the Mobile-Alabama River systems (MARS) in order to try to understand the influence of various geologic provinces on the REE geochemistry of these waters. Even though REE concentrations are in the low ngL^{-1} range for fresh water, the detection limits of the Perkin-Elmer Elan 6000 allows direct determination without preconcentration of samples. (4) With the development of direct detection, the analysis of REEs has become more commonplace. REE distribution in rivers can provide important information regarding both the aquatic and landscape processes. (7) In general, there is an inverse relationship between pH and REE concentrations in natural waters. (1, 5) Additionally, alkalinity can play a major role in increasing the effective solubility of heavy rare earths (HREEs) via the formation the strong carbonate solution complexes. (2, 3, 8, 9)

1.1 Geologic Provinces and Drainage Basins

The Mobile River basin drains the majority of the State of Alabama. Having an area of over 85,800 square km in Alabama and more than 119,000 square km total, including areas from contiguous states, the Mobile River basin comprises one of the largest drainage systems east of the Mississippi River. The Black Warrior River and the Upper and Lower Tombigbee Rivers drain the western basin, while the eastern section is drained by the Alabama River and its three major tributaries, the Coosa, Tallapoosa and Cahaba Rivers. They join just north of Mobile, Alabama to form the Mobile River, which then divides into a complex delta system that flows into Mobile Bay. The four major provinces are the Appalachian Plateau, the Valley and Ridge, the Piedmont, and the Coastal Plain. The Coosa and Cahaba Rivers drain the Valley and Ridge, joining the Alabama River in the Coastal Plain. The Piedmont province is drained by the Tallapoosa River, which also joins

The Mobile-Alabama River System (MARS)

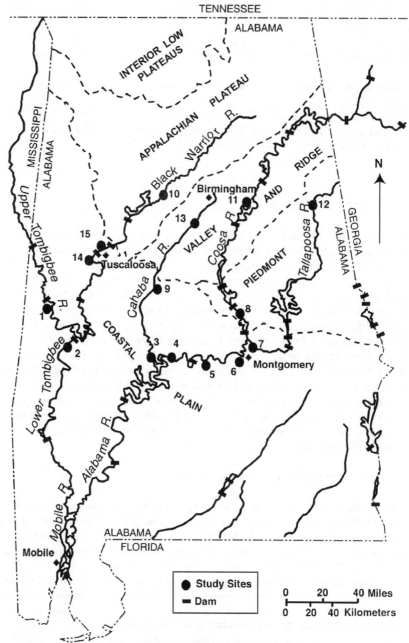

Figure 1 *Mobile-Alabama River System Sampling and Dam Sites*

the Alabama River in the Coastal Plain. The Black Warrior River arises in the Appalachian Plateau and joins the Tombigbee River in the Coastal Plain. The Upper and Lower Tombigbee Rivers are almost exclusively in the Coastal Plain (Figure 1). Each of these geologic provinces has distinct rock types and geochemical signatures. One of the goals of this investigation was to ascertain if rock type differences were important to REE concentrations in these waters.

2 METHODS

Strict adherence to ultra clean sampling techniques was followed in our study. These techniques include the collection and storage of samples in Teflon^{TM} bottles that have been precleaned by soaking for at least seven days in 10% reagent grade nitric acid, thoroughly rinsed using 18 MΩ water, and dried under a class 100 laminar flow hood. Bottles were double-bagged in clean polyethylene bags before being taken into the field. Samplers wore polyethylene gloves, used the "clean hands-dirty hands" technique, and rinsed the sampling bottles three times with sample prior to collection. (8) Samples were placed on ice until returning to the laboratory, usually within 12 hours. Upon returning, the samples were filtered using 0.4 μm Nuclepore^{TM} filters in filtering apparatus that had been precleaned by soaking in 18 MΩ water. The filtrate was acidified with Ultrex^{TM} nitric acid to ~2%.

Table 1 Limits of detection and quantification

	lod	loq		
La	139	0.160	0.532	ngL^{-1}
Pr	141	0.246	0.821	
Ce	140	0.121	0.402	
Nd	143	2.042	6.805	
Nd	146	0.745	2.482	
Sm	148	0.760	2.535	
Sm	149	3.785	12.616	
Sm	152	0.419	1.395	
Eu	151	0.288	0.962	
Eu	153	0.315	1.050	
Gd	157	0.613	2.045	
Gd	158	0.454	1.514	
Tb	159	0.227	0.757	
Dy	161	0.282	0.939	
Dy	163	0.439	1.463	
Dy	164	0.540	1.800	
Ho	165	0.205	0.683	
Er	166	0.345	1.149	
Er	167	0.530	1.766	
Tm	169	0.185	0.618	
Yb	173	1.259	4.197	
Yb	174	0.672	2.241	
Lu	175	0.069	0.231	

At least seven blanks (18 MΩ water acidified to ~2% with Ultrex™ nitric acid) were analyzed to determine limits of detection and quantification for the REEs. They are shown in Table 1. Limit of detection was defined as 3 times the standard deviation of the average of the blanks plus the average of the blanks, while the limit of quantification was defined as 10 times the standard deviation of the average of the blanks plus the average of the blanks. BaO^+ formation for Ba135 and Ba137 was monitored daily, as before, to calculate the effect on mass 151 and mass 153. During the REE analyses, BaO^+151 and BaO^+153 were monitored and appropriate correction equations were entered in the method for Eu151 and Eu153. (4)

3 RESULTS

3.1 Alabama River System

Water from each of the sampling sites in the Alabama River System shows a higher shale-normalized (SN) concentration of heavy rare earths (HREE) than of light rare earths (LREE) as expected from previous works (Figure 2). (1, 5, 9) The relative enrichments of the HREE compared to LREE in the river waters of the Mobile-Alabama River System are consistent with the clay rich provinces these rivers drain. Previous investigators have demonstrated that clay minerals can readily adsorb the positively charged LREE, leading to relative lower concentrations of the LREE in solution (10, 11) The other significant data illustrated by these graphs are the consistent positive Eu anomalies, perhaps caused by plagiocase weathering. BaO^+ interference may also be a contributing factor, but as much care as possible was used to avoid that interference.

3.1.1   Tallapoosa River

Site 12 is in the Talledega Block of the Northern Piedmont Province (Figure 1). (12) Figure 2 indicates that river water from site 12 not only has the smallest positive Eu anomaly, but is also the only sample that shows a positive Ce anomaly. In addition, site 12 has one of the highest La/Er_{SN} ratios at 0.586 and thus shows one of the lowest HREE enrichments (Table 2). Most of the REEs are trivalent in low-temperature aqueous solutions. (13, 14) However, Ce can exist as Ce(IV) and Eu as Eu(II). It is possible that the redox conditions of the waters may affect the relative concentrations of these redox sensitive REEs, leading to observed "anomalies." For example, if the river waters were strongly reducing, we would expect that all of the dissolved Ce would be in the trivalent form, as opposed to the less soluble (CeIV) which forms CeO_2 and there would be no negative Ce anomalies, or perhaps positive Ce anomalies. It is less clear if strongly reducing natural waters would facilitate reduction of Eu(III) to Eu(II), with most evidence pointing to this being unlikely. (13) Although we only have two actual sampling sites along the Tallapoosa River, there is a positive correlation ($r^2 = 0.84$) between the Fe concentration and the La concentration (Figure 3). Site 12 is relatively Fe-rich compared to the Alabama River sites in the Coastal Plain (~0.8 mgL^{-1} vs. ~0.3 mgL^{-1}). One explanation of the higher Fe value might suggest the presence of Fe^{2+} which would lead to the desorption of LREEs. This is supported somewhat by the positive Ce anomaly and the lower Eu anomaly. Another explanation might be the higher La/Er_{SN}, the higher overall REE concentrations and higher Fe concentrations of the Tallapoosa River may reflect greater weathering in general and the colloidal Fe stabilizing the LREE in solution.

Colloidal Fe would still be measured in our samples because of the filter size used: 0.4 μm.

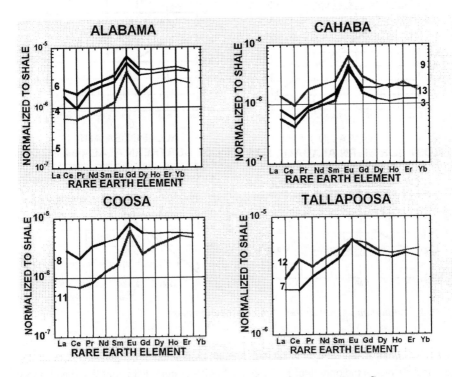

Figure 2 *Rare Earth Element Concentrations for the Alabama River System*

Table 2 *Ratios of Lanthanum:Erbium in the Alabama River System*

River	Site	Ratio
Alabama	6	0.3707
	5	0.2269
	4	0.4118
Coosa	11	0.1445
	8	0.5145
Tallapoosa	12	0.5856
	7	0.4898
Cahaba	13	0.4088
	9	0.5877
	3	0.4652

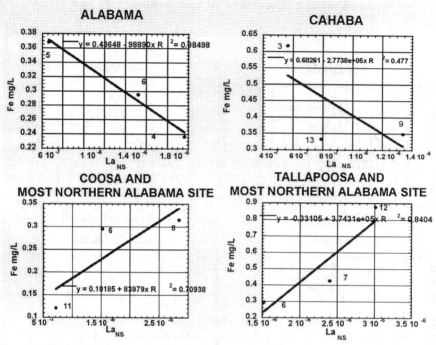

Figure 3 *Lanthanum vs Iron in the Alabama River System*

3.1.2   Cahaba River

The Cahaba River is the only river in the entire MARS system that is not dammed. Many previous investigators have shown that the HREEs form more stable carbonate complexes in solution than the LREEs, which in effect leads to greater solubility of the HREEs compared to the LREEs (Figure 4). (3, 15, 16) The greater alkalinities of the Cahaba River compared to others in the MARS system may reflect weathering of shale and carbonates along the valley floors of the Valley and Ridge Province through which the river flows. Interestingly, site 13 shows a greater enrichment in HREE than the other sites along the Cahaba River within the Coastal Plain region (i.e., sites 9 and 3); La/Er_{NS}=0.4088, 0.5877 and 0.4652, for sites 13, 9, and 3, respectively (Table 2). (12) However, the alkalinities increase with flow downstream in the Cahaba, and especially as it enters the Coastal Plain (site 9) (Figure 1), where the measured alkalinity values are the highest of all the sites along the river (Figure 4). Higher HREE concentrations correspond with the higher alkalinities. Because the Fe concentration in the Coastal Plain sites are higher, the same mechanism leading to the elevated Fe values probably also leads to the enhanced LREE concentrations, which becomes the driving factor in the La/Er ratios. The Cahaba River shows a strong *positive* correlation (r^2=0.9993) between the ratio of La/Er and pH (Figure 5). The highest pH and REE concentrations are reported for the river water sample collected where the Cahaba leaves the Valley and Ridge Province and enters the Coastal Plain (i.e., site 9).   At this point, the Cahaba River leaves the Pottsville Formation, which consists primarily of sandstone, siltstone, shale, conglomerate and coal, and enters the Tuscaloosa Group, which is chiefly composed of clay, sand and gravel (12).

We interpret this change of REE chemistry as being due in part to the change of the geology within the basin.

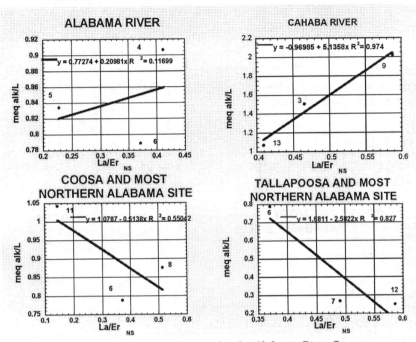

Figure 4 *Alkalinity vs Lanthanum/Erbium for the Alabama River System*

3.1.3 Coosa River

Site 11 from the region of the Coosa River located within the Valley and Ridge Province shows the greatest enrichment in HREEs compared to LREEs for our samples collected from this river, La/Er$_{NS}$=0.1445. In addition, the site 11 sample has a lower dissolved Fe concentration and higher alkalinity compared to site 8 on this river. Farther downstream, the Coosa River flows through the Piedmont Province and into the Coosa Block consisting of feldspathic garnet-quartz-muscovite schist, where the La/Er$_{NS}$ ratio rises to 0.5145. (12) Along this reach of the river, Fe concentrations increase and alkalinities decrease.

3.1.4 Alabama River

For most rivers and streams, a decrease in pH will lead to an increase in REE concentrations. (1, 3, 9, 15, 17) However, the Alabama River shows no correlation between pH and REE concentrations. Instead, the weak inverse correlation between La and Fe points to a possible change in redox conditions along the basin that may affect REE concentrations. If for some reason the Fe here exists not as colloidal Fe but as dissolved Fe, then the Fe concentration would have less of an effect on the LREE concentrations. There is a strong positive correlation between alkalinity, [HCO$_3^-$], and the ratio of La/Er$_{NS}$ (r^2=0.97367). (Figure 3) A very similar correlation exists between Ca (in mg/L) and La/Er$_{NS}$ (r^2=0.97985). When the northern most sample site along the Alabama River is included in plots with the data from the Tallapoosa and Coosa Rivers (Figure 3), the expected fairly strong positive correlation between La and Fe concentrations is observed, perhaps illustrating that the unexpected observations in the Alabama River are caused by changes farther downstream.

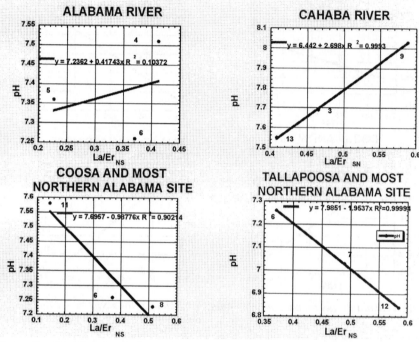

Figure 5 *pH vs Lanthanum/Erbium for the Alabama River System*

3.2 Black Warrior-Tombigbee Rivers

The city as well as other anthropogenic influences such as coal mining, coal bed methane effluents and several dams (Figures 1, 6 and 7) have impacted sites 14 and 15 around the city of Tuscaloosa. Site 10 on the Black Warrior River is located upstream from the city of Tuscaloosa within a rural setting of the Appalachian Plateau. Sampling sites 1 and 2 are along the Tombigbee River and are located on the Coastal Plain. Sites 10, 1 and 2 show very similar REE$_{SN}$ concentrations and patterns. Their shale-normalized La/Er ratios are more similar to each other than to sites 14 and 15. The normalized La/Er ratios of these river waters are directly related to alkalinity, and slightly less so to pH, as demonstrated in Figure 7. The data also reveal a dramatic decline in Fe concentration in the downstream direction from site 10 (0.325 mg/L) to sites 14 and 15 (0.052 and 0.080mg/L respectively) although the correlation between, for example, La and Fe concentrations is relatively weak. These relationships could reflect differences in redox conditions along the rivers that may have developed in response to the urban setting. Once again, the alkalinity and

Ca plots mimic each other showing the influence of limestone weathering within the catchment basins of the rivers.

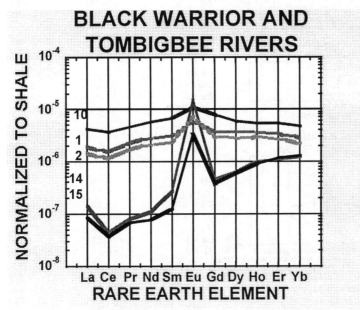

Figure 6 *Rare Earth Element Concentrations for the Black Warrior and Tombigbee Rivers*

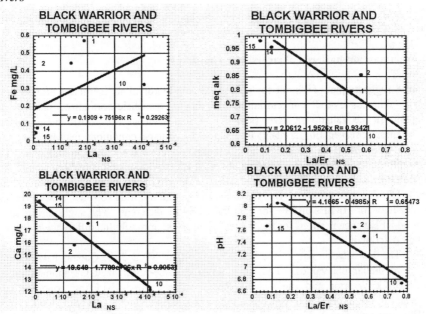

Figure 7 *Correlation of Fe vs La, Alkalinity vs La/Er, Ca vs La and pH vs La/Er for the Black Warrior and Tombigbee Rivers.*

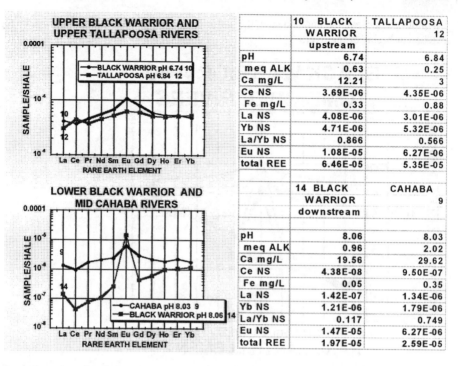

	10 BLACK WARRIOR upstream	TALLAPOOSA 12
pH	6.74	6.84
meq ALK	0.63	0.25
Ca mg/L	12.21	3
Ce NS	3.69E-06	4.35E-06
Fe mg/L	0.33	0.88
La NS	4.08E-06	3.01E-06
Yb NS	4.71E-06	5.32E-06
La/Yb NS	0.866	0.566
Eu NS	1.08E-05	6.27E-06
total REE	6.46E-05	5.35E-05

	14 BLACK WARRIOR downstream	CAHABA 9
pH	8.06	8.03
meq ALK	0.96	2.02
Ca mg/L	19.56	29.62
Ce NS	4.38E-08	9.50E-07
Fe mg/L	0.05	0.35
La NS	1.42E-07	1.34E-06
Yb NS	1.21E-06	1.79E-06
La/Yb NS	0.117	0.749
Eu NS	1.47E-05	6.27E-06
total REE	1.97E-05	2.59E-05

Figure 8   *Comparisons of Rivers with Similar Ph*

4 CONCLUSIONS

4.1 Fractionation patterns of rivers with similar pH

Figure 8 suggests that pH is not the primary driving factor of REE fractionation in these rivers. Whereas overall concentrations are similar, the shale-normalized patterns are different. In the Tallapoosa River (site 12) we find evidence (i.e., positive Ce anomaly and smallest observed Eu anomaly) that suggests reducing conditions may be important for REE concentrations (although no Eh measurements were taken) in this river. The major differences in the REE_{SN} patterns observed for the Tallapoosa site 12 and the Black Warrior site 10 are in the Ce and Eu behaviors. The HREE concentrations are similar, as are the LREE concentrations. But the usual factors contributing to these concentrations, i.e. alkalinity for HREEs and Fe for LREEs, are quite different. The slight difference in the total REE concentrations for these sites seems to be primarily the difference in the size of the Eu anomaly.

The downstream part of the Black Warrior (site 14) and the Cahaba River (site 9) also have similar pH values. The large negative, shale-normalized Ce anomaly of the lower Black Warrior and the Cahaba Rivers suggests that oxidizing conditions are important in the rivers. River water from the lower Black Warrior River is also the most HREE- enriched reported in this study, and is even more HREE-enriched than Cahaba River waters despite the higher alkalinity of water from the Cahaba. The La/Yb$_{NS}$ ratios of these waters seem to be driven more by the loss of Fe and Mn particulates (precipitation as oxyhydroxides) with the accompanying loss of LREEs, than by solution complexation of HREEs with carbonate ions that results in greater stability and solubility of the HREEs. Sites 14 and 15 on the Black Warrior River exhibit the lowest Fe concentration in the Black Warrior-Tombigbee system. These data suggest that there has been a progressive removal of REEs with flow downstream, with more LREEs being removed by precipitation or coprecipitation with Fe/Mn oxyhydroxides, and a lesser loss of HREEs.

4.2 The system as a whole

As shown by previous works, solution chemistry has a major influence on dissolved REE concentration in terms of total aqueous concentrations and the fractionation between the LREEs and the HREEs. (2, 9, 18, 19) The behavior of the REEs in the Cahaba was completely unexpected based on the observations of previous studies, as noted before. The Alabama River, which lies wholly within the Coastal Plain, is also an enigma. The Alabama River does not follow the expected patterns in terms of pH, alkalinity, and even exhibits an inverse relationship between La and Fe concentrations. The Alabama River observations may be due to the Alabama River being the most downstream end member of that system, representing a mix of waters from the Cahaba, Coosa and Tallapoosa Rivers.

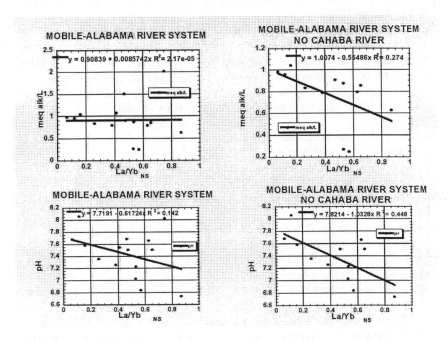

Figure 9 *The Mobile-Alabama River System as a Whole*

The system as a whole shows no correlation between alkalinity and La/Yb (Figure 9). However, if the Cahaba River samples are not considered, a correlation between the La/Yb$_{NS}$ ratio and alkalinity is apparent. Moreover, the correlation between the pH and the La/Yb$_{NS}$ ratio increases by a factor of 3 when the Cahaba River data are not included in the calculation. There is a stronger correlation between LREE concentrations and pH than between the HREE concentrations and pH. In the system as a whole, no strong correlation exists between Fe concentrations and La concentrations, or between alkalinity and Yb concentrations.

Clearly our data show that state of the art REE measurements can be made by the Perkin-Elmer system without preconcentration. The data demonstrate that both solution chemistry and geologic province affect the REE distributions in this system.

5 REFERENCES

1. H. Elderfield, R. Upstill-Goddard, and E. R. Sholkovitz, *Geochim. Cosmochim. Acta*, 1990, **54**, 971.
2. S. J. Goldstein and S. B. Jacobsen, *Earth Planetary Sci. Letters*, 1988, **89**, 35.
3. K. H. Johannesson, W. B. Lyons and D. B. Bird, *Geophysical Res. Letters*, 1994, **21**, 773.
4. E. Y. Graham, L. A. Ramsey, W. B. Lyons and K. A. Welch, in *Plasma Source Mass Spectrometry*, ed. G. Holland and S. D. Tanner, Royal Society of Chemistry, Cambridge, 1997, 253.
5. S. J. Goldstein and S. B. Jacobsen, *Earth Planet. Sci. Letters*, 1988, **88**, 241.
6. E. R. Sholkovitz, *Earth Planetary Sci. Letters*, 1992, **114**, 77.
7. C. A. Nezat, E. Y. Graham, N. L. Green, W. B. Lyons, K. Neumann, A. E. Carey and M. Hicks, *The Physical and Chemical Controls on Rare Earth Element Concentrations in Dissolved Load, Suspended Load, and River Channel Sediments of the Hokitika River, New Zealand*, GSA Annual Meeting, 1999, **31**, A-256.
8. K. A. Welch, W. B. Lyons, E. Graham, K. Neumann, J. M. Thomas and D. Mikesell, *J. Chromatography A*, 1996, **739**, 257.
9. E. R. Sholkovitz, *Aquatic Geochemistry*, 1995, **1**, 1.
10. I. R. Duddy, *Chem. Geol.*, 1980, **30**, 363.
11. J-J. Braun, M. Pagel, A. Herbillon and C. Rosin, *Geochim. Cosmochim. Acta*, 1993, **57**, 4419.
12. D.E. Raymond, W. E. Osborne, C. W. Copeland and T. L. Neathery, *Alabama Stratigraphy*, Geological Survey of Alabama, 1988, Circular **140**.
13. D. A. Sverjensky, *Earth Planet. Sci. Letters*, 1984, **67**, 70.
14. S. A. Wood, *Chemical Geology*, 1990, **82**, 159.
15. K. H. Johannesson and M. J. Hendry, *Geochim. Cosmochim.*, 2000, **64**, 1493.
16. K. H. Johannesson and W. B. Lyons, *Limnol. Oceanogr.*, 1994, **39**, 1141.
17. K. M. Keasler and W. D. Loveland, *Earth and Planetary Letters*, 1982, **61**, 68.
18. W. B. Lyons, K. H. Johannesson, B. Y. Graham, B. C. Astor and J.-C. Bonzongo, *Are Geological Differences in Watersheds Reflected in Trace Element Geochemistry of Riverine Systems?*, Proceedings AGU Spring Meeting, 1999, **80**, S107.

19.   K. Neumann, E. Y. Graham, A. E. Carey, W. B. Lyons and E. Callender, *Historic Monitoring of Trace Metal Concentrations in Rivers: Backcasting Using Sediment Cores and Present-day Water Data*, National Water Quality Monitoring Council Meeting, Austin, TX, 2000.

6 ACKNOWLEDGEMENTS

We wish to thank Dr. Kevin H. Johannesson for his thoughtful review of this manuscript.

APPLICATION OF ICP-MS TO A TRACE ELEMENTS MASS BALANCE STUDY IN A POWER PLANT

M. Bettinelli, S. Spezia, A. Fiore, N. Pastorelli, C. Terni

ENEL Produzione Spa – Laboratory of Piacenza, Via Nino Bixio 39, 29100 Piacenza, Italy.

1 INTRODUCTION

To generate electricity, fuels are combusted at high temperatures in utility boilers. This process releases in the atmosphere many organic and inorganic pollutants, both in gaseous and solid form (absorbed on suspended particulate).

Limiting the polluting emissions in the environment is therefore one of the main goals for the management as well as being a precise requirement of the national regulation: in Italy (1) 19 inorganic elements (As Be Cd Co Cr Cu Hg Mn Ni Pb Pd Pt Rh Sb Se Sn Te Tl and V), are under regulation to the emission in the environment. ENEL, the principal electricity company in Italy, regularly monitors the amount of organic and inorganic pollutants emitted by power plants, but, to gain a better knowledge of all the processes involved, it is necessary to establish a broad analytical approach in order to determine, using a mass balance approach, the elemental distribution in the various flow of the plant process.

In this study, all the matrices (fuels, gaseous emissions, waters etc.) that are involved in the power process were detected for all the elements before cited. A power plant was chosen in which it was possible to burn different types of fuel (coal, fuel oil and orimulsion, an emulsion of 30% of water in bitumen), and equipped with a desulphurization (DeSOx) and a denitrification (DeNOx) plant for the abatement of sulphur and nitrogen oxides emissions. The DeSOx plant is designed in order to reduce the amount of sulphur oxides in the gaseous emissions, by mixing the fumes with a suspension of limestone. Calcium sulphite is by this way produced and subsequently oxidized to gypsum ($CaSO_4 \cdot 2H_2O$) by fluxing air. DeNOx plant, a series of Vanadium based catalysts, reduces the N-oxides by means of the following reaction: $6 NO + 4 NH_3 \Rightarrow 5 N_2 + 6 H_2O$.

A schematic design of the power plant, is reported in Fig. 1, showing the extreme wide variety of matrices to be considered, while DeNOx scheme is omitted, because of it doesn't produce waste waters and solids and the produced Nitrogen is dispersed in the stack gas. It is difficult to formulate a unified analytical approach for the characterization of such complex matrices. In the literature, multi-element approaches to the analysis of the emission products have been already reported (2-4). In this case an analytical method was required which able to determine the elements in a wide range of concentrations (see, as an example, the case of vanadium varying from 2 ppm in the limestone to 5% in oil fly ashes) and to reach, at the same time, very low detection limits. In addition, good analytical precision was required, in order to perform mass balances as much as possible reliable, assuring contemporarily an high throughput of samples.

ICP-MS has been recognised as the technique most suited to satisfying all these requirements, having the possibility to couple different and complementary devices, such as flow injection (FI), hydride generation (HG), and microconcentric nebulizer (MCN).

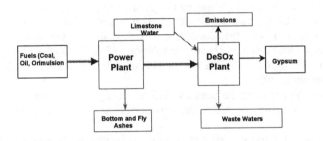

Figure 1 *Scheme of input and output flows in a Power plant equipped with a desulphurization system (DeSOx)*

2   EXPERIMENTAL

2.1 Instrumentation

An ELAN 5000 ICP-MS (Perkin Elmer Sciex Instruments, Concord Ontario, Canada) equipped with two mass flow controllers was used and Perkin Elmer   AS-60 autosampler (Perkin Elmer, Norwalk, CT USA).   Depending on the required analysis, the instrument was coupled to a Perkin Elmer Gem-tip cross-flow nebulizer, a Perkin-Elmer FIAS 400 flow injection system with or without hydride generation device or a MCN-1 microconcentric nebulizer (Cetac Technologies Inc. Nebraska USA).

For the acid dissolution of fuels and solid samples, a Milestone MLS 1200 MEGA microwave oven (FKV, Sorisole, BG, Italy), operating at 1200 W and 2450 MHz was adopted. The digestion vessels were the MDR 100 high pressure vessels (maximum pressure 100 Bar) in teflon TFM.

2.2 Reagents

Nitric acid 65% (m/v), hydrochloric acid, 37% (m/v) hydrofluoric acid 40% (m/v) and hydrogen peroxide 30% (m/v) were Suprapur reagents (Merck, Darmstad, Germany). High purity water was produced by passing distilled water through a Milli-Q deionizing system (Millipore, Bedford, MA, USA)

2.3 Standard and Reference Materials

Multielemental standard solutions were prepared from 10 mg/L multielemental atomic absorption standards, "ICP-MS Calibration Standard n. 1, 2, 3, 4" (PE Pure, Perkin-Elmer), while the standard reference materials used for the validation of the method were NIST SRM 1633a "Trace elements in coal fly ash"; NIST SRM 1648 "Urban Particulate Matter" and NIST 1634b "Trace elements in fuel Oil".

2.4 Analytical Protocol

In this campaign, we collected samples of fuels, water, limestone, as well as gypsum, waste waters, sludges, bottom and fly ashes and emissions samples, taken at the chimney of the plant. In particular, the gases emitted by the plant were gathered according to the requirements of VDI 3868 German regulation (5) that requires the collection of the particulate on a quartz filter, the condensable gases in refrigerators and the uncondensable gases in three different absorption solutions. (Sol. A with 10% (v/v) *aqua regia*; sol. B with 6% HNO_3 (v/v) and 6% H_2O_2 (v/v) and sol. C with 2% (w/v) $KMnO_4$ for the mercury determination).

Depending on the nature of the samples, different analytical problems can occur during the determination, mainly related to the presence of many ICP-MS interferences such as:

- High salinities, that can affect the efficiency of the nebulization (being the plant fed with sea water, the final waste waters were concentrated up to a 50,000 ppm of chlorides and 20,000 ppm of sulphates)
- Chlorides with the well known spectral interferences, principally on As and V
- Organics and Acids giving non-spectral interferences
  In addition, at times, only small amounts of samples were available.

All these difficulties were overcome by adopting, for every specific problem, the appropriate instrumental configuration:

- The liquid samples were directly analyzed after filtration and, if necessary, appropriate dilution. In the case of analysis of high saline solutions, the flow injection system was used, in order to avoid the clogging of the nebulizer and the cones. Moreover, when few volumes of samples were available, the microconcentric nebulizer was adopted instead of the cross-flow device.
- The solid samples (as coals, limestone, ashes, gypsum, particulate collected on filters,) were digested in the microwave oven by means of a aqua regia/HF mixture (Table 4), while the fuel oils and orimulsion samples were digested with a HNO_3/H_2O_2 solution (Table 5) . Both these procedures were previously reported in literature by Bettinelli et al. (6,7)
- Se and Hg were determined by means of the hydride generation system in all the liquid and solid samples.

The instrumental parameters are summarized in Tables 1 and 2. Linear working range from three points (10, 50 and 100 μg/L) for multi-elemental standards were used for the quantification. Standard solutions were prepared at the same acid or salt concentration as the samples.

When analyzing trace elements, low and stable blanks values are important. So, blank samples were included as part of the sampling and analysis procedure, in order to estimate the method detection limit (MDL) in a real way. Figure 2 shows the MDL values in the case of the determination of trace elements in the particulate matter collected on quartz filters.

These MDL were determined as three times the standard deviation of several (n=5) blanks, obtained in different days and are expressed as absolute µg of single elements: for all the elements the MDL was lower than 0.05 µg, with the exception of Cr and Cu that had respectively an 0.3 and 0.15 µg MDL.

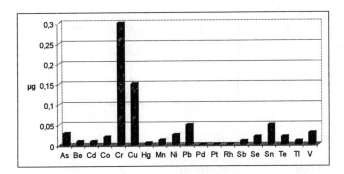

Figure 2 *Method Detection Limits (MDL) for particulate analysis*

Table 1 *Analytical conditions for PN and MCN ICP-MS analysis*

Parameters	
PF power	1150 W
Plasma argon	15.0 L/min
Nebulizer flow	0.85 L/min
Auxiliary flow	1.0 L/min
Resolution	Normal
Sample flow rate	1 ml/min (PN) *or* 90 µL/min (MCN)
Scanning Mode	Peak hop transient
Dwell time	200 ms
Sweeps / replicates	3
Number of replicates	5
Curve fit	Linear thr. zero
Nebulizer	Cross-flow *or* MCN-1 CETAC

Inter-element corrections:

$$^{75}As = {}^{75}As - 3.087 \, {}^{77}Se + 2.619 \, {}^{82}Se \qquad {}^{51}V = {}^{51}V - 3.09 \, {}^{53}Cr + 0.353 \, {}^{52}Cr$$

Table 2 *Operating conditions for FI-HG-ICPMS analysis*

Element	Se	Hg
Isotope	82	202
Sample Volume (μL)	200	200
Linearity (μgL^{-1})	100	50
Detection Limit (μgL^{-1})	0.03	0.01
Carrier Solution	HCl 3% (w/v)	
Reducing Solution	NaBH4 0.2% in NaOH 0.05% (w/v)	
Argon Auxiliary Flow	1 Lmin^{-1}	
Dwell time	20 s	
Readings / replicate	60	
N° of Replicates	3	

Table 3 *Operating Conditions for MW Dissolution of Solid Samples*

Power (Watts)	Time (min)
250	5
600	1
0	1
300	3
Vent	2

0.25 g of sample with 6 mL HCl, 2 mL HNO_3, and 2 mL HF

Table 4 *Operating Conditions for MW Dissolution of Liquid Fuels*

Power (Watts)	Time (min)
250	6
400	4
500	3' 30"
0	2
500	4
Vent.	2

nd 2 mL H_2O_2

3 RESULTS

3.1 Method validation

The proposed methods of analysis have been already validated in previous works (3, 7) by analyzing both standard reference materials (as NIST SRM 1633a "Trace elements in coal fly ash"; NIST SRM 1648 "Urban Particulate Matter" and NIST 1634b "Trace elements in fuel Oil") and comparing the ICP-MS results obtained on a series of real samples with those obtained by other techniques as GFAAS, FI-HGAAS and NAA.

The analytical campaign required a very high number of analyses, with an high throughput of samples. (About 150 samples were characterized per more than 2700 parameters in a three weeks long period). The quality of the measures had then to be continuously monitored in terms of accuracy, precision and sensitivity. To do this, the cited reference materials were periodically analyzed during the whole measuring campaign. In Table 5 the recovery percentages obtained in the case of NIST 1633a, with the corresponding RSD values, are reported as example.

Table 5 *Analysis of NIST SRM 1633a "Trace Elements in Coal Fly Ashes"*

Element	Recovery (%) (RSD n= 5)	Element	Recovery (%) (RSD n= 5)
As	105 (4)	Pb	109 (4)
Cd	102 (4)	Sb	92 (5)
Cr	102 (5)	Se	95 (4)
Cu	99 (2)	Tl	89 (2)
Ni	104 (5)	V	99 (1)
Non certified elements			
Be	100 (2)	Co	98 (3)
Sn	110 (7)		

3.2   Elemental distribution and mass balance

In a mass balance, the analytical data are related to the total amount of the matrices involved in the plant process, in order to calculate the effective quantity of element present in the plant in the same unit of time. On the basis of these calculations, the distribution pattern of the elements in the output flow from the power plant was calculated, relative to the three different combustion processes.

In this situation it is necessary to count only those elements that had, in the fuel, concentrations significantly different from the respective analytical detection limits, in order to avoid calculations affected by too high uncertainties. For this reason the results obtained for Pd, Pt, Rh, Te and Tl were not considered.

In Table 6 some results are reported in order to evidence the different behaviors that the elements have depending on the combustion process. As an example, in the case of coal combustion, the mercury is scattered between ashes (30.7%), waste waters (40.5%) and emissions (28.0%), while it is mainly present in the emissions in the case of orimulsion and fuel oil combustion (64.5% and 54.4% respectively).

The other elements are concentrated in the ashes in the case of coal combustion, while a more complex situation is shown for orimulsion and fuel oil combustion: Cr and Sn are present in the gypsum for both these fuels, while Mn, Ni and V are distributed between the various matrices. In order to graphically express the mass balance calculations, the differences between the element concentrations in the input and output flows of the power plant have been reported in Figure 3 for the more representative elements in the case, as example, of coal combustion.

The reported bias, defined as [|IN-OUT|/IN * 100], were in the order of 20 – 30 %, showing a more than acceptable result, considering all the uncertainties deriving both from the analysis, and especially from sampling operations that are involved in the process. For the other elements that are present at very low concentration, it wasn't possible to make any mass balance consideration, being their determinations affected by too high uncertainties.

Finally, the concentrations of the inorganic elements emitted by the power plant were compared to the limits of the Italian legislation. For almost all the elements analysed emissions were under 1% of the limits, with the only exceptions of Cr (for coal and fuel oil), Se (for coal) and V (for orimulsion and fuel oil) that are in the range of 5-10% and Ni for which emission levels in the order of 15-20% of the Italian limits were observed.

Table 6 *Elemental distribution in output flows of power plant*

		Ash %	Waste Water %	Gypsum %	Emissions %
Orimulsion	Cr	1.8	3.7	91.9	2.5
	Hg	0.1	41.0	4.5	54.4
	Mn	8.2	37.7	52.0	2.2
	Ni	43.2	49.5	3.6	3.8
	Sn	1.4	0.3	94.9	3.4
	V	27.4	68.6	1.2	2.7
Coal	Cr	92.9	0.4	2.6	4.2
	Hg	30.7	40.5	0.9	28.0
	Mn	98.9	0.4	0.2	0.5
	Ni	95.5	0.5	0.5	3.5
	Sn	94.7	0.1	2.9	2.3
	V	99.1	0.3	0.1	0.5
Fuel Oil	Cr	7.0	0.1	89.2	3.7
	Hg	0.1	2.6	32.8	64.5
	Mn	2.5	63.5	27.3	6.7
	Ni	33.4	44.1	17.5	4.9
	Sn	0.6	0.5	95.0	4.0
	V	27.1	59.1	7.6	6.1

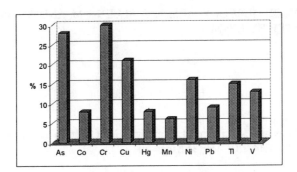

Figure 3 *Mass Balance calculation - Differences between IN and OUT contributions in the case of Coal Combustion*

4 CONCLUSIONS

In this paper, the results obtained in a campaign of measurements made in a power plant equipped with a DeSOx and DeNOx plant, testing three different fuels, coal, fuel oil and orimulsion, have been presented. The use of different configurations of ICP-MS allowed to overcome the interferences due to the presence of salts, chlorides, organics, acids, etc., making a precise and accurate determination on a very wide variety of samples. The analyses showed good detection limits (generally lower than 0.1 µg) and precision (lower than 5% as RSD%) for all the elements.

The results obtained from this campaign gave first of all indications about the pollutants emissions in this power plant, showing that the absolute values at the emissions are very low respect to the Limits established by the Italian regulation.

On a second hand this campaign gave the plant management important information about the element distributions in the industrial process, allowing a more precise knowledge of the chemical and physical factors that regulate the whole process.

5 REFERENCES

1. Ministero dell'Ambiente DM del 12/7/90, Linee guida per il contenimento delle emissioni inquinanti degli impianti industriali e la fissazione dei limiti di emissione, Rome, Italy, (30 July, 1990)
2. D. E. Kimbrough, I. H. 'Mel' Suffet, *Analyst*, 1996, **121**, 309-315.
3. M. Bettinelli, S. Spezia, U. Baroni, G. Bizzarri, *Microchemical Journal* 1198, **59**, 203-218.
4. L. Morselli, S. Zappoli, S. Militerno, *Toxicological and Env. Chem.* 1993, **37**, 139-145.
5. VDI 3868 " Determination of total emission of metals, metalloid, and their compounds - Manual measurements in flowing, emitted gases - Sampling system for particulate and filter-passing matter " Dusseldorf, Germany, 1994.
6. M. Bettinelli, U. Baroni, N. Pastorelli, *Anal. Chim. Acta* 1989, **225**, 159-174.
7. M. Bettinelli, S. Spezia, U. Baroni, G. Bizzarri, *J. Anal. At. Spectrom.* 1995, **10**, 555-560

DETERMINATION OF MERCURY, ARSENIC, SELENIUM, AND ANTIMONY IN POTABLE WATER BY ICP-MS

J. Allibone, E. Fatemian and P. J. Walker
Thames Water Utilities Ltd, Millharbour Laboratory, Great Eastern Enterprise,
3 Millharbour, London E14 9XP, United Kingdom

1 INTRODUCTION

A method for the routine determination of Mercury, Arsenic, Selenium and Antimony in potable water samples has been developed that includes addition of gold to samples to preserve mercury and the subsequent analysis for all four metals concurrently by Inductively Coupled Plasma-Mass Spectrometer (ICP-MS) . The method has been validated using the guidelines of the NS30[1] document and shows compliance with the analytical performance criteria required by the UK's Drinking Water Inspectorate. This work forms part of a project to enable the routine determination of all the low concentration metals in potable water samples to be achieved in a single run using the ICP-MS.

In the European Commission area the four metals Hg, Se, As and Sb have to be monitored in potable water to ensure compliance with an EC Directive, this being enacted in the UK by The Water Supply (Water Quality) Regulations 1989. This statutory analysis is regulated by the Drinking Water Inspectorate, who define analytical performance criteria for all analysis concerned with potable water. The guidelines for the validation process for analytical methods is given in the document NS 30, published by WRc[1.]

Antimony, Arsenic and Selenium analysis usually involves the preparation of the samples in the appropriate oxidation state. They are then reduced to the hydride form and analysed using atomic absorption, emission or fluorescence techniques. The preparation for selenium is different from that of As and Sb.

The analysis of mercury has commonly been carried out using either cold vapor fluorescence or ICP-MS to detect elemental mercury vapor following the reduction of inorganic mercury using stannous chloride or other reducing agents. This technique was originally developed by Thompson and Godden[2] in 1975, and further refined by others including Stockwell and Godden[3], 1989. These techniques will certainly meet the requirements of the DWI analytical performance criteria, but they have the disadvantage that the sample for mercury analysis has to be separately prepared and analysed, while the other metals are analysed on an ICP.

There is a need for a technique that will allow mercury, arsenic, antimony and selenium to be analysed at the same time as the other potable water metals as this will make more effective use of staff and instrument time. Mercury has another problem in that it has a very low vapour pressure and will be easily lost from environmental samples if allowed to undergo reduction to the elemental species via natural biochemical pathways. It has been know for a long time that gold will amalgamate mercury and in doing so will effectively stop its loss.

Inductively coupled plasma would be an ideal technique for the determination of these metals including a gold/mercury amalgam. Unfortunately the use of optical emission spectrometry (ICP-OES) for mercury analysis is not feasible for potable waters because its response is too low to be practical at the concentrations expected. However the use of an ICP-MS will enable the low concentrations to be detected successfully, as shown by others in differing matrices[4-7]. One problem that does occur is the long memory effect between samples caused by the contact of the mercury within samples with the materials that comprise the sample introduction system. This is usually overcome by extended washing periods between samples.

It was our contention that we could make the routine analysis for mercury practical by using gold addition to the samples to preserve them and to use an ICP-MS for the simultaneous determination of mercury with other low level metals, overcoming the problem of memory by addition of gold to the instrument wash water. The use of gold as a preservative has definite advantages over the nitric acid/dichromate preservation mixture used currently, particularly with PET (polyethylene teraphthalate) containers. This work will be described in a future paper.

In this paper we describe in detail a method for the routine analysis of potable water utilising the addition of gold to samples and standards. The results of a validation exercise to DWI criteria is given to show that the method fulfills the performance criteria of this UK regulatory body.

2 EXPERIMENTAL

The PE Sciex Elan 6000 ICP-MS (Perkin-Elmer, Beaconsfield, UK) was used for the analysis of Mercury in potable water samples. The ICP-MS was equipped with PE-AS91 auto sampler. Samples were introduced via a cross–flow nebuliser with Scott-type spray chamber. The Elan 6000 and auto sampler and peristaltic pump are controlled by the Elan Windows NT software and are fully automated. The operational criteria for the ICP-MS are given in Table 1.

The Elan 6000 instrument was optimised daily to ensure that it was functioning correctly, involving checking the tuning, mass calibration and system performance. Optimisation was carried out according to manufacturer's instructions using various multi-element solutions in 2% nitric acid (Aristar grade). Pre-prepared single element 1000 mg l^{-1} metal standards (Spectrosol – BDH, Poole, Dorset, UK.) were mixed to give the solutions used in the optimisation process. Serial dilutions of mixed intermediate solutions were used to obtain the concentrations stated. Where necessary these optimisation solutions were matrix matched by the addition of calcium, sodium and potassium as describe. Tuning was carried out using the solution described in Table 2, to cover the mass range for routine use, which included that of mercury at 202. After tuning was completed a full check on the whole system performance was then carried out using the solution in Table 3. This gave performance figures for the rhodium intensity, for the formation of oxides (CeO) and doubly charged ions (Ba^{++}). The performance of the analogue section of the MS detector was then optimised using the solution shown in Table 4. No analysis was carried out unless all the manufacturer's performance criteria for the instrument were satisfactory.

Table 1 *Instrument Setting*

Sweeps/Reading	65
Reading/Replicate	1
Number of replicate	3
Tuning File	Default.tun
Optimisation File	Default.dac
Scan Mood	Peak hopping
MCA	1
Dwell time	25.0 ms
Integration time	1625.0 ms
Detector Mode	Dual
Auto Lens	On
Spectral Peak processing	Average
Signal profile processing	Average
Curve type	Linear through Zero
Rh 103	Internal Std
Std 1	Blank
Std 2	0.6 μg l^{-1}
Std 3	1.20 μg l^{-1}
Sample flush time	30s
Sample flush speed	48rpm
Read Delay	60s
Read delay between stds	60s
Wash time between stds	1s, 60s, 120s
Wash speed	48rpm
Nebuliser gas flow	1.0 l/m
RF power	1025

Table 2 *Tuning solution and Mass Range*

Element	Mass	Concentration µg l^{-1}
Be	9	10 µg l^{-1}
Mg	24	10 µg l^{-1}
Rh	102	10 µg l^{-1}
Pb	208	10 µg l^{-1}
U	238	10 µg l^{-1}
Ca	NA	100 mg l^{-1}
Na	NA	30 mg l^{-1}
K	NA	10 mg l^{-1}

NA - Not appropriate

Table 3 *System Optimisation Solution*

Element	Mass	Concentration
Li	3	10 µg l^{-1}
Be	9	10 µg l^{-1}
Mg	24	10 µg l^{-1}
Co	59	10 µg l^{-1}
Ni	59	10 µg l^{-1}
Rh	103	10 µg l^{-1}
In	114	10 µg l^{-1}
Ba	137	10 µg l^{-1}
Ce	140	10 µg l^{-1}
Tl	204	10 µg l^{-1}
Bi	213	10 µg l^{-1}
Pb	208	10 µg l^{-1}
U	238	10 µg l^{-1}
Ca		100 mg l^{-1}
Na		30 mg l^{-1}
K		10 mg l^{-1}

3 REAGENTS

3.1 Rhodium

A 10 µg l^{-1} Rhodium (Rh) stock internal standard solution was prepared from single element 1000 mg l^{-1} Rh (BDH - Spectrosol) by serial dilution. All the standard solutions including calibration standards were spiked to final concentration of 10µg l^{-1} with Rh , as were the samples. All data is referenced to the internal standard.

3.2 Gold

Gold was obtained in the form of an ICP standard, 1000 mg l^{-1} Au concentration from Fisher Scientific International Co, Loughborough, UK. and diluted as necessary using pipettes and volumetric flasks.

Table 4 *Detector (analogue) Optimisation Solution*

Element	Mass	Concentration
Mg	24	200 µg l^{-1}
Cu	63	200 µg l^{-1}
Cd	114	200 µg l^{-1}
Rh	103	200 µg l^{-1}
Pb	208	200 µg l^{-1}
Ca	NA	100 mg l^{-1}
Na	NA	30 mg l^{-1}
K	NA	10 mg l^{-1}

NA - Not Appropriate

3.3 Calibration standards

The standards used for calibration were prepared in 1% nitric acid (Aristar) in deionised water with the addition of calcium, sodium and potassium as described in Table 5, to match the concentrations of these elements found in potable water samples.

Table 5 *Calibration Solutions*

	Blank	STD 1	STD 2
Hg	----	0.6 µg l^{-1}	1.02 µg l^{-1}
Ca	100 mg l^{-1}	100 mg l^{-1}	100 mg l^{-1}
Na	30 mg l^{-1}	30 mg l^{-1}	30 mg l^{-1}
K	10 mg l^{-1}	10 mg l^{-1}	10 mg l^{-1}
Rh	10 µg l^{-1}	10 µg l^{-1}	10 µg l^{-1}
Au	5 mg l^{-1}	5 mg l^{-1}	5 mg l^{-1}

The range of calibration, 0 - 1.2 µg l^{-1} Hg was chosen to slightly exceed the PCV limit of 1 µg l^{-1} Hg applied to potable water. Similarly the calibration ranges for arsenic, selenium and antimony were chosen to exceed the PCV limits of 50, 10, 10 µg l^{-1}. All standards were spiked with 10 µg l^{-1} rhodium (Rh) as an internal standard and 5 mg l^{-1} gold (Au) to mirror the samples and to act as a preservative.

3.4 Validation standards

The validation standards were prepared to the concentrations shown in Table 6, in 1% nitric acid (Aristar). The sodium, calcium and potassium were added to provide matrix matching with the samples. Rhodium was added as an internal standard, while the gold was added to preserve the mercury.

3.5 Rinse water and Samples

The rinse water used between samples was 2% nitric acid (Aristar) in deionised water. In keeping with the guidelines in NS 30, real samples of potable water were used in the experimental work and validation. For the validation studies a large volume of locally obtained potable water was sub-divided into two batches of which one was spiked with mercury to give the PCV concentration of 1 µg l^{-1} Hg. Other elements of interest were spiked into the solution at the appropriate PCV concentrations. The spiked and unspiked solutions had gold solution added to give a concentration of 5mg l^{-1} Au, and were used throughout the validation exercise.

Table 6 *Solution details used for validation*

	As µg l^{-1}	Se µg l^{-1}	Sb µg l^{-1}	Hg µg l^{-1}
Blank	----	----	----	----
10%	6.0	1.2	1.2	0.12
90%	54	10.8	10.8	1.08
Tap	----	----	----	----
Spiked tap	50	10	10	1.0

	Na mg l^{-1}	Ca mg l^{-1}	K mg l^{-1}	Rh µg l^{-1}	Au mg l^{-1}
Blank	30	100	10	10	5
10%	30	100	10	10	5
90%	30	100	10	10	5
Tap	30	----	----	10	5
Spiked tap	30	-----	----	10	5

NB all other elements of interest were present at the appropriate concentrations.

Table 7 *Results from the Direct Analysis of Spiked Tap Water*

	Measured Hg µg l^{-1}	Actual Hg µg l^{-1}
Spiked	1.41	1.0
Spiked	1.47	1.0
Spiked	1.34	1.0
Spiked	1.28	1.0
Spiked	1.18	1.0
Spiked	1.12	1.0
Spiked	1.02	1.0
Spiked	0.977	1.0
Spiked	0.766	1.0
Spiked	0.757	1.0

4 PROCEDURE

Initial work on the ICP-MS using the operating conditions for normal metals analysis showed that the results for direct mercury analysis without any pre-treatment were highly erratic, see Table 7. These results were obtained within a single run of identical samples taken from a bulk sample of potable water spiked to the PCV of 1 μg l^{-1} Hg, and acidified to 1% with nitric acid. It is apparent that the mercury is being lost with time, indicating that it is being retained somewhere within the system between the auto-sampler cup and the ICP torch. No indication is given of the gradual reduction in this loss with time, which would indicate that any active sites were being saturated. In fact the results indicate that the loss increases with time which could indicate losses in many different areas occurring at differing rates. The imprecision of the results was attributed to the loss of mercury on the materials of the sample delivery system of the ICP-MS, e.g. pump tubing, nebuliser plastics and glass components of the torch. There is also the possibility that having lost this mercury onto these materials it could be released gradually over time, leading to contamination of subsequent samples and poor calibration. It was obvious from this work that there was no possibility that this technique would satisfy the analytical performance criteria of the DWI, and that some form of stabilising/preserving agent would be required to make this method of analysis successful. The work on the addition of gold as a preservative for mercury in real samples, the subject of a later paper, prompted the addition of gold at a concentration of 5mg l^{-1} Au, to the wash solution. The results from these studies were so promising that it was decided to go straight into a full NS 30 validation, using gold in samples and standards.   The analysis of arsenic, selenium and antimony was far less problematical than that of mercury.   With the ICP-MS conditions for the analysis established previously by the mercury study it was then necessary to establish the best way to include arsenic, selenium and antimony into this scheme.   Initial trials showed which atomic masses were the most appropriate. (Arsenic being monoisotopic gave no choice).   The software inter element corrections were shown to function correctly by analysing a number of spiked samples. Once the conditions were set to give optimum results it was then possible to carry out validation studies for each metal following the NS30 guidelines.

5 VALIDATION STUDIES

To be shown to be suitable for the analysis of potable water the validation exercise must be carried out under the protocol described in the document NS30[1] . This is the standard that is specified by the DWI to ensure that the validation of methods for the analysis of potable water in the UK is carried out in a uniform manner which will allow comparisons to be made between different laboratories and methods. The protocol normally involves the analysis of a batch of duplicate samples comprising blanks, a standard solution with a concentration of the determined about 10% of the PCV, a standard solution with a concentration of 90% of the PCV, a sample of real potable water and a real water sample spiked to the PCV concentration. The samples are randomised before analysis to ensure that the data obtained truly reflects what would be happening with a mixture of real samples of unknown and varying concentrations, where a low concentration sample may have been preceded by one with a high concentration. Any carry-over from one sample to another would be highlighted by the comparison between duplicates, as would any system instability. The NS30[1] document covers not only the validation protocol but all major aspects of laboratory Analytical Quality Control for use within the water industry in great detail, including statistical operations (including worked examples), design of tests and inter-laboratory quality control. To ensure that the validation exercise mimics normal operation only a maximum of two batches can be analysed daily, with full re-calibration of the instrument occurring between batches. Following daily optimisation and calibration under the described operating conditions (Tables 1-5), the Elan 6000 instrument was used to analyse the following samples for mercury, arsenic, selenium, antimony and other elements of interest.

1 Blank - a deionised water blank
2 Low standard 10% of calibration range
3 High standard 90% of calibration range
4 Real potable sample
5 Potable sample (4 above), spiked with mercury to 1 µg l^{-1} Hg, 50 As µg l^{-1}, 10 µg l^{-1} Sb, 10 µg l^{-1} Se and other elements of interest at the appropriate PCV values.

All samples including Blank had matrix matching elements, internal standard and gold added as specified in Table 7, and were acidified to 1% with nitric acid. The full NS30 protocol was followed with two batches only being analysed daily each having a full calibration prior to batch analysis. Samples were randomised on the auto-sampler.

6 VALIDATION REPORTING

The DWI acceptance criteria for potable water analysis is given in the booklet "Guidance on Safeguarding the Quality of Public Water Supplies"[8] and other supplementary documentation available from DWI. The Section 4 deals with the performance required of analytical systems, specifying for most for non-microbiological parameters. The maximum tolerable total error of individual results should not exceed C or 20% of the result, whichever is the greater; the maximum tolerable total standard deviation of individual results should not exceed C/4 or 5% of the result, whichever is the greater; and the maximum tolerable systematic error (or bias) of individual results should not exceed C/2 or 10% of the result, whichever is the greater. In most cases C = one-tenth of the prescribed concentration or value (PCV for Hg 1µg l^{-1}). The limit of detection implied from this specification is 4.645 times the within-batch standard deviation of results for blanks , and must at least equal 10% of the PCV, i.e. for mercury the LOD should be at least 0.1 µg l^{-1} Hg, arsenic 5 µg l^{-1}, selenium 1 µg l^{-1}, antimony 1 µg l^{-1}. The AQCsoft (Published by WRc) data evaluation package has been specifically designed for use on validation exercises carried out to the NS30 protocol. The results of this validation study are suitably arranged into a paired logical sequence, with the statistical analysis calculated by this performance evaluation

Table 8 *Validation Results for Mercury Stabilised with Gold using ICP-MS*

Results in µg l^{-1} Hg , solution composition as in Table 7

Batch	Blank	0.12 µg l^{-1} (10% PCV)	1.08 µg l^{-1} (90% PCV)	Tap	Spiked Tap to 1.0 µg l^{-1} (PCV)
1a	0.003	0.122	1.070	0.010	0.993
1b	0.001	0.124	1.080	0.008	0.987
2a	0.000	0.107	1.050	0.009	0.975
2b	0.005	0.115	1.060	0.001	0.970
3a	0.010	0.121	1.040	0.017	0.937
3b	0.001	0.124	1.030	0.012	0.932
4a	0.025	0.117	1.050	0.015	0.944
4b	0.002	0.118	1.060	0.008	0.944
5a	0.019	0.115	1.040	0.009	0.939
5b	0.005	0.113	1.050	0.001	0.948
6a	0.012	0.119	1.120	0.015	1.020
6b	0.011	0.118	1.100	0.011	1.040
7a	0.007	0.114	1.060	0.016	0.997
7b	0.004	0.114	1.070	0.009	0.990
8a	0.005	0.106	1.030	0.009	0.980
8b	0.001	0.105	1.050	0.008	0.989
9a	-0.009	0.107	1.090	0.006	0.988
9b	0.000	0.116	1.090	-0.001	1.010
10a	-0.002	0.103	1.060	0.002	1.010
10b	0.004	0.107	1.070	-0.002	1.010
11a	0.001	0.118	1.100	0.001	1.040
11b	0.000	0.120	1.080	0.001	1.060
12a	-0.005	0.100	1.080	0.005	1.040
12b	-0.008	0.106	1.080	-0.001	1.030
Mean	0.0038	0.1137	1.0671	0.0070	0.9905
M1	0.0001	0.0001	0.0010	0.0001	0.0028
M0	0.0000	0.0000	0.0001	0.0000	0.0001
F value	1.8800	9.8573	13.0287	3.3602	40.0015
Sw (within)	0.0064	0.0030	0.0089	0.0039	0.0084
Sb (between)		0.0064	0.0218	0.0043	0.0370
St (total)		0.0071	0.0236	0.0058	0.0379
Rel SD		6.2	2.2	82.7	3.8
Significance	N.S	***	***	*	***
Calculated F		0.0800	0.1951	0.0542	0.5854
Est Degs F		13	13	17	12
LOD	0.032 µg l^{-1}				
Recovery	99.05 %				

N.S   Not significant   *, **, *** Significant at 0.05, 0.01 and 0.001 respectively

software package. This data passes the DWI acceptance criteria. The results of the individual batch analysis are ordered by sample type by batch and subjected to statistical analysis (Tables 8,9,10,11). The means for each sample type, within and between batch standard deviations, the bias, the limit of detection and the recovery are all calculated and compared to the guideline values specified by DWI for analytical method performance. The estimated number of degrees of freedom must not be less than 11 for the exercise to be valid.

Table 9 *Validation results for As using ICP-MS*

Batch	Blank	0.12 µg l^{-1} (10% PCV)	1.08 µg l^{-1} (90% PCV)	Tap	Spiked Tap to 1.0 µg l^{-1} (PCV)
1	0.002	5.870	53.0	2.09	54.0
	0.002	5.880	52.6	2.07	53.5
2	-0.013	5.740	51.9	2.18	55.1
	-0.011	5.770	51.9	2.18	55.3
3	0.005	5.890	52.5	2.11	53.4
	-0.008	5.890	52.7	2.11	53.7
4	0.432	6.060	54.1	2.35	55.0
	-0.026	6.030	54.1	2.33	53.3
5	-0.008	5.970	53.2	2.12	55.1
	-0.003	5.930	52.9	2.11	54.9
6	-0.003	6.070	53.7	2.16	55.2
	-0.001	5.980	53.8	2.14	54.9
7	0.013	6.090	54.1	2.37	55.2
	0.008	6.100	53.7	2.42	54.5
8	0.342	6.120	54.2	2.30	54.9
	-0.015	6.090	54.5	2.30	55.6
9	0.011	6.020	53.7	2.33	55.4
	-0.006	5.880	53.2	2.28	55.1
10	0.003	5.940	52.9	2.13	54.3
	-0.007	5.820	53.1	2.08	54.7
11	0.008	5.930	53.1	2.13	53.8
	0.008	5.890	53.2	2.14	53.8
Mean	**0.033**	**5.95**	**53.27**	**2.20**	**54.57**
M1	0.0112	0.0227	1.0394	0.0249	0.8604
M0	0.0154	0.0022	0.0386	0.0004	0.2105
F value	1.3686	10.349	26.901	61.552	4.0881
Significance	N.S	***	***	*	*
Rel SD		**1.9**	**1.4**	**5.1**	**1.3**
Calculated F		0.0080	0.0760	0.0081	0.0719
Est Degs F		12	11	10	15
LOD	**0.629 µg/l**				
Recovery	**104.78 %**				

N.S	Not significant	* **, *** Significant at 0.05, 0.01 and 0.001 respectively
M1	Between batch mean square,	
M0	Within batch mean square	
F	Value M1/M0	

Rel SD   Relative Standard Deviation
LOD   Limit of Detection.

Table 10 *Validation results for Se using ICP-MS*

Batch	Blank	$0.12\ \mu g\ l^{-1}$ (10% PCV)	$1.08\ \mu g\ l^{-1}$ (90% PCV)	Tap	Spiked Tap to 1.0 μg l^{-1} (PCV)
1	-0.032	1.03	10.6	2.06	12.6
	-0.064	1.00	10.6	2.14	12.5
2	0.042	1.07	10.6	2.03	12.3
	-0.139	0.947	10.5	2.25	12.7
3	0.020	1.31	12.08	1.47	13.23
	-0.010	1.37	12.20	1.67	13.51
4	0.321	1.40	11.40	2.02	12.60
	0.293	1.36	11.20	2.12	12.80
5	0.078	1.35	12.1	2.53	14.10
	0.120	1.33	12.10	2.63	14.20
6	-0.003	1.12	10.90	2.06	12.70
	-0.020	1.12	10.70	2.21	12.80
7	0.107	1.39	12.10	2.05	13.50
	-0.070	1.44	12.20	2.12	13.60
8	0.128	1.45	10.90	2.42	13.30
	0.117	1.37	10.60	2.64	13.10
9	0.150	1.28	11.13	1.97	12.42
	0.086	1.21	10.89	2.16	12.44
10	0.096	1.30	11.10	1.86	12.50
	0.080	1.27	11.30	1.95	12.50
11	0.005	1.25	10.90	2.42	12.30
	0.051	1.08	10.60	2.74	12.50
12	0.076	1.17	11.10	1.68	12.60
	0.065	1.20	11.30	1.88	12.60
Mean	0.0624	1.242	11.212	2.128	12.891
M1	0.0191	0.0415	0.7251	0.1932	0.6111
M0	0.0031	0.0028	0.0180	0.0157	0.0166
F value	6.060	15.069	40.281	12.292	36.775
Significance	**	***	***	***	***
Rel SD	0.0561	0.0525	0.1342	0.1254	0.1289
Calculated F		0.3542	1.1821	1.6713	0.7554
Est Degs F		12	13	13	12
LOD	0.2828 $\mu g/l$				
Recovery	104.88				

N.S   Not significant $*$ $**$, $***$ Significant at 0.05, 0.01 and 0.001 respectively
M1   Between batch mean square,
M0   Within batch mean square
F   Value M1/M0
Rel SD   Relative standard Deviation
LOD   Limit of Detection.

Table 11 *Validation results for Sb using ICP-MS*

Batch	Blank	0.12 µg l⁻¹ (10% PCV)	1.08 µg l⁻¹ (90% PCV)	Tap	Spiked Tap to 1.0 µg l⁻¹ (PCV)
1	0.064	1.42	11.20	0.652	11.30
	0.106	1.38	11.20	0.657	11.20
2	0.403	1.26	10.60	0.487	10.50
	0.424	1.20	10.50	0.492	10.20
3	0.020	1.44	11.20	0.676	11.20
	-0.013	1.36	11.20	0.653	11.10
4	0.150	1.14	10.70	0.567	10.30
	0.053	1.42	10.80	0.529	10.10
5	0.321	1.45	11.40	0.663	11.20
	-0.014	1.37	11.30	0.638	11.10
6	-0.011	1.33	11.00	0.678	11.10
	-0.031	1.33	10.90	0.643	10.90
7	0.173	1.36	11.00	0.692	11.10
	0.004	1.32	10.80	0.694	11.30
8	0.010	1.36	10.90	0.672	11.10
	0.179	1.34	10.80	0.679	11.40
9	0.020	1.38	10.80	0.685	11.20
	-0.013	1.29	10.80	0.681	11.40
10	0.024	1.41	11.30	0.697	11.50
	-0.004	1.36	11.30	0.691	11.60
11	0.031	1.38	10.90	0.664	10.90
	0.001	1.34	10.80	0.647	11.10
12	0.008	1.33	11.00	0.640	11.10
	-0.019	1.30	10.90	0.625	11.20
Mean	**0.078**	**1.344**	**10.97**	**0.641**	**11.04**
M1	0.0283	0.0056	0.1177	0.0076	0.3022
M0	0.0077	0.0046	0.0046	0.0002	0.0179
F value	3.6510	1.2063	25.677	40.362	16.868
Significance	*	N.S	***	***	***
Rel SD	-----	**5.3**	**2.3**	**9.7**	**3.6**
Calculated F		0.0820	0.2032	0.0619	0.5248
Est Degs F	23	12	12	12	12
LOD	0.4437 µg/l				
Recovery	104.105 %				

N.S   Not significant * **, *** Significant at 0.05, 0.01 and 0.001 respectively
M1   Between batch mean square,
M0   Within batch mean square
F   Value M1/M0
Rel SD   Relative Standard Deviation
LOD   Limit of Detection.

7 DISCUSSION

The initial work undertaken on the analysis of mercury using the ICP-MS gave results on standard solutions that were indicative of major problems with the retention of mercury within the sample introduction and torch system. It was apparent that it would not be possible to directly analyse mercury on this system unless a method was developed to retain the mercury in solution throughout its passage from sample cup to torch. The validation exercise for mercury has fulfilled the expectations given in the preliminary work using gold addition to retain the mercury in solution and prevent it being lost within the sample introduction process to the ICP-MS. It has also shown that the addition of the gold to both samples and standards will effectively preserve the metal in solution for at least a three week period – that being the length of time taken to complete the validation study. The use of gold means that now mercury can be co-analysed with the normal low level metals using the ICP-MS without using extended wash out times and thus improving sample throughput and efficiency. The results of the validation show that the method performs very well over the range required for potable water analysis. The limit of detection at $0.032\mu g\ l^{-1}$ Hg and the recovery of 99% at the PCV concentration using real potable water matrix fully justifies the future use of this method for the routine determination of mercury and all the other metals of interest (Table 12). The bias and standard deviation of all samples that contain mercury are acceptable to the Drinking Water Inspectorate, fulfilling the national analytical performance criteria. Gold is shown to be a suitable preservation agent for both standards and samples as no significant loss of the metal is shown from any solution and the results are more precise than when run without the gold being present.

Table 12 *Recovery of multi-element in solution containing Au and Hg*

Element	Spiked µg/l	Recovered µg/l
Al	200	198
Fe	200	193
Cr	50	47.9
Mn	50	49.2
Ni	50	48.5
Pb	50	51.6
Ag	10	10.1
Cd	5	5.03
As	50	50.3
Se	10	9.85
Sb	10	9.16
Cu	3000	2920
Zn	5000	4850

The results from the validation studies on arsenic, selenium and antimony show full compliance with the analytical standards set by DWI. The limits of detection are comparable or better that those obtained using the fluorescence techniques currently in place. Table 12 is a summary of a small investigation into the effects of the inclusion of gold in a solution of all the low concentration metals of interest to the DWI in potable water. There is no significant affect on the analytical results obtained, proving that the presence gold has no detrimental effects on the analysis using ICP-MS.

8 CONCLUSION

This work shows that it is possible to determine mercury, arsenic, selenium, antimony in potable water to DWI analytical criteria using the ICP-MS, under a regime of operating conditions usually applied to the analysis of other low concentration metals. This has been achieved by the addition of gold to amalgamate the mercury within the samples and standard solutions, causing it to be retained within solution instead of being lost onto the materials of the ICP-MS sample introduction system. This allows simultaneous determination of mercury with these metals and will result in an increased efficiency within the laboratory.

9 REFERENCES

1. NS 30 A Manual on Analytical Quality Control for the Water Industry - R. V. Cheeseman and A. L. Wilson (Revised M. J. Gardner 1989) ISBN 0 902156 85 3 Water Research Centre plc, Marlow, Bucks, UK.
2. K. C. Thompson and R. G. Godden, Analyst 1975, 100, 544 R. G. Godden and P. B. Stockwell, *Journal of Analytical Atomic Spectrometry*, 1989, **Vol 4**, 301.
3. I. B-A. Razagui and S. J. Haswell, *Journal of Analytical Toxicology*, **Vol 21**, March/April 1997.
4. B. Passariello, M. Barbaro, S. Quaresima, A. Casciello and A. Marabini, *Microchemical Journal 54*, 1996, 348-354.
5. A. Woller, H. Garroud, F. Martin, O. Donard, P. Fodor, *Journal of Analytical Atomic Spectrometry*, January 1997, **Vol 12**, 53-56.
6. J. Yoshinaga and M. Morita, *Journal of Analytical Atomic Spectrometry*, April 1997, **Vol 12**, 417-420.
7. Guidance on Safeguarding the Quality of Public Water Supplies, HM Stationary Office, 2nd Impression 1990, ISBN 0 117522627.

DETERMINATION OF METALS IN SEWAGE AND INDUSTRIAL WASTE WATERS BY ICP-MS IN A SINGLE RUN

E.Fatemian, J.Allibone, and P.J.Walker

Millharbour Laboratory, Thames Water Utilities Ltd, Great Eastern Enterprise, 3 Millharbour, London E14 9XP, United Kingdom

1 INTRODUCTION

The analysis of industrial waste water, crude sewage and treated final effluent form an important part of the laboratory routine analysis in addition to drinking water. This type of environmental analysis is not unique and it is becoming important in all environmentally conscious societies. A method has been developed for the determination of Cr, Mn, Mo, Ni, Pb, Zn, Sn, Co,Cu, Be, Sn Cd, V,Ag, Se, Sb, ,Hg and As in industrial waste waters, crude sewage and final effluent using the ICP-MS instrument. There are no clear and well-defined regulations for validation of methods related to these sample types unlike potable water, where the validation is regulated by Drinking Water Inspectorate and defined in the NS30 document. In the absence of clear guidelines the method demonstrated here has been validated using NS-30 criteria with the self-imposed condition that the assigned PCV (proscribed concentration value) for the determinand must be better or equal to that obtained for potable waters.

After successful development of methods for Hg, As, Se and Sb and other metals of interest in a single run using an ICP-MS for drinking water analysis [1,2], the application was expanded to very different matrices, Crude Sewage, final treated effluent and industrial waste waters. Our objective was to develop a method that would allow the analysis all the metals of interest including As, Se, Sb and Hg, in one pass through the instrument for these matrices. The 17 metals of concern in these matrices are Chromium (Cr), Manganese (Mn), Molybdenum (Mo), Nickel (Ni), Lead (Pb), Zinc (Zn), Tin (Sn), Cobalt (Co), Copper (Cu), Beryllium (Be), Cadmium (Cd), Vanadium (V),Silver (Ag), Selenium (Se), Antimony (Sb), Mercury (Hg) and Arsenic(As). These elements are usually analysed by ICP, AAS, fluorescence and hydride generation techniques, requiring several passes and a variety of instrumentation to complete the analytical suite.

2 INSTRUMENTATION

The PE Sciex Elan 6000 ICP-MS (Perkin- Elmer, Beaconsfield, UK) was used for the analysis of metals in industrial waste water, crude sewages and final effluents samples. The ICP-MS was equipped with PE-AS91 Auto Sampler. Samples were introduced via a cross–flow nebuliser with Scott-type spray chamber. The Elan 6000 and auto sampler and peristaltic pump are controlled by the Elan Windows NT software and are fully automated. The operational criteria for the ICP-MS are given in Table 1. The Elan 6000 instrument was optimised daily by checking the tuning, mass calibration and system performance to ensure that it was functioning correctly. Optimisation was carried out according to manufacturers instructions using various multi-element solutions in 2% nitric acid(Aristar grade- BDH, Poole, Dorset,UK).   Pre-prepared single element, 1000 mg l^{-1} metal standards (Spectrosol – BDH, Poole, Dorset,UK.) were mixed to give the solutions used in the optimisation process. Serial dilution's of mixed intermediate solutions were used to obtain the concentrations stated. Tuning was carried out using the solution described in Table 2, to cover the mass range for routine use. After tuning was completed a full check on the whole system performance was then carried out using the solution in Table 3. This gave performance figures for the rhodium intensity, for the formation of oxides (CeO) and doubly charged ions (Ba^{++}). The performance of the analogue section and dual and mass calibration of the MS detector was carried out using the solution shown in Table 4. No analysis was carried out unless all the manufacturers performance criteria for the instrument were satisfactory.

3 REAGENTS

3.1 Internal standard

A 10 μg l^{-1} Rhodium(Rh) stock internal standard solution was prepared from single element 1000 mg l^{-1} Rh (BDH -Spectrosol) by serial dilution. All the standard solutions including calibration standards were spiked to final concentration of 10μg l^{-1} with Rh , as were the samples. All data is referenced to the internal standard.

3.2 Gold

Gold was obtained in the form of an ICP standard, 1000 mg l^{-1} Au concentration, (Fisher Scientific International Co, Loughborough, UK) and diluted as necessary using pipettes and volumetric flasks. Gold was used in all samples including calibration standards at a concentration of 5mg l^{-1} Au as preservative for Hg.

3.3 Rinse water

The rinse water used between samples was 2% nitric acid (Aristar) in deionised water.

Table 1 *ELAN 6000 Instrument settings for the industrial waste water and sewage effluent method*

Sweeps/Reading	25
Reading/Replicate	1
Number of replicate	3
Tuning File	Default.tun
Optimisation File	Default.dac
Scan Mood	Peak hopping
MCA	1
Dwell time	100ms
Integration time	2500 ms
Detector Mode	Dual
Auto Lens	On
Spectral Peak processing	Average
Signal profile processing	Average
Curve type	Linear through Zero
Rh 103	Internal Std
Sample flush time	30s
Sample flush speed	-35rpm
Read Delay	60s
Delay analysis speed	-24
Wash time between stds	1s, 60s, 120s
SamplesWash speed	-35rpm
Samples wash time	90s
Nebuliser gas flow	1.0 l/m
RAF power	1075

Table 2 *Tuning and optimisation solution*

Element	Mass	Concentration μg l⁻¹
Be	9	10 μg l⁻¹
Mg	24	10 μg l⁻¹
Fe	54	10 μg l⁻¹
Ni	60	10 μg l⁻¹
Rh	103	10 μg l⁻¹
Sn	120	10 μg l⁻¹
Ce	140	10 μg l⁻¹
Pb	208	10 μg l⁻¹
U	238	10 μg l⁻¹
Ca	na	100 mg l⁻¹
Na	na	30 mg l⁻¹
K	na	10 mg l⁻¹

Table 3 *System optimisation solution*

Element	Mass	Concentration
Li	3	10 μg l^{-1}
Be	9	10 μg l^{-1}
Mg	24	10 μg l^{-1}
Co	59	10 μg l^{-1}
Ni	59	10 μg l^{-1}
Rh	103	10 μg l^{-1}
In	114	10 μg l^{-1}
Ba	137	10 μg l^{-1}
Ce	140	10 μg l^{-1}
Tl	204	10 μg l^{-1}
Bi	213	10 μg l^{-1}
Pb	208	10 μg l^{-1}
U	238	10 μg l^{-1}
Ca	na	100 mg l^{-1}
Na	na	30 mg l^{-1}
K	na	10 mg l^{-1}

na - Not applicable

3.4 Mixed digest acid

430ml Hydrochloric acid (Aristar) and 320ml of nitric acid (Aristar) diluted to 2L
Figure 4 explores how the Ir addition causes the offset in Pd-corrected Ag isotopic composition. Addition of Ir causes no resolvable variation in measured Ag isotopic composition, but does change the fractionation experienced by Pd as shown by the offset in measured $^{108}Pd/^{105}Pd$ with Ir addition. This "extra" fractionation induced in Pd is then translated into the Ag isotopic composition through the mass fractionation correction. Surprisingly, although the Ir addition is most clearly expressed in the isotopic composition of Pd and not at all in Ag, Ir addition at all concentration levels studied suppresses the Ag ion current by approximately 25% while only slightly (6%) enhancing that of Pd. If this matrix effect were the result of charge exchange between Ir and Pd-Ag in the plasma, one might expect that the magnitude of change of mass fractionation would track that of signal suppression/enhancement, yet this is not the case in the Ir-Pd-Ag experiment with de-ionised water.

Table 4 *Detector (analogue) optimisation solution*

Element	Mass	Concentration
Mg	24	200 µg l^{-1}
Al	27	200 µg l^{-1}
Cr	50	200 µg l^{-1}
Fe	54	200 µg l^{-1}
Mn	55	200 µg l^{-1}
Ni	58	200 µg l^{-1}
Cu	63	200 µg l^{-1}
Zn	64	200 µg l^{-1}
As	75	200 µg l^{-1}
Se	82	200 µg l^{-1}
Rh	103	200 µg l^{-1}
Cd	114	200 µg l^{-1}
Sb	121	200 µg l^{-1}
Pb	208	200 µg l^{-1}
Ca	Na	100 mg l^{-1}
Na	Na	30 mg l^{-1}
K	Na	10 mg l^{-1}

na - Not applicable

3.5 Calibration standards

The standards used for calibration were prepared in 1% mixed hydrochloric and nitric acid (Aristar) in deionised water The range of calibration, was chosen to exceed the assigned PCV limit applied (Table 8). All standards were spiked with 10 µg l^{-1} rhodium (Rh) as an internal standard and 5 mg l^{-1} gold (Au) to act as a preservative for Hg. The calibration range for industrial waste water, final effluent and crude sewage are presented in Tables 5 and 6.

Table 5 *Calibration ranges*

Elements	µg l^{-1}
Ag	0-10
Al	0-250
As	0-60
Be	0-50
Cd	0-12
Co	0-50
Cr	0-50
Cu	0-50
Fe	0-50
Hg	0-1.2

Mn	0-50
Mo	0-50
Ni	0-50
Pb	0-50
Sb	0-12
Se	0-12
Sn	0-50
V	0-120
Zn	0-50

Table 6 *Mass selection and correction formula*

Elements	Mass	Correction formula
Be	9	
Al	27	
Cr	50	-0.739726 * Ti47 –0.002506 * V51
V	51	-3.127*(ClO-(0.113*Cr52))
Mn	55	
Ni	58	-0.003053 * Fe56
Co	59	
Zn	64	-0.035313 * Ni 60
Cu	65	
As	75	-3.127 (mass77- (0.874mass 82))
Se	82	-1.008696 * Kr83
Mo	98	-0.110588 * Ru 101
Ag	107	
Cd	114	-0.06826 * Sn118
Sn	120	-0.127189 * Te125
Sb	121	
Hg	202	
Pb	208	

Mass selection and correction formula selected for industrial waste waters, final effluent and crude sewage are presented in Table 7.

Table 7 *Components of industrial waste water*

Source of sample	Determinant
Aerospace	Cr,Cu,Zn,Ni,Cd,Pb
Agro Chemicals	Cr,Cu,Zn,Ni,Cd,Pb, Hg
Car Production Plant 1	Hg, Cd
Car Production Plant 2	Cr,Cu,Zn,Ni,Cd,Pb
Food Production 1	Al

Food Production 2	Cr,Cu ,Zn, Ni, Cd, Pb
Laundry	Cr,Cu,Zn,Ni,Cd,Pb
Metal Plating 1	Cr,Cu, Zn, Ni, Cd, Pb, Hg
Metal Plating 2	Cr, Cu ,Zn, Ni,Cd ,Pb
Metal Plating 3	Cr,Cu,Zn,Ni,Cd,Pb Hg
Metal Plating 4	Cr,Cu,Zn,Ni,Cd,Pb
Paint Production	Cr,Cu,Zn,Ni,Cd,Pb Hg
Photographic	Ag

4 VALIDATION

To demonstrate that the procedure was suitable for the analysis of industrial waste water, final effluent and crude swages a validation exercise was carried out under the protocol described in the document NS30 (Ref. [3]). This protocol involves the analysis of a batch of duplicate samples comprising: blanks, standard solutions with concentrations of the determinant at approximately 10% and 90% of the PCV, a real sample and that real sample spiked to the PCV concentration. The samples are randomised before analysis to ensure that the data obtained truly reflects what would be happening with a mixture of real samples of unknown and varying concentrations. Any carry-over from one sample to another would be highlighted by the comparison.

To ensure that the validation exercise mimics normal operation, only a maximum number of two batches can be analysed daily, with full re-calibration of the instrument occurring between batches.

4.1 Samples

In keeping with the guidelines in NS 30, genuine samples of industrial waste, crude sewage and final effluent were used in the experimental work and validation. A representative sample of industrial waste water was produced by bulking equal volumes of every type of industrial waste water analysed in the laboratory (Table 7). This solution would then include all types of industrial samples that we might normally analyse via ICP-MS. Representative samples of final effluent and crude sewage were prepared by mixing equal portions of each obtained from a number of sewage treatment works.

Table 8 *Industrial waste water spiked concentration (assigned PCV)*

Elements	Concentration $\mu g\ l^{-1}$
As	50
Be	10
Cd	10
Co	10
Cr	10
Cu	10

Hg	1
Mn	10
Mo	10
Ni	10
Pb	10
Sb	10
Se	10
Sn	10
V	10
Zn	10

Table 9 *Final effluent and crude sewage spiked concentration (assigned PCV)*

Elements	$\mu g\ l^{-1}$
As	50
Cd	10
Cr	10
Cu	10
Hg	1
Mn	10
Ni	10
Pb	10
Sb	10
Se	10

Table 11 *Results obtained*

Element	LOD ugl^{-1}	%Recovery
Aluminium	2.08	104.1
Antimony	0.22	104.5
Arsenic	0.12	102.9
Beryllium	0.015	99.1
Cadmium	0.0065	99.5
Chromium	0.25	98.9
Cobalt	0.008	100.9
Copper	0.53	99.3
Lead	0.1	104.8
Manganese	0.21	99.5
Mercury	0.05	103.6
Molybdenum	0.25	98.9
Nickel	0.44	98.1

Selenium	0.19	97.7
Silver	0.36	101.3
Tin	0.31	101.7
Vanadium	0.58	104.6
Zinc	1.09	94.6

Table 10 *Example of the final analytical performance report from AQC-94 Precision Results for Hg in industrial waste water*

Batch	Blank	(10% PCV) µg l⁻¹	(90% PCV) µg l⁻¹	Real Sample µg l⁻¹	Spiked sample at (PCV) µg l⁻¹
1a	0.025	0.117	1.215	-0.010	1.075
1b	0.009	0.111	1.105	-0.023	1.075
2a	0.016	0.126	1.104	0.014	1.104
2b	0.002	0.116	1.124	-0.005	1.124
3a	0.005	0.111	1.075	-0.001	1.055
3b	-0.015	0.097	1.075	-0.015	1.115
4a	-0.001	0.140	1.051	0.028	1.051
4b	-0.006	0.126	1.051	0.015	1.061
5a	0.007	0.146	1.093	0.017	1.043
5b	0.004	0.124	1.103	0.006	1.043
6a	0.026	0.099	1.074	-0.003	1.024
6b	0.009	0.106	1.094	-0.013	1.034
7a	0.000	0.105	1.040	-0.005	0.987
7b	-0.015	0.101	1.050	-0.016	1.010
8a	0.001	0.131	1.079	0.023	1.069
8b	-0.008	0.128	1.119	0.009	1.099
9a	0.001	0.107	1.069	-0.002	1.009
9b	-0.002	0.102	1.069	-0.017	1.029
10a	0.004	0.110	1.066	0.002	1.016
10b	-0.011	0.104	1.076	-0.013	1.016
11a	0.017	0.110	1.103	-0.002	1.073
11b	-0.001	0.109	1.113	-0.015	1.073
Mean	**0.0030**	**0.1148**	**1.0885**	**-0.0012**	**1.0539**
M1	0.0002	0.0003	0.0021	0.0003	0.0026
M0	0.0001	**0.0001**	**0.0007**	**0.0001**	**0.0003**
F value	1.6977	6.1195	3.1097	3.3421	9.3138
Sw (within)	0.0096	0.0072	0.0260	0.0097	0.0166
Sb (between)		0.0116	0.067	0.0105	0.0338
St (total)		0.0136	0.0373	0.0142	0.0376
Rel SD		11.9	3.4	0.0	3.6
Significance	N.S.	**	*	*	***
Calculated F		0.2972	0.4698	0.3240	0.5090

Est Degs F		13	16	16	12
LOD	0.0489				
Recovery	103.6				

N.S   Not significant   *, **, *** Significant at 0.05, 0.01 and 0.001 respectively

4.2 Sample Preparation

The three sample types were acid digested using our normal procedure. This consists of the addition of hydrochloric/nitric acid reagent and boiling for one hour, the samples being made up to volume with deionised water. The only modification required was the dilution of the samples to bring the determinands within the calibration ranges of the ICP-MS. This was achieved by diluting the samples 1:99 with deionised water prior to digestion. The spiked sample was prepared by the addition of a multi-standard, dilution 1:99, followed by digestion.

5 DISCUSSION

This work is part of an ongoing project to fully utilise ICP-MS instrumentation for the analysis of all matrices of interest to the water industry. Previous papers (Ref.[1,2]) have dealt with the determination of mercury in drinking waters, the co-determination of mercury/hydride forming elements and other metallic elements routinely analysed at low levels in potable waters. In this work the scope of the analysis using an ICP-MS has been extended to cover waste waters. These are potentially more prone to a variety of interference's [4,5,6,7] which affect the precision of the results. Interference's can arise from presence of other isotopes / elements with the same mass number as the analyte of interest or molecular combinations which give the same mass number. Matrix effects can arise due to sample introduction and are common in ICP-MS. The matrix can influence arousal formation, ion transport into plasma through variation in surface tension and viscosity. A complex matrix may also effect the focusing of ions into sampler and skimmer cone orifices, the transportation of the ions through ion lenses and quadruple. The main reason for including all types of industrial waste in the method development was to provide for as many possible different types of interference. Most interference's would be revealed by the spiked recovery values showing excessive or low recovery values.

In the absence of any national or standard values, appropriate arbitrary values were assigned. These values were kept as low as possible. Some were set at potable water levels (As, Se. Sb, Hg) most of the remainder at lower levels, e.g. Be, Ni, Cu, Cr, Mn, Pb and Zn. The only exception is Cadmium, this was not because of any problems associated with cadmium but only to ease the preparation of multi-standards.

It may seem unusual to set industrial and effluent discharge limits around or below drinking water limits. However when the disposal route for both discharges end in a river which may have raw water abstracted downstream for drinking water purposes then such a policy may become necessary.

Validation of the method was carried out according to the NS-30 protocol used for potable waters. A 1% solution of nitric acid in deionised water was used as the blank solution, 10% and 90% of the top standard solution, were used as low and high standards. The mixture of industrial waste waters was run as a 'real sample', and the industrial waste spiked to the assigned value was the 'spiked real sample'.The requirements of NS30 are:- to have at least 11 degree of freedom, the samples to be run randomly, in duplicate. Not more than 2 batches to be run on a single day, with full calibration between the batches

The calculation of the results has been carried out using AQC94 software with manual entry of the results. The recoveries (Table. 11) are within the range of 95% to 105% limit which is acceptable for this analysis. Addition of gold at 5mg l^{-1} again proved successful for the determination of mercury even in the complex industrial waste water mixture, giving a method detection limit of 0.048ug $^{l-1Hg}$. Similar results were achieved for the hydride forming elements, As, Se, and Sb with method detection limits of 0.120, and 0.219 ug l^{-1} respectively.

6 CONCLUSIONS

All of the elements of interest passed the requirements of NS30, the limits of detection and recovery are satisfactory for the routine analysis of industrial waste waters, final effluent and crude sewage.

The ability to run all the elements of interest with low detection limits in a single run makes the use of ICP-MS extremely cost effective. The savings in time and operational /material costs using a single pass ICP-MS analytical run are very significant when compared to the techniques it replaces.

This paper also demonstrates that the statistical techniques used to validate analytical methods for drinking water can be successfully transferred to other matrices. To demonstrate the effectiveness some assumptions were made for appropriate limits for these matrices. This analytical method is now in routine use within the laboratory.

7 REFERENCES

1.   J.Allibone, E.Fatemian and P.J.Walker, Determination of mercury in potable water by ICP-MS using gold as a stabilising agent., *J.Anal.At.Spectrom*, **14**, 235-239 (1999)
2.   E.Fatemian, J.Allibone and P.J.Walker, Use of gold as a routine and long term preservative for mercury in potable water, as determined by ICP-MS, *Analyst*, 1999, **124**, 1233-1236
3.   R.V Cheeseman and A.L Wilson (revised M.J Gardner, 1989), NS-30. A manual on analytical quality control for the water industry. Water Research Centre, Marlow Bucks.
4.   D.J. Douglas and J.B French, Importance of interference's Gas dynamics of ICP-MS interface. *J.Anal.At.spectrom*. **3**,743-747 (1988)

5. G.R. Gillson, D.J Douglas, J.E Fulford, K.W Halligan and S.D Tanner Matrix induced interference non spectroscopic
    inter element interference's in ICP-MS. *Anal.Chem.* **60**, 1472-1474 (1988)
6. S.H Tan and G Horlick, Background spectral features in ICP-MS, *Appl.Spectrosc.* **40**, 445-460 (1986)
7. Y.S Kim, et.al., Non-spectroscopic matrix interference's in ICP-MS *Spectrochim. Acta* **45B**, 333-339 (1990)

Section 5

Isotope Ratio Measurement

THE PERFORMANCE OF COMMERCIAL MONO-ELEMENTAL SOLUTIONS FOR HIGH ACCURACY MASS SPECTROMETRY

P.Evans[1*], B. Fairman[1], C. Wolff-Briche[1], S. Merson[1], C. Harrington[2]

[1]LGC (Teddington) Ltd., Queens Road, Teddington, Middlesex, TW11 0LY.
[2] Dept. Chemistry, De Montfort University, The Gateway, Leicester LE1 9BA.

1 INTRODUCTION

The availability of reliable, cost-effective calibration standards is pivotal to modern analytical services. The accuracy, and the uncertainty associated with the certificate values play a major role in the reliability of reported measurements. Improvements in analytical precision, combined with a greater emphasis on accounting for combined measurement uncertainty, means that the need for these high accuracy standards has never been greater.

Despite the key role that such solutions play, there is minimal independent verification of certificate values. A recent European survey of calibration standards revealed that differences of 4-5% between certificate and actual abundances were not uncommon[1]. Here we present the analysis of four 1000 ppm mono-elemental solutions (Fe, Mg, Cu and Al) purchased from four commercial suppliers. Each solution is supplied with a certificate value and an uncertainty. These measurements are designed to assess the fitness of these standards for high accuracy calibration. Following *Eurachem* guidelines[2], the combined uncertainty in these measurements is calculated. Using this approach, a full assessment can be made of whether the standards contain the certified concentration and whether similar precision can be attained within the laboratory. Furthermore, this study acts as the bench-mark by which the increased uncertainty associated with more complex systems, where matrix effects or sample preparation is required, can be judged.

Fe, Mg and Cu were analysed by isotope dilution inductively coupled plasma mass spectrometry (ID-ICP-MS) in which the samples were 'spiked' with an isotopically enriched solution of the same element. Isotope dilution represents the purest type of internal standard, as there is no difference in ionisation efficiency between isotopes of the same element. Isotope dilution mass spectrometry is recognised by the BIPM (Bureau International des Poids et Measures) as a primary ratio method for the determination of the amount of substance[3]. Al, which is effectively mono-isotopic, cannot be analysed by isotope dilution and was therefore analysed using a high accuracy inductively coupled plasma optical emission spectroscopy (ICP-OES) method. In order to compare these two analytical techniques Fe was analysed by both methods.

For each solution a combined uncertainty calculation was performed following *Eurachem* guidelines[2]. This considers all of the realistically quantifiable uncertainties from the preparatory stages and precision of the measured isotope ratio or spectral data.

The four elements chosen (Fe, Mg, Cu & Al) represent some of the most commonly measured elements in both commercial and academic laboratories, covering a broad spectrum of environmental, biological and industrial applications. In addition, certified reference materials obtained from NIST were analysed as quality control solutions. Each commercial standard was purchased exclusively for this assessment exercise and was stored and handled according to the manufacturer's instructions.

Despite the increase in the use of gravimetry (e.g. $\mu g\ g^{-1}$) as the units of choice in sample preparation, manufacturers continue to report primarily the volumetric assay ($\mu g\ ml^{-1}$) introducing the need for a density conversion. Two of the manufacturers investigated do not provide a density value for their standards. Here, we have performed an additional density calculation and the effects of this added stage of analysis to the uncertainty are considered.

2 ANALYTICAL

2.1 Instrumentation

All analyses were carried out at LGC (Teddington) Ltd. Specialised Techniques division. ID-ICP-MS analyses of Cu and Mg were made using a conventional quadrupole ICP-MS (Elan 5000A, Perkin-Elmer, Beaconsfield, UK), Fe was analysed using a double focusing magnetic sector ICP mass spectrometer (Element, Finnigan MAT, Bremen, Germany). The Element operated under 'cold plasma' conditions using a Pt guard electrode to minimise isobaric interferences on ^{56}Fe from $^{40}Ar-^{16}O$. ICP-OES analyses of Al were made using an Optima 3300RL (Perkin-Elmer, Beaconsfield, UK). Details of normal operating conditions are given in Table 1.

For the commercial standards that were not supplied with density measurements conversion from $\mu g\ ml^{-1}$ to $\mu g\ g^{-1}$ was performed using a DMA 55 digital density meter (Paar scientific, London, UK) that permits accurate density readings to 4 decimal places. Each analysis was made at 20 °C and calibrated relative to a series of HNO_3 dilutions. An average of two replicates was used in the final calculation for each sample.

2.2 Reagents

Analytical grade Ultrex II acids (J.T.Baker, Berks. UK) were used throughout the procedure diluted with deionized water (18MΩ) taken from an Elga Maxima water purification unit (Elga, Marlow, UK).

2.3 Standards

'In-house' primary standard solutions of each element, with natural isotopic compositions, were prepared by dissolving pure metals (Alfa Aesar, Karlsruhe, Germany) and diluting them to reach the desired concentration. Grade 1 Mg and Fe metal and Cu rod (99.98%, 99.9985%, 99.999 % pure respectively) were dissolved in 5% HNO_3. Grade 1 Al wire (99.9999% purity) was dissolved in 20 % HCl. A relative uncertainty of 0.2% was estimated for each of the metals from the certificate of analysis. These stock solutions were then further diluted to the desired concentration.

As the isotopic composition of Cu, Mg and Fe do not vary significantly in nature the values from IUPAC were used to calculate the natural isotopic ratios for the different elements[4] as illustrated in Table 2.

2.4 Spikes

Isotopically-enriched elemental solutions are required for ID-ICP-MS. These· were obtained from Teknolab A/S (Dröbak, Norway) for Fe and from AEA technology (Harwell, UK) for Cu and Mg. The abundances of the spike isotopes are also presented in Table 2.

Table 1 *Operating conditions for the three instruments used for the analysis of mono-elemental solutions.*

Perkin-Elmer Elan 5000A Quadrupole ICP-MS		Perkin Elmer Optima 3300RL ICP-OES	
Forward power (W)	1040	Forward power (W)	1300
Ar Coolant gas flow rate I min⁻¹	15	Ar Plasma gas (L/min)	15
Ar Auxiliary gas flow rate I min⁻¹	0.8	Ar Auxiliary gas (L/min)	0.5
Ar nebulizer gas flow rate I min⁻¹	0.9	Ar Nebulizer gas (L/min)	0.8
Nebulizer	cross flow	Nebulizer	cross-flow
nebulizer uptake (ml/min)	0.7	nebulizer uptake (ml/min)	1.0
Dwell time (ms)	35	integration time (s)	0.1
No. sweeps per reading	200	Read time (s)	7
No. replicates per pass	10	No. replicates	10
Points per peak	3	Points per peak	5
Finnigan MAT Element magnetic sector ICP-MS		Wavelengths analysed:	
		Al	308.215 nm
Forward power (W)	750	In (drift monitor)	325.603 nm
Ar Coolant gas flow rate I min⁻¹	12.6	Fe	259.939 nm
Ar Auxiliary gas flow rate I min⁻¹	0.9	Mn (drift monitor)	259.372 nm
Ar nebulizer gas flow rate I min⁻¹	1.3		
Nebulizer	glass expansion		
nebulizer uptake ml min⁻¹	0.05		
Samples per peak	20		
Settling time (ms)	5		
No. sweeps	5 x 1000		
Mass window	90%		
Intergration window	70%		

The units I min⁻¹ above should read $l\ min^{-1}$.

Isotopic abundances as %			
Isotope	^{54}Fe	^{56}Fe	^{57}Fe	^{58}Fe
Natural	5.8	91.72	2.2	0.28
Spike	0	7.05	92.74	0
Isotope	^{24}Mg	^{25}Mg	^{26}Mg	
Natural	78.99	10	11.01	
Spike	0.963	98.814	0.223	
Isotope	^{63}Cu	^{65}Cu		
Natural	69.17	30.83		
Spike	8.8	91.2		

Table 2 *Natural and spike isotopic abundances used for isotope dilution analysis.*

3 PROCEDURES

3.1 Isotope dilution mass spectrometry

The analysis of Fe, Mg and Cu were carried out by isotope dilution inductively coupled plasma mass spectrometry (ID-ICP-MS) in which samples and standards are 'spiked' with a known amount of an isotopically enriched source of the element under investigation. The ID-ICP-MS procedure adopted has previously been reported[5]. It is based on the 'matching' method recommended by Henrion[6], but has been modified from a time and effort consuming iterative method to a single step analysis. The stages of the matching method are graphically illustrated in Figure 1. In the "exact matching" method the isotope abundance ratio [spike/sample] is adjusted to close to 1. In doing so, problems associated with isotope ratio measurements by ICP-MS, such as the need for dead-time correction, are minimised.

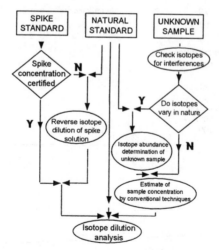

Figure 1 *A graphical illustration of the ID-ICP-MS procedure*

The sample solutions and 'in house' metal standards were diluted to achieve a concentration of ~ 10 $\mu g \cdot g^{-1}$ in dilute nitric acid. A known amount of spike solution was then added to the sample (B) and the primary standard (B_c) to achieve isotope ratios $(n(^{24}Mg)/n(^{25}Mg))$ for Mg analysis, $n(^{63}Cu)/n(^{65}Cu)$ for Cu analysis and $n(^{56}Fe)/n(^{57}Fe)$ for Fe analysis close to 1 (± 2%). The sample blend was measured bracketed by a spike/natural standard calibration blend of known abundances. An average isotope abundance ratio from the 2 bracketing measurements was used to correct the sample blend for mass discrimination effects arising from the plasma interface and ion optics. The concentration (c'_X) in the spiked solution was calculated according to equation 1.

$$c'_X = c_Z \cdot \frac{m_Y}{m_X} \cdot \frac{m_{Zc}}{m_{Yc}} \cdot \frac{R_Y - R'_B \cdot \dfrac{R_{Bc}}{R'_{Bc}}}{R'_B \cdot \dfrac{R_{Bc}}{R'_{Bc}} - R_X} \cdot \frac{R_{Bc} - R_X}{R_Y - R_{Bc}} \qquad \text{Equation 1}$$

Where,

c'_X mass fraction of analyte in sample X obtained from the measurement of one aliquot

c_Z mass fraction of analyte in primary standard Z

m_Y mass of spike Y added to the sample X to prepare the blend B (X+Y)

m_X mass of sample X added to prepare the blend B (X+Y)

m_{Zc} mass of primary standard solution Z added to the spike Y to make calibration blend Bc (=Y+Z)

m_{Yc} mass of spike Y added to the primary standard Z to make calibration blend Bc (=Y+Z)

R'_B measured isotope ratio of sample blend (X+Y)

R'_{Bc} measured isotope ratio of calibration blend (Bc = Z+Y)

R_{Bc} gravimetric value of the isotope ratio of calibration blend (Bc = Z+Y)

R_Y isotope ratio of spike Y (value from certificate)

c'_X average mass fraction of the different blends measured

The final concentration was then recalculated back to the original sample, using the average concentration calculated from several aliquots' measurements, the dilution factor D, and a recovery factor R, which was calculated by performing the same measurement procedure on a NIST SRM certified reference solution according to equation 2.

$$c_X = \frac{\overline{c'_X} \cdot R_{re\,cov\,ery}}{D_{dilution}} \qquad \text{Equation 2}$$

Where,

$D_{dilution}$ dilution factor

$R_{recovery}$ recovery factor calculated from a reference material

This isotope ratio of the calibration blend is calculated from the data of the spike and standard that are gravimetrically mixed as described in equation 3.

$$R_{Bc} = \frac{R_Z \cdot m_{Zc} \cdot c_Z \cdot \sum R_{Yi} + R_Y \cdot m_{Yc} \cdot c_Y \cdot \sum R_{Zi}}{m_{Zc} \cdot c_Z \cdot \sum R_{Yi} + m_{Yc} \cdot c_Y \cdot \sum R_{Zi}}$$ Equation 3

R_Z is the isotope ratio of the primary standard.

Fe, Mg and Cu isotope ratios do not vary significantly in nature, unlike for example Pb, and therefore are taken from IUPAC tabulations[3].

3.2 Inductively coupled plasma optical emission spectroscopy

Inductively coupled plasma – optical emission spectroscopy was employed for the analysis of Al. Al is effectively mono-isotopic, all isotopes other than ^{27}Al being radiogenic with short half lives, and thus excluding it from isotope dilution analysis. Analysis of the Fe solutions was also performed using ICP-OES. Fe represents one of the more difficult elements to analyse by ICP-MS due to isobaric interference from ^{40}Ar-^{16}O. Duplication of the Fe results assesses the relative methods of the techniques for such problematic elements.

The sample solution was first diluted by a factor of ~30 and analysed using a single element calibration to obtain an indicative concentration. Based on this result a standard solution of the 'in-house' pure metal was prepared whose concentration was closely matched to that of the sample dilution. A NIST SRM solution was diluted to a concentration that closely matched that of the standard solution. Both the NIST SRM and the sample dilution were analysed by bracketing with the 'in-house' standard solution. Instrumental drift was corrected for by on-line addition of In as internal standard. For each analysis the results were calculated from the mean of seven bracketed runs.

4 DATA EVALUATION AND UNCERTAINTY CALCULATION

4.1 Combined uncertainty of ID-ICP-MS

All the variables present in equation 1 have associated standard uncertainties and are contributing to the combined standard uncertainty of the concentration c_X. Combined standard uncertainties were calculated applying the uncertainty propagation law to Equation 1 as described in the *Eurachem* guide 'quantifying uncertainty in analytical measurement'[2].

The concentration of the 'in house' primary standard of metal in solution c_Z, and associated standard uncertainty, was calculated from the data of its gravimetric preparation. The concentration of the spike was determined by performing reverse isotope dilution mass spectrometry analysis. The uncertainty of the weighings, as performed on a 4-figure balance (Sartorius A200S), were based upon previously calibrated weight measurements and the balance certificate. Standard uncertainties are 6×10^{-5} g, 1.2×10^{-4} g and 2.4×10^{-4} g for 1, 10 and 100 g weighing respectively. The standard uncertainty of the measured isotope ratios was calculated from the relative standard deviation given by the instrument and calculated from repeated measurements. For the measured isotope amount ratio of the calibration blend, R'_{Bc}, a contribution from the drift of the instrument was combined to the relative standard deviation of the measurement.

The combined standard uncertainty was then calculated by combining the largest uncertainty of the individual ICP-MS measurements, the uncertainty in the dilution factors, the standard deviation of the mean of the aliquots and the uncertainty associated with the recovery factor.

4.2 Calculation of ICP-OES uncertainty

The parameters that contribute to the combined uncertainty of ICP-OES measurements were determined to be the method recovery, method precision, dilution factors and the concentration of the calibration standard.

A method recovery factor from the NIST SRM was calculated for each run. It has nevertheless not been applied in the calculation of the uncertainty budgets as the ratio $C_{observed}/C_{SRM}$ was close to 1. Furthermore, $C_{observed}$ was within the certified concentration range.

For each sample, 3 runs each consisting of a number of replicates (7) were made. Analysis of variance (ANOVA) was used to compare the sources of variation for each sample (i.e., the within run variability and the between run variability).

For all samples, the ANOVA showed that both within and between group variation contribute to the overall variability. The precision uncertainty is therefore taken as the standard deviation of the mean of the means. The uncertainty associated with dilution and weighing of the samples and quality control solutions are the same as presented for the ID-ICP-MS measurements.

4.3 Comparison between results and certificate values

Combined standard uncertainties were calculated, following *Eurachem* guidelines, for the commercial solutions to include their certificate values (typically $1000 +/- 3$ µg ml^{-1}) and, where applicable, the uncertainty associated with conversion to gravimetric values. In the absence of sufficient data, the uncertainty in density measurements for all of the commercial solutions is assumed to be equal to that of the NIST measurements. A value of 0.2% is used, which is greater than the standard quoted uncertainty of the procedure used in this study and therefore considered to be a suitable coverage factor.

The significance of the difference between the certificate and experimental results are assessed by considering the certificate value, the mean of the experimental results and their associated combined uncertainties as outlined in equation 4. If t <2 (the standard coverage factor) then the difference between results is considered not to be significant i.e the analysis and certificate are indistinguishable.

$$t = \frac{1-r}{u(r)} \qquad \qquad \text{Equation 4}$$

Where

r   ratio difference between the experimental mean and certificate value
u(r)   combined standard uncertainty of the experimental and certificate values

2 degrees of freedom are used in this analysis as the majority of uncertainty in plasma spectrometry results from the analytical precision for which very large numbers of replicate analyses are obtained. For example the ID-ICP-MS protocol employed measures 5000 repeat isotope ratios.

5 RESULTS AND DISCUSSION

5.1 Results

Both ID-ICP-MS and ICP-OES analyses give values for the NIST certified reference materials that are in good agreement with certificate values. Consequently no recovery factors have been applied to the respective analyses. A full compilation of the results is presented in Table 3, including standard uncertainties and comparisons.

In general the results show that the majority of commercial calibration solutions can be analysed to a value that is statistically indistinguishable from the certificate and with combined uncertainty that is <1% (at the 95% confidence level). Deviations from the certificate values are represented in Figures 2 & 3.

Fe analysis by ID-ICP-MS gave the poorest results relative to certificate values with two of the four commercial solutions yielding results that were significantly different to the expected values. However, due to the high precision capability of the of the HR-ICP MS (Finnigan Element), the combined uncertainty for Fe analysis is smaller than for Mg or Cu thereby creating a smaller tolerance for overlap between the certificate and experimental values.

All Mg and Cu analyses by ID-ICP-MS lie within the reported margin of uncertainty of the manufacturer (t < 2) and hence there is no significant difference between the analyses and certificates. Fe and Al analyses by ICP-OES also lie within the uncertainty of their certificate values and with combined uncertainty comparable to the ID-ICP-MS results.

5.2 Sources of uncertainty in ID-ICP-MS

An assessment of the sources of the uncertainty of the measurements is shown graphically in Figure 4. Here, it is evident that the uncertainty budget is dominated by contributions from the ratio measurements of the sample and calibration blends.

5.3 Comparison of ID-ICP-MS and ICP-OES measurements

Both ID-ICP-MS and ICP-OES analyses can provide equal levels of accuracy when handling these relatively high concentration samples which raises the issue of 'fitness for purpose' when selecting an analytical technique. Given the added time and expense of ID-ICP-MS measurements compared to ICP-OES there is minimal improvement in precision and possible reductions in accuracy when analysing high concentrations of problematic elements such as Fe. The advantages of ID-ICP-MS only become apparent at low concentrations or small sample sizes. For example, the ICP-OES technique consumed approximately 50 g of a 50 µg g^{-1} solution per analysis whereas ID-ICP-MS required ~10 g at ~50 ng g^{-1}, a 5000-fold reduction in analyte consumption.

5.4 Problems associated with density conversion

Poor correlation between commercial standard certificates and analytical results are most frequently encountered when an additional density reading has been necessary to convert from volumetric to gravimetric values. Whilst it cannot unequivocally be proven to be the cause of this difference, the need for density values on calibration solutions intended for high accuracy analysis is highlighted as density meters are not readily available to many laboratories.

An additional problem with the need for density conversion is the effect that it has on the combined uncertainty of the calibration standard. Consider, for example, a 1000 µg ml^{-1} solution with a certificate uncertainty of 0.3% and a density of 1.059. If an uncertainty of 0.2 % is ascribed to the density measurement then the gravimetric value for the solution is 944 µg g^{-1} and a significantly increased combined uncertainty of 0.4%.

6 CONCLUSION

In conclusion, the certificate values of commercially available calibration solutions can be verified within the laboratory with an acceptable degree of precision and uncertainty, although not without considerable effort and time. Problems such as the absence of density values, which are easily resolved by the manufacturer, can lead to significant problems when left to the individual laboratory to overcome.

Element	CERTIFICATE VALUES				ANALYTICAL RESULTS					
	Certificate concentration	Density g cm^{-3}	Recalculated as µg g^{-1}	Expanded uncertainty % k=2 density corrected	LGC Technique	No. samples	LGC Value	Expanded uncertainty as a % K=2	% difference	Significant Difference t
Fe	10 mg g^{-1}	N/A	10000.0	0.3	ID-ICP-MS	5	9983.5	0.3	-0.2	0.8
	1000 µg ml^{-1}	1.059 (20 °C) LGC	944.0	0.4	ID-ICP-MS	3	938.1	0.3	-0.6	2.3
	1003 µg ml^{-1}	1.015 (24.2 °C)/1.0114 (20 °C)	988.0	0.6	ID-ICP-MS	3	992.1	0.3	0.4	1.2
	1000 µg ml^{-1}	1.002 (21 °C)/ 1.002 (20 °C)	998.0	0.4	ID-ICP-MS	3	996.7	0.3	-0.1	0.5
	1000 ppm	1.018 (20 °C) LGC	982.8	0.6	ID-ICP-MS	3	966.7	0.3	-1.6	4.8
	10 mg g^{-1}	N/A	10000.0	0.3	ICP-OES	5	10006.0	0.7	0.1	0.4
	1000 µg ml^{-1}	1.059 (20 °C) LGC	944.0	0.4	ICP-OES	3	942.5	0.7	-0.2	1.2
	1003 µg ml^{-1}	1.015 (24.2 °C)/1.0114 (20 °C)	988.2	0.6	ICP-OES	3	993.9	0.7	0.6	0.4
	1000 µg ml^{-1}	1.002 (21 °C)/ 1.002 (20 °C)	998.0	0.4	ICP-OES	3	999.7	0.7	0.2	0.3
	1000 ppm	1.018 (20 °C) LGC	982.8	0.6	ICP-OES	3	981.4	0.7	-0.1	0.16
Mg	9.62 mg ml^{-1}	1.083 +/- 0.002 (22 °C)	8882.7	0.4	ID-ICP-MS	3	8914.2	0.7	0.4	0.9
	1000 µg ml^{-1}	1.024 (20 °C) LGC	976.6	0.4	ID-ICP-MS	3	982.3	0.6	0.6	1.5
	1001 µg ml^{-1}	1.016 (24.2 °C)	985.2	0.6	ID-ICP-MS	3	987.4	0.7	0.2	0.5
	1000 µg ml^{-1}	1.007 (21 °C)	993.1	0.4	ID-ICP-MS	3	997.8	0.7	0.5	1.1
	1000 ppm	1.016 (20 °C) LGC	984.2	0.4	ID-ICP-MS	3	989.7	0.6	0.6	1.5
Cu	9.99 mg ml^{-1}	1.074 +/- 0.002 (22 °C)	9301.7	0.4	ID-ICP-MS	3	9325.9	0.9	0.3	0.6
	1000 µg ml^{-1}	1.020 (20 °C) LGC	980.0	0.4	ID-ICP-MS	3	978.9	1.0	-0.1	0.2
	1002 ppm	1.011 (24.5 °C)	991.1	0.6	ID-ICP-MS	3	990.3	0.8	-0.1	0.16
	1000 µg ml^{-1}	1.004 (21 °C)	996.0	0.4	ID-ICP-MS	3	998.2	1.1	0.2	0.37
	1003 ppm	1.016 (20 °C) LGC	987.5	0.3	ID-ICP-MS	3	986.0	0.9	-0.2	0.3
Al	9.67 mg ml^{-1}	N/A	9670.0	0.3	ICP-OES	3	9656.7	0.9	-0.1	0.8
	1000 µg ml^{-1}	1.012 (20 °C) LGC	988.1	0.4	ICP-OES	3	991.0	0.7	0.3	0.7
	1000 µg ml^{-1}	1.007 (24.9 °C)	993.1	0.6	ICP-OES	3	999.2	0.7	0.6	1.3
	1000 µg ml^{-1}	1.011 (21 °C)	989.1	0.4	ICP-OES	3	994.1	0.7	0.5	1
	1004 ppm	1.011 (20 °C) LGC	993.5	0.4	ICP-OES	3	997.1	0.7	0.4	0.9

Table 3 *A comparison of commercial mono-elemental solutions of Fe, Mg, Cu and Al*

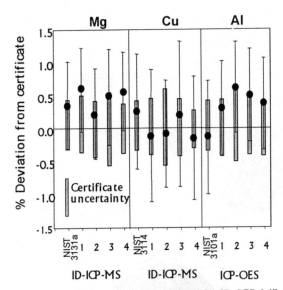

Figure 2 *A comparison of Mg, Cu and Al analyses by ID-ICP-MS &*
ICP-OES for commercial mono-elemental solutions with their
certificate values. Error bars show combined uncertainty k=2

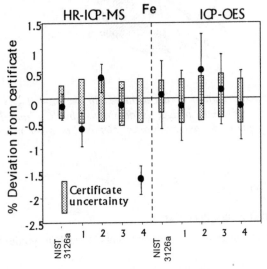

Figure 3 *A comparison of four Fe analyses of commercial mono-elemental solutions by*
ID-ICP-MS and ICP-OES and their certificate values. Results shown with combined
uncertainty at k=2

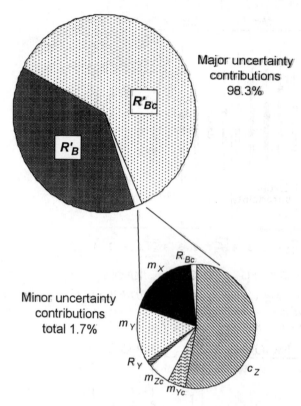

Figure 4 *The relative contribution of uncertainties in the ID-ICP-MS method, as illustrated for Fe. The uncertainty budget is dominated by R'_{BC} and R'_B the ratio measurements of the sample and mass calibration blends. For other notations refer to text.*

7   REFERENCES

1   Laboratoire National D'essais. Pers. Comm.

2   Quantifying uncertainty in analytical measurements. Eurachem guide. ISBN 0-948-926-08-2

3   Comite consultatif pour la quantite de matiere. Rapport de la 4 session (1998)

4   T.B. Coplen. *Pure & Appl. Chem.* 1996 **68** 2339 – 2359.

5   T. Catterick, B. Fairman, C.J. Harrington. *J. Anal. At. Spectrom.* 1998 **13** 1109

6   A. Henrion. *Fresenius J. Anal. Chem.* 1994 **350** 657 - 658

ICP-MS APPLIED TO ISOTOPE ABUNDANCE RATIO MEASUREMENTS: PERFORMANCE STUDY AND DEVELOPMENT OF A METHOD FOR COMBINING UNCERTAINTY CONTRIBUTIONS FROM MEASUREMENT CORRECTION FACTORS

Christophe R. Quétel[1], Thomas Prohaska[1,2], Simon M. Nelms[1,3], Jürgen Diemer[1] and Philip D. P. Taylor[1]

[1]European Commission, Joint Research Centre, Institute for Reference Materials and Measurements, Retieseweg, B-2440 Geel, Belgium.

[2]Now at University of Agricultural Sciences,Institute of Chemistry, Muthgasse 18, A-1190 Vienna, Austria.

[3]Now at TJA Solutions, Ion Path, Road Three, Winsford, Cheshire, CW7 3BX, UK.

1 INTRODUCTION

The principle of element amount content determination by isotope dilution mass spectrometry (IDMS) was established more than 40 years ago.[1, 2] There is a well-defined mathematical expression to describe "direct" IDMS, based on the measurement of isotope amount ratios, that can be written as follows:[3]

$$C_X = C_Y \cdot \frac{m_Y}{m_X} \cdot \left(\frac{R_Y - R_B}{R_B - R_X} \right) \cdot \frac{\sum (R_i)_X}{\sum (R_i)_Y} \tag{1}$$

in which C_X and C_Y are the element amount contents in the sample and the spike solutions, m_X and m_Y are the respective masses of the sample and the spike, R_X, R_Y, and R_B are the sample, spike and blend ratios and ΣR_{ix} and ΣR_{iy} are the sum of the ratios in the sample and the spike solutions (all referenced to the same isotope of the element). This equation links, in a clear way, what is believed to have been measured to what was actually measured. The Comité Consultatif pour la Quantité de Matière (CCQM) recognized IDMS as a potentially primary method of measurement.[4] *"A primary method of measurement is a method having the highest metrological qualities, whose operation can be completely described and understood, for which a complete uncertainty statement can be written down in term of SI units [...]*".[4] The CCQM also stated that *"Measurements of amount of substance, to be considered primary, must be made using a method [...] for which the values of all parameters, or corrections which depend on other species or the matrix, are known or can be calculated with appropriate uncertainty"*.[4] Differing correction factors can be implemented into equation 1 to account for the various sources of measurement biases. In the case of inductively coupled plasma mass spectrometry (ICP-MS) measurements, sources of bias include the sample introduction systems and the plasma flickering, matrix effects, instrumental background and the way ions are focused in energy, transmitted and detected. If these effects influence individual isotope signal intensities in a cumulative way, they can be partially or totally removed (depending on their nature and their frequency) by calculating ratios.[5] Instrument parameters, and particularly the time parameters on sequential instruments, can be optimized to filter out more of these instabilities.[6] However, some effects are "additive" and hence cannot cancel out mutually when ratios of isotope signal intensities are calculated. These include the instrumental background, any isobaric interference and, to a certain extent, the dead-time effect. Correcting for these effects represents additional sources of uncertainty to the final measurement result. Although these corrections are often performed correctly in the literature, their influence on the measurement uncertainty budget is frequently ignored.

A performance study of the three main types of commercially available ICP-MS instruments was conducted in our laboratory[7, 8] by means of certified synthetic uranium isotope mixtures (same samples, different dilutions). Differences in measurement combined uncertainties achievable per instrument could be illustrated. Possible biases affecting single detector magnetic sector ICP-MS measurements were identified. These studies will be summarized. A method to calculate uncertainties arising from different corrections applied to ICP-MS measurements was developed.[9] This method and the application to the calculation of ID-ICP-MS results, to enhance their "metrological qualities", will be presented.

2 EXPERIMENTAL

The experimental results introduced during this lecture are based on uranium measurements published elsewhere[7, 8] and on lead measurements performed during a study described elsewhere.[10] Experimental conditions are described in detail in these references and will only be summarised here.

2.1 Instrumentation and measurement procedures

Measurements were performed on two quadrupole sector ICP-MS (Q-ICP-MS) and two magnetic sector ICP-MS (S-ICP-MS) instruments: respectively a "PQ2+" (VG Elemental, Winsford, England) equipped with turbo pumps, an

"Elan 6000" (Perkin-Elmer Sciex, Norwalk, USA), an "Element2" (Finnigan MAT, Bremen, Germany) and a "Nu Plasma" (Nu Instruments, Wrexham, Wales).

Routine experimental settings were applied throughout the measurements. Careful rinsing followed by a blank analysis was performed prior to every sample measurement in order to monitor and correct for sample to sample memory effects as closely as possible.

Only digital isotope abundance ratio measurements were performed on the "PQ2+", the "Elan 6000" and the "Element2", even when a dual mode detector was available. Measurements with these three instruments are sequential, i.e. only the signal of one isotope is measured at a time and the same single detector is used for all isotopes. This is opposite to the simultaneous detection mode of the "Nu Plasma", where the ion beams are focussed into a fixed array of 12 Faraday cups (analogue acquisition) via variable dispersion optics. All instruments were operated at low resolution. Where necessary (i.e. first three instruments), dead time effects[11] were corrected for using eqn. (2):

$$I_{\tau\_corr} = I_{obs} / (1 - \tau \cdot I_{obs}) \qquad (2)$$

in which I_{obs} is the measured counting rate and I_{τ_corr} is the counting rate corrected for dead-time effects (both in counts $\cdot$ s^{-1}), and τ is the dead time (s) of the detector. Dead time values were determined by means of a method described elsewhere.[12]

2.2 Reagents and solutions

The uranium experiments were performed using the IRMM-072 series, a unique suite of isotopic reference material (IRM) solutions. These solutions, which can be used in particular for checking the linearity of isotope mass spectrometers,[13, 14, 15] have been described in detail previously.[16] IRMM-072/1 to IRMM-072/10 (except IRMM-072/5 and IRMM/072/8) were measured. Samples were diluted in HNO$_3$ 2%, and dilution level was adjusted depending on instrument sensitivity and mode of detection: ~ 10 ng U $\cdot$ g^{-1} for the Q-ICP-MS experiments, ~ 1 ng U $\cdot$ g^{-1} for the single detector S-ICP-MS experiments and ~ 1000 ng U $\cdot$ g^{-1} for the multiple collector S-ICP-MS experiments.

Similarly to the uranium experiments, same samples at different dilution levels were used for the lead experiments: ~ 5 ng Pb $\cdot$ g^{-1} for the Q-ICP-MS and the single detector S-ICP-MS experiments and ~ 50 ng Pb $\cdot$ g^{-1} for the multiple collector S-ICP-MS experiments. Sample preparation (incl. spiking and digestion) is explained in more details elsewhere.[10]

3 MASS-DISCRIMINATION EFFECT

The various possible or demonstrated sources of mass-discrimination during ICP-MS measurements were reviewed recently.[8,17] Common to all types of ICP-MS instruments are the effects induced by electrostatic interactions between isotopes and by space charge effects due to the presence of the plasma and the matrix ions. Other effects, more specific of the design and the way of operating the mass spectrometer, might also have a significant impact on the measurement results, as indicated later for the single detector S-ICP-MS.

Two ways of correcting for mass-discrimination will be illustrated in this paper. In both cases, a ratio (K factor) between the certified (R_{Cert}) and the measured (R_{Meas}) values of a known isotope amount ratio is calculated (eqn.3):

$$K = R_{Cert} / R_{Meas} \qquad (3)$$

The first approach is an internal correction. The known isotope amount ratio is measured together with the unknown isotope amount ratio to be corrected (same run). A K factor is calculated according to eqn.3 and normalised to the mass difference between the isotopes of the known ratio. The normalised value is then used to correct for mass-discrimination the measured value of the unknown ratio. Ideally, the isotopes of the known and the unknown ratios belong to the same element ($^{146}Nd/^{144}Nd$ ratio to correct other Nd ratios for instance[18]). Otherwise, an element of known isotopic composition and close in mass to the element of interest is artificially added to the sample (if not already present at measurable levels): $^{203}Tl/^{205}Tl$ ratio to correct Pb isotope amount ratio measurements.[19] For normalisation to one mass unit difference, mostly three models (eqns.4 to 6) are applied:[8, 15, 20]

$$K_{lin} = 1 + (m_2 - m_1) \cdot \varepsilon_{lin} \qquad (4)$$

$$K_{pow} = (1 + \varepsilon_{pow})^{(m_2 - m_1)} \qquad (5)$$

$$K_{exp} = (m_2 / m_1)^{\varepsilon_{exp}} \qquad (6)$$

in which m_1 and m_2 are the respective atomic masses of the two isotopes ^1E and ^2E involved in the ratio $n(^1E) / n(^2E)$, ε_{lin}, ε_{pow} and ε_{exp} are the mass-discrimination per mass unit factors for the three models respectively. Clearly, the internal approach is valid only if the instrument does not induce any significant mass or element dependency of the mass-discrimination factors (at least for a limited mass-range). Internal correction for mass-discrimination was applied to uranium measurements, using the linear model (eqn.4) for the three single detector ICP-MS and the exponential model (eqn.6) for the multiple collector S-ICP-MS.

The second approach consists in correcting externally for mass-discrimination using an IRM measured before (and) or after the unknown sample. The same pair of isotopes is measured for both solutions and, ideally, both ratio values are almost identical. Element concentration and sample matrix conditions in the unknown sample must also be matched in the IRM. The unknown ratio is corrected by multiplication with the K factor (eqn.3) calculated using the IRM measurement result. Obviously, applying this approach will depend on the availability of an appropriate IRM, adapted to the characteristics of the unknown sample measured. Moreover, mass-discrimination and any other effect monitored by the IRM must remain as stable as possible over the time spent at measuring the IRM and the unknown sample (including rinsing times, blank acquisitions etc). Despite these possible limitations, the external calibration remains an attractive approach particularly when it is difficult or impossible to apply internal calibration and/or the mass-discrimination effects are not sufficiently mass independent. For "direct" IDMS (eqn.1), correction for mass-discrimination is usually external. In most cases only R_B, the blend isotope amount ratio, must be corrected (all R_y values are certified and all R_x values can be found in column 9 of the IUPAC table[21]):

$$C_X = C_Y \cdot \frac{m_Y}{m_X} \cdot \left(\frac{R_Y - (R_{Cert}/R_{Meas}) \cdot R_B}{(R_{Cert}/R_{Meas}) \cdot R_B - R_X} \right) \cdot \frac{\sum (R_i)_X}{\sum (R_i)_Y} \qquad (7)$$

It must be noted that in case the isotopic composition of the element of interest in the unknown sample varies in nature (Pb for instance) it must be measured instead of using IUPAC data. Hence, every measured R_x value must also be corrected for mass-discrimination effects specifically.

4 COMBINED UNCERTAINTY CALCULATIONS

Combined uncertainties attached to the measurement results were calculated according to the ISO/GUM and Eurachem guides[22, 23] by applying an uncertainty propagation procedure to individual uncertainty contributions. In practice, a dedicated software program[24] was used, based on the numerical method of differentiation described by Kragten.[25] All calculated combined uncertainties are expanded with a coverage factor k = 2.

5 RESULTS AND DISCUSSION

5.1 Uranium "isotopic measurements" using ICP-MS: a comparison

From IRMM-072/1 to IRMM-072/10 the $^{233}U/^{235}U$ and $^{233}U/^{238}U$ isotope abundance ratios varied from ~ 1 to ~ $2 \cdot 10^{-3}$ respectively (certified to within 0.03%, k = 2). For each sample the $^{235}U/^{238}U$ ratio was always close to unity (certified to within 0.02%, k = 2) and was measured to correct internally for mass discrimination effects the measurements of the other two ratios. Identical calculations were performed for all four instruments (only the model differed, cf. eqns. 4 to 6). Three uncertainty components were combined together to build up the final uncertainty on the corrected measurement results. For the $^{233}U/^{235}U$ ratio and for the $^{233}U/^{238}U$ ratio, the repeatability of their respective measurements was combined together with the $^{235}U/^{238}U$ ratio measurement repeatability and the standard uncertainty (k = 1) on the certified value of the same $^{235}U/^{238}U$ ratio. As illustrated with figures 1 to 4 for the $^{233}U/^{235}U$ ratio, all certified values could be matched experimentally within the expanded uncertainty of the measurements (similar results for the $^{233}U/^{238}U$ ratio).

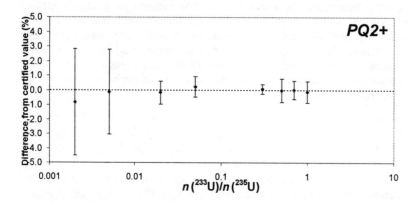

Figure 1 (reproduced from ref. 8). *Compare comments in figure 4 below*

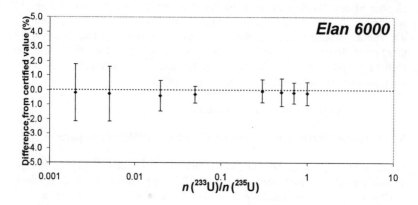

Figure 2 (reproduced from ref. 8) *Compare comments in figure 4 below*

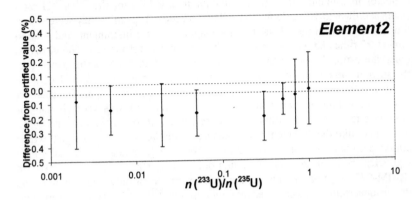

Figure 3 (reproduced from ref. 7) *Compare comments in Figure 4 below*

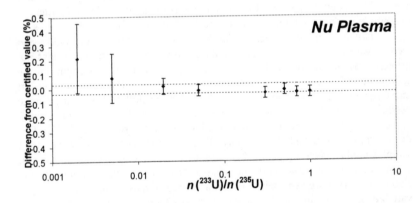

Figure 4 (reproduced from ref. 8). *Comments figures 1 to 4. Difference between the certified and the experimental $^{233}U/^{235}U$ isotope abundance ratio values for IRMM-072 samples in the range ~ 1 to ~ $2 \cdot 10^{-3}$. Measurement results are corrected for mass-discrimination effects, using the linear law model for the "PQ2+", the "Elan 6000" and the "Element2", and using the exponential law model for the "Nu Plasma". The vertical bars and the horizontal dotted lines represent the expanded uncertainties (k = 2) for the experimental and the certified values respectively.*

Expanded uncertainties on measurements performed with the multiple collector instrument varied from ± 0.04% to ± 0.24% for the $^{233}U/^{235}U$ ratio (from ± 0.08% to ± 0.27% for the $^{233}U/^{238}U$ ratio). Typically, they were ~ 1 to 5 times larger with the single detector magnetic sector instrument, and ~ 10 to almost 25 times larger with both quadrupole sector instruments. The closer to unity the ratio, the better the multiple collector S-ICP-MS performance (i.e. the smaller the combined uncertainties of the measurements and the larger the relative difference with other instruments' performance). For the smallest ratios, the combined uncertainty of the measurements was very similar for both magnetic sector instruments. These results illustrate the capacity of an instrument like the "Element2" to measure a ratio between a ^{235}U signal which equivalent concentration in solution is ~ 500 pg • g^{-1} and a ^{233}U signal which equivalent concentration in solution is only ~ 1 pg • g^{-1} ($^{233}U/^{235}U$ ~ $2 \cdot 10^{-3}$ in IRMM-072/10 sample). For all the instruments, with only three components in the uncertainty budgets, repeatability of the $^{233}U/^{235}U$ ratio measurement was the main source of uncertainty. For close to unity ratios, the counting statistics represented only 33% of the single detector S-ICP-MS measurement repeatability. For the smallest ratios the same theory could explain 100% of the observed repeatability (0.16%), thus indicating an optimum performance of this instrument and leading to an estimated combined uncertainty between 0.30%-0.35%. Therefore, this range could be considered to be the best achievable uncertainty values for measurements of ratios of nearly 3 orders of magnitude performed with the single detector S-ICP-MS instrument.

However, it can be seen in Fig. 5 that the results were not random and were always negative. This indicates the presence of a measurement bias not properly corrected for, and thus not properly taken into account in the estimation of the combined uncertainty budget. Possible sources of measurement bias were examined.[7,8] Calculation to simulate 1 ns change in the dead time value (15 ns instead of 16 ns) showed upward shifts on ratios (up to 0.1% on ~ $2 \cdot 10^{-3}$ ratios). Contamination could be ruled out but it was demonstrated that the addition of only 100 pg natural uranium to the measured samples could induce up to 0.2% difference on the measured ratios. Careful examination of the variation of the mass-discrimination per mass unit factors also pointed at more than only one major source of mass-discrimination, with variable size depending on the ratios measured. More sources of measurement bias could not be identified. However, these results illustrate why uncertainty on isotope abundance ratio measurement by ICP-MS, when performed with instruments of different design and characteristics in particular, cannot systematically be reduced to measurement repeatability only.

In the following section, a method for combining uncertainty contributions from corrections for "additive" measurement biases is presented.

5.2 Combination of uncertainty contribution from correction of isotope signal intensity for "additive" measurement biases

The three principal "additive" biases on ion current measurement are the correction for dead time effects (eqn.2), the correction for instrumental background (eqn.8) and the correction for isobaric interference (eqn.9).

$$I_{bkg\_corr} = I_{obs} - I_{bkg} \tag{8}$$

$$I_{in\_corr} = I_{obs} - I_{in} \tag{9}$$

in which I_{bkg} and I_{in} are the measured counting rates for background and isobaric interference signals respectively, and I_{bkg_corr} and I_{in_corr} are the counting rates corrected for instrumental background and isobaric interference respectively (all counting rates in counts $\cdot$ s^{-1}).

These corrections apply to intensities and do not cancel out mutually when calculating isotope ratios. Combining uncertainties including contributions from these corrections would require propagating the uncertainty contribution from the measured intensities (i.e. experimental repeatability) and not from the ratios directly. It is well known that the repeatability and the reproducibility of the intensities of a ratio are significantly worse than those of the ratio itself.[5] There is therefore a risk of exaggerating the combined uncertainty of the evaluated measurements. In order to avoid this risk, a method is proposed for the combined uncertainty calculation that "translate" "additive" corrections on intensities into multiplicative corrections on ratios. In eqn.10, each measured ratio ($R_{effect\_corr\_I}$) is multiplied by unity factors (k_{effect}) carrying standard uncertainties (SU_{effect}) representative of the uncertainty contributions arising from the "additive" corrections applied to the ratio intensities.

$$R_{Meas} = R_{effect\_corr\_I} \cdot k_{effect} \tag{10}$$

with $k_{effect} = 1 \pm SU_{effect}$

In eqn.7 the experimentally measured ratios R_{Meas} and R_B are almost systematically corrected for the effects described in eqns.2, 8 and 9. Application to "direct IDMS" of the method proposed above is obtained by combining eqn.7 with eqn.10 for the three effects (eqn.11).

$$C_X = C_Y \cdot \frac{m_Y}{m_X} \cdot (..) \cdot \frac{\sum (R_i)_X}{\sum (R_i)_Y} \tag{11}$$

$$\text{with } (\ldots) = \left(\frac{R_Y - \left[\dfrac{R_{Cert}}{(k_{\tau\_R_{Meas}} \cdot k_{bkg\_R_{Meas}} \cdot k_{in\_R_{Meas}} \cdot R_{Meas})} \cdot k_{\tau\_R_B} \cdot k_{bkg\_R_B} \cdot k_{in\_R_B} \cdot R_B \right]}{\left[\dfrac{R_{Cert}}{(k_{\tau\_R_{Meas}} \cdot k_{bkg\_R_{Meas}} \cdot k_{in\_R_{Meas}} \cdot R_{Meas})} \cdot k_{\tau\_R_B} \cdot k_{bkg\_R_B} \cdot k_{in\_R_B} \cdot R_B \right] - R_X} \right)$$

in which $k_{\tau\_R_{Meas}} = 1 \pm SU_{\tau\_R_{meas}}$, $k_{bkg\_R_{Meas}} = 1 \pm SU_{bkg\_R_{meas}}$, $k_{in\_R_{Meas}} = 1 \pm SU_{in\_R_{meas}}$, $k_{\tau\_R_B} = 1 \pm SU_{\tau\_R_B}$, $k_{bkg\_R_B} = 1 \pm SU_{bkg\_R_B}$ and $k_{in\_R_B} = 1 \pm SU_{in\_R_B}$.

The way to calculate these SU_{effect} is now illustrated with the correction for dead time effects. Typically, ratios measured with ICP-MS instruments are obtained by calculating the average value of individual ratios measured for n successive replicate measurements.

First, the "additive" correction is applied to the intensities measured for isotopes ^1E and ^2E ($I_{Ri\_}{}^1E_{obs}$ and $I_{Ri\_}{}^2E_{obs}$) of the ratio during each replicate i: eqn. 12, with dead time correction (eqn. 2) applied to $n = 6$ replicate measurements.

$$\tag{12}$$

Repl. 1: $(I_{R1\_}{}^1E_{obs} / (1 - I_{R1\_}{}^1E_{obs} \cdot \tau)) / (I_{R1\_}{}^2E_{obs} / (1 - I_{R1\_}{}^2E_{obs} \cdot \tau)) = R_{1\_\tau}$
   2: $(I_{R2\_}{}^1E_{obs} / (1 - I_{R2\_}{}^1E_{obs} \cdot \tau)) / (I_{R2\_}{}^2E_{obs} / (1 - I_{R2\_}{}^2E_{obs} \cdot \tau)) = R_{2\_\tau}$
   3: $(I_{R3\_}{}^1E_{obs} / (1 - I_{R3\_}{}^1E_{obs} \cdot \tau)) / (I_{R3\_}{}^2E_{obs} / (1 - I_{R3\_}{}^2E_{obs} \cdot \tau)) = R_{3\_\tau}$
   4: $(I_{R4\_}{}^1E_{obs} / (1 - I_{R4\_}{}^1E_{obs} \cdot \tau)) / (I_{R4\_}{}^2E_{obs} / (1 - I_{R4\_}{}^2E_{obs} \cdot \tau)) = R_{4\_\tau}$
   5: $(I_{R5\_}{}^1E_{obs} / (1 - I_{R5\_}{}^1E_{obs} \cdot \tau)) / (I_{R5\_}{}^2E_{obs} / (1 - I_{R5\_}{}^2E_{obs} \cdot \tau)) = R_{5\_\tau}$
   6: $(I_{R6\_}{}^1E_{obs} / (1 - I_{R6\_}{}^1E_{obs} \cdot \tau)) / (I_{R6\_}{}^2E_{obs} / (1 - I_{R6\_}{}^2E_{obs} \cdot \tau)) = R_{6\_\tau}$

Second, an average ratio (R_{Meas}) is calculated out of the n replicate ratios corrected for dead time effects (R_{i_τ}).

Third, the standard uncertainty of each measured intensity is set equal to 0 (thus only remains standard uncertainty on τ).

Fourth, the combined uncertainty u_{c_τ} on R_{Meas} is calculated by numerical differentiation. Fifth, SU_τ is calculated in relative (i.e. in %): eqn. 13.

$$SU_\tau = (u_{c\_\tau} / R_{Meas}) \cdot 100 \tag{13}$$

Same reasoning is applied to the other "additive" corrections. A performance comparison study between Q-ICP-MS and multiple-collector S-ICP-MS for element amount content certification by IDMS in isotopically enriched materials conducted in our laboratory is reported in these proceedings.[26] In this study, the method described above was used to evaluate comprehensive measurement combined uncertainties. For one element (Pb), the results obtained with the Q-ICP-MS indicate that the main uncertainty component arises from the blend sample measurement, and nearly half of the remaining uncertainty contribution (28 %) arises from the correction for dead time effects. In the following example (below section), Pb was certified in a fly-ash material by IDMS using Q-ICP-MS.[10] Correction for mass-discrimination was external. Same samples (different dilution levels) were also run on a single detector S-ICP-MS and a multiple collector S-ICP-MS.

5.3 Comparison of uncertainty contributions for Pb certification by IDMS in fly ash material using different types of ICP-MS

For such a certification exercise, the uncertainty from the sample preparation step is a major contribution and will have an overruling effect and minimize the differences between the final combined uncertainties evaluated for the various types of ICP-MS instruments. When looking only at the combined uncertainty on the respective blend ratio measurements, some significant differences can be observed (Figure 5).

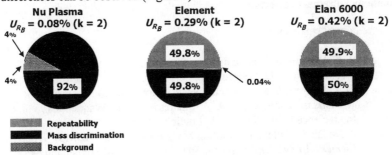

Figure 5 *Combined uncertainty on near to unity Pb blend ratio measurement (U_{Rb}) showing distribution of the various uncertainty contributions.*

Since the blend ratio in these experiments was close to unity there is no significant contribution of the correction for dead time effects to the combined uncertainty, neither for the "Element" nor for the "Elan 6000". Consistently with the results obtained for the uranium experiments, the measurements performed with the "Nu Plasma" produced the smallest combined uncertainty. Distribution of the various uncertainty contributions change radically between the results observed for both single detector ICP-MS instruments and those observed for the multiple collector ICP-MS instruments. In one case there is nearly equal distribution between the repeatability of the blend ratio measurement and the external correction for mass-discrimination, whereas in the other case the contribution from the correction for external mass-discrimination dominates largely. For the "Nu Plasma", it is worth noting the non-negligible contribution of the correction for the instrumental background. This is due to the fact that only a 50 ng Pb • g^{-1} solution was available for these experiments (i.e. 20 times less concentrated than for the uranium measurements), and thus the Pb signal intensities measured were more sensitive to the magnitude of the instrumental background signal intensities.

6 CONCLUSIONS

Combined uncertainty on isotope ratio measurements over almost 3 orders of magnitude performed with a magnetic sector multi-collector ICP-MS can be ~ 10-25 times smaller than with a quadrupole ICP-MS, and ~ 1-5 smaller than with a magnetic sector single collector ICP-MS. However, these results might differ depending on the ratio and the element considered and depending on the number and the magnitude of the measurement uncertainty contributions considered. A series of "additive" corrections, such as the ones for dead time effects or for instrumental background signal or for isobaric interference signal(s), might play a non negligible role. A method, applicable to the IDMS equation, was proposed to combine uncertainty contributions arising from these measurement corrections.

7 REFERENCES

1   R. K. Webster *in* Isotope dilution analysis, *Advances in mass spectrometry*; J. D. Waldron ed., Pergamon Press, London, 1958.
2   R. K. Webster *in* Mass spectrometric isotope dilution analysis, *Methods in geochemistry*; A. A. Smales and L. R. Wagner eds., Interscience Publ., New-York, London, 1960.
3   P. De Bièvre *in* Isotope Dilution Mass Spectrometry (IDMS), *Trace Element Analysis in Biological Specimens*; R. F. M Herber and M. Stoeppler eds., Elsevier, Amsterdam, 1994.
4   Comité Consultatif pour la Quantité de Matière *in* Rapport de la 1ère session; BIPM ed., Pavillon de Breteuil, F-92312 Sèvres Cedex, France, 1995
5   F. Vanhaecke, L. Moens and P. D. P. Taylor *in* Use of ICPMS for isotope ratio measurements, *ICP Spectrometry and its applications*; S. J. Hill ed., Sheffield Academic Press, England, 1999.
6   C. R. Quétel, B. Thomas, O. F. X. Donard and F. E. Grousset, *Spectrochim. Acta Part B*, 1997, 52 (2), 177.
7   C. R. Quétel, T. Prohaska, M. Hamester, W. Kerl and P. D. P. Taylor, *J. Anal. At. Spectrom.*, 2000, 15, 353.
8   C. R. Quétel, J. Vogl, T. Prohaska, S. Nelms, P. D. P. Taylor and P. De Bièvre, *Fresenius J. Anal.Chem.*, 2000, 368, 148.
9   C. R. Quétel, T. Prohaska, *IRMM Internal report GE/R/IM/44/99*, 1999
10   T. Prohaska, C.R. Quétel, C. Hennessy, D. Liesegang, I. Papadakis and P.D.P. Taylor, J. Env. Monitoring, in press
11   G. F. Knoll *in* Radiation detection and measurement; Wiley and Sons, New York, 3rd ed., 2000.

12   S. Nelms, C. R. Quétel, T. Prohaska, J. Vogl and P. D. P. Taylor, *J. Anal At. Spectrom.*, accepted for publication

13   W. Lycke, P. De Bièvre, A. Verbruggen, F. Hendrickx and K. Rosman, *Fresenius J. Anal. Chem.*, 1988, 331, 214.

14   A. S. Cohen, N. S. Belshaw and R. K. O'Nions, *Int. J. Mass Spectrom. and Ion Processes*, 1992, 116, 71.

15   P. D. P. Taylor, P. De Bièvre, A. J. Walder and A. Entwistle, *J. Anal. At. Spectrom.*, 1995, 10, 395.

16   K. J. R. Rosman, W. Lycke, R. Damen, R. Werz, F. Hendrickx, L. Traas and P. De Bièvre, *Int. J. Mass Spectrom. and Ion Processes*, 1987, 79, 61.

17   K. G. Heumann, S. M. Gallus, G. Radlinger and J. Vogl, *J. Anal. At. Spectrom.*, 1998, 13, 1001.

18   I. T. Platzner *in* Modern isotope ratio mass spectrometry; J. D. Winefordner ed.; Wiley and Sons, New-York, 1997.

19   H. P. Longerich, B. J. Fryer and D. F. Strong, *Spectrochim. Acta Part B*, 1987, 42, 39.

20   C. N. Maréchal, P. Télouk and F. Albarède, *Chemical Geol.*, 1999, 156, 251.

21   K. Rosman and P. D. P. Taylor *in* IUPAC – Isotopic compositions of the elements 1997, *Pure Appl. Chem.*, 1998, 70, 217.

22   Guide to the Expression of Uncertainty in Measurement, ISBN 92-67-10188-9, © International Organization for Standardization, Genève, 1995

23   EURACHEM, *Quantifying Uncertainty in Analytical Measurement*, ISBN 0-948926-08-2, © Crown Copyright, 1995.

24   GUM Workbench®, Metrodata GmbH, D-79639 Grenzach-Wyhlen, Germany

25   J. Kratgen, *Analyst*, 1994, 119, 2161.

26   S. Nelms, T. Prohaska, C. R. Quétel and P. D. P. Taylor *in* Plasma Source Mass Spectrometry: the new Millenium, *Proceedings of the Durham Conference 2000*

COMPARISON OF MC-ICP-MS WITH QUADRUPOLE ICP-MS FOR THE CERTIFICATION OF THE AMOUNT, CONTENT AND ISOTOPIC COMPOSITION OF ENRICHED ^{206}PB AND ^{203}TL MATERIALS USING ISOTOPE DILUTION

Simon M. Nelms[1], Thomas Prohaska[2], Christophe R. Quétel and Philip D. P. Taylor

European Commission Joint Research Center, Institute for Reference Materials and Measurements, Retieseweg, B-2440 Geel, Belgium

[1]current address:- TJA Solutions, Ion Path, Road Three, Winsford, Cheshire, CW7 3BX, UK
[2]current address:- University of Agricultural Sciences, Institute of Chemistry, Muthgasse 18, A-1190 Vienna, Austria

1 INTRODUCTION

Multi-collector ICP-MS is becoming an accepted technique for making high precision determinations of isotope ratios. Compared to multi-collector thermal ionisation mass spectrometry, MC-ICP-MS has a higher sample throughput, does not sustain a time based change in mass discrimination during the measurement cycle and can be applied to elements whose first ionisation potentials are above 7eV, which cannot be ionised by a conventional thermal source, without resorting to the addition of chemical modifiers.

Improved precision reduces uncertainty in the overall measurement, and this has been demonstrated recently by a study into the comparative performance of ICP mass spectrometers using uranium isotopic measurements.[1] By constructing a comprehensive uncertainty budget that contains all the individual uncertainty components of the measurement, the contribution of each component to the overall uncertainty can be evaluated. Using this approach, a direct comparison of measurements made by MC-ICP-MS and quadrupole ICP-MS can be made and also the influence of parameters other than just measurement repeatability on the measurement uncertainty can be elucidated.

In the current work, a double focusing MC-ICP-MS was compared to quadrupole ICP-MS for the certification of the amount content and isotopic composition of both a ^{206}Pb enriched and a ^{203}Tl enriched spike material, using identical mother solutions diluted to the appropriate concentration for each instrument. The amount content certification was achieved using isotope dilution ICP-MS as a primary method of measurement. In this paper, the results from each measurement will be described with particular emphasis on how the data compared in terms of the total combined uncertainty of the result, and which parameters had the strongest effect on the uncertainty of the final result.

2   THEORETICAL BACKGROUND OF ISOTOPE DILUTION

Isotope dilution as an accurate and precise method of quantitation for mass spectrometry is well known.[2] In its simplest form, the method involves addition of a known quantity of an enriched elemental spike of known concentration to a known weight of sample, followed by isotopic equlibration of the mixture. The concentration of the element in the sample is calculated from isotope ratio measurements of the blend and the unspiked sample, together with given values for the corresponding ratio in the spike material, (referenced to one isotope of the element) and the sample and spike weights, using the equation:

$$C_x = C_y \cdot \frac{m_y}{m_x} \cdot \left(\frac{R_y - R_b}{R_b - R_x} \right) \cdot \frac{\Sigma R_{ix}}{\Sigma R_{iy}}$$

where:
C_x and C_y are the element concentrations in the sample and spike
m_y and m_x are the spike and sample weights
R_y, R_b and R_x are the spike, blend and sample ratios and
ΣR_{iy} and ΣR_{ix} are the sum of the ratios (all referenced to one isotope of the element e.g. ^{208}Pb).

For elements whose isotopic composition varies in nature, such as Pb, it is necessary to measure R_x and ΣR_{ix}. For other elements, reference values such as those from IUPAC,[3] can be used. Isotope dilution has a number of advantages as a method of quantitation. Since it is based on isotope ratio measurements, which can be measured with higher precision than absolute signal intensities, the precision of the determination is improved over external calibration methods. This feature also means that quantitative recovery of the analyte of interest in sample preparation steps is not necessary, although the recovery should be sufficient to ensure that a good quality measurement can be made. Since the analyte concentration is determined by mixing the sample with an enriched spike of the same element, excellent correction for non-spectral matrix effects is achieved as the measurement is internally standardised in the best way possible. The method does have some disadvantages, in that a limited number of enriched spikes are available, mono-isotopic elements can only be quantified if a non-natural, sufficiently long-lived isotope of the element is available and the spike materials are expensive. Contamination of the sample during the preparation procedure can be problematic if the blanks are not adequately controlled. Highly enriched spike materials must also be handled carefully as contamination will invalidate subsequent measurements unless the spike amount content and isotopic composition are re-certified.

3 INSTRUMENTATION

3.1 Quadrupole ICP-MS

An ELAN 6000 (PE Sciex), equipped with a quartz low-flow glass concentric nebuliser and an air-cooled quartz mini-cyclonic spray chamber (Glass Expansion Pty. Ltd., Hawthorn, Victoria, Australia) was used. The instrument was operated with plasma, auxiliary and nebuliser gas flows of 14.8, 0.95 and 0.93 L min^{-1} respectively. The autolens option was set to off for the measurements. Ions were detected with a simultaneous pulse counting / analog ion discrete dynode ion multiplier (ETP, Sydney, Australia).

3.2 Magnetic sector double focusing multi-collector ICP-MS

A Nu Plasma (Nu Instruments, Wrexham, Wales, UK), equipped with a quartz low-flow glass concentric nebuliser (Glass Expansion Pty. Ltd.) and a quartz peltier cooled Scott-type spray chamber, was used. The instrument was operated with plasma, auxiliary and nebuliser gas flows of 13, 0.96 and 0.97 L min^{-1} respectively. Signals were collected using Faraday cup detectors.

3.3 Experimental

3.3.1 Sample and isotope dilution spike description   Two enriched element spike solutions were supplied (Merck Ltd, Darmstadt, Germany). The first was a ^{206}Pb solution of nominal concentration 10 µg g^{-1} and the second was a ^{203}Tl solution, also of nominal concentration 10 µg g^{-1}. For the Pb study, "natural-like" Pb NBS-981 (NIST, Gaithersburg, USA) was used as the spike material. This was prepared as a 2.4 mmol kg^{-1} solution in 0.4 mol dm^{-3} nitric acid by metrological weighing and dissolution of the parent metal wire.[4] For the Tl study, a natural Tl material, IRMM-648, (certified for Tl amount content and isotopic composition) was used.

3.3.2 Sample preparation for isotope dilution measurements   Six blends of the enriched ^{206}Pb sample and the NBS-981 spike material were prepared by metrological weighing of both materials under Class 100 clean room conditions. The sample and spike weights were selected to yield a $^{206}Pb/^{208}Pb$ isotope ratio of around 2:1 in the blends. Samples of the parent blends (around 10 µg g^{-1}) were diluted with 0.4 mol dm^{-3} HNO$_3$, prepared from sub-boiled quality acid and Reagent Grade I water (18.2MΩ cm resistivity) to 5 and 200 ng g^{-1} for the ELAN 6000 and Nu Plasma measurements respectively. For the Tl measurements, six blends of the enriched ^{203}Tl sample and the IRMM-648 spike material were prepared by metrological weighing of both materials also under Class 100 clean room conditions. In this case, the sample and spike weights were selected to produce a $^{203}Tl/^{205}Tl$ isotope ratio of around 1:1 in the blends. Samples of the parent blends were diluted to 5 and 500 ng g^{-1} for the ELAN 6000 and Nu Plasma measurements respectively.

3.3.3   Sample preparation for isotopic composition measurements
Six solutions of the ^{206}Pb and ^{203}Tl materials were separately prepared by serial dilution of the parent 10 µg g^{-1} materials with 0.4 mol dm^{-3} HNO$_3$ to 5 ng g^{-1} and 500 ng g^{-1} for the ELAN 6000 and Nu Plasma instruments respectively.

3.3.4 Analytical procedure   For the Pb blend measurements, Pb NBS-982 was measured at intervals (every two blends) throughout the run to determine the mass discrimination correction factor needed to correct the measured blend ratios. NBS-982 has a near unity $^{206}Pb/^{208}Pb$ ratio so the effect of dead time on the mass discrimination factor measurement for the quadrupole results was negligible. NBS-982 was also used to measure the mass discrimination factor for the isotopic composition measurements, at intervals of three samples. From the factor derived for the $^{206}Pb/^{208}Pb$ a linear law (ELAN 6000) and an exponential law (Nu Plasma) was applied to derive the corresponding mass discrimination factors for the $^{207}Pb/^{208}Pb$ and $^{204}Pb/^{208}Pb$ ratios. For the latter ratio, a correction was made for the ^{204}Hg interference on ^{204}Pb, using the signal response at ^{202}Hg and calculating the ^{204}Hg response via the IUPAC Hg isotope abundances. All samples were blank subtracted before the ratios were calculated.

3.3.5 Uncertainty calculations   In the following section, reference is made to the terms accuracy of measurement, standard uncertainty, coverage factor and precision. According to the Guide to the Expression of Uncertainty in Measurement[5] these words are defined as follows. Accuracy of measurement is the closeness of the agreement between the result of a measurement and a true value of the measurand, standard uncertainty is the result of a measurement expressed as a standard deviation, coverage factor (k) is a numerical factor used as a multiplier of the combined standard uncertainty in order to obtain an expanded uncertainty and precision is the experimental relative standard deviation of a series of measurements of the same measurand.

Uncertainty contributions to the final result of a measurement can be derived by applying an error propagation procedure, based on using partial differentiation of the components that contribute to the measurement result. In the case of an isotope dilution measurement this means deriving the partial differentials for each component of the isotope dilution equation. In practice, a numerical approach, based on inserting each input quantity to the equation in turn and evaluating the effect of the uncertainty of each of these quantities on the uncertainty of the final result, is easier to apply. The method that is usually followed is the spreadsheet approach described by Kragten.[6] Guidelines for categorising and correctly evaluating the uncertainty components of a measurement are given by the ISO and Eurachem organisations. In this work, a dedicated software program, GUM Workbench[®7] was used to calculate the total combined uncertainty of each measurement result. This software uses the numerical approach described above to perform the uncertainty calculations. All uncertainties reported herein are given as expanded uncertainties with a coverage factor k of 2.

4 RESULTS AND DISCUSSION

The results for the total amount content measurements of the Pb and Tl materials, together with their total combined expanded uncertainties are given below (Table 1).

Table 1 *Pb and Tl amount content results for both instruments*

Instrument	Total Pb (μg g^{-1})	Relative uncertainty (%) k = 2	Total Tl (μg g^{-1})	Relative uncertainty (%) k = 2
Elan 6000	9.971(39)	0.4	9.963(60)	0.6
Nu Plasma	9.942(20)	0.2	9.994(10)	0.1

The total Pb and Tl amount content results were close to the given nominal value of 10 μg g^{-1}. The results for both materials on both instruments agree well and there is no statistically significant difference between the mean values found using either instrument (at the 95% confidence limit, Students t-test), for either material.[8] The total combined uncertainty of the Pb amount content result using the Nu Plasma was found to be only half that achieved with the ELAN 6000, despite the fact that the isotope ratio measurement precision of the former instrument was better by typically a factor of 5 to 10. For the corresponding Tl results, there is a factor of 6 difference, compared to a factor of 10 difference in isotope ratio measurement precision. This clearly illustrates that measurement uncertainty in isotope dilution mass spectrometry is not simply a function of isotope ratio precision alone and other factors must be considered. The uncertainties reported above were derived by thorough investigation of each of the parameters that could contribute to the uncertainty of the final result. The main contributors to the amount content uncertainty for the Pb and Tl measurements are shown in Table 2 below.

Table 2 *Uncertainty contributions to the total Pb and Tl amount content result for both instruments*

Parameter	ELAN 6000		Nu Plasma	
	Contribution (%)		Contribution (%)	
	Pb	Tl	Pb	Tl
Blend ratio measurement precision	72.1	29.4	3.3	3.5
Dead time	13.3	0.1	---	---
Spike purity	6.6	---	25.0	---
Mass discrimination correction	4.2	69.3	56.9	63.5
Metrological weighings	2.6	---	9.2	---
Blank	0.7	---	1.2	0.1
Certified ratios	0.5	0.5	2.2	18.9
Spike concentration	---	0.3	---	9.6
Spike weighings	---	0.3	---	4.1
Sample weighings	---	---	---	0.2
Isotopic composition	---	0.3	2.1	0.1

For the ELAN 6000 Pb results, the majority of the total uncertainty does arise from the measurement precision of the blends (72%). Nearly half of the remaining 28% arise from uncertainty in the detector dead time. The effect of the uncertainty of this parameter on the measured isotope ratios was calculated using the method proposed by Quétel *et al.*[9] for evaluation of uncertainty components in ICP-MS isotope ratio measurements. The effect of blank measurement uncertainty, where applicable, was determined in a similar way. Experiments in our laboratory have shown that measurement of the dead time using any of the methods reported in the literature leads to a relatively large uncertainty (around ± 10%).[10,11] Including dead time uncertainty in the evaluation is therefore essential for measurements of non-unity ratios. Of the remaining uncertainty, mass discrimination correction uncertainty (also derived from ratio measurement imprecision), metrological weighing uncertainty (limited by the resolution of the balance readings) and even uncertainty in the spike purity, make significant contributions to the final result uncertainty. For the Nu Plasma results, the improved isotope ratio measurement precision leads directly to a smaller contribution from this parameter to the uncertainty. A slight drift in the mass discrimination of the instrument during the course of the measurements increased the relative influence of this parameter compared to the ELAN 6000 results. Finally, the effect of the metrological weighing and spike purity uncertainties were much larger for these measurements, illustrating that the uncertainty of the final result was in fact limited by these parameters and not the ratio measurement precision.

For the ELAN 6000 Tl results, almost all of the total uncertainty arises from the measurement precision of the blends and the effect of the mass discrimination correction (98.7%). The reasons for this are firstly, the instrumental Tl blank was very low (compared to the Pb blank) so had an insignificant effect and secondly, the blend ratios were all close to 1:1 and so the effect of dead time uncertainty was also negligible. Nonetheless, small contributions to the uncertainty were observed from the certificate ^{203}Tl/^{205}Tl ratio uncertainty and the measured isotopic composition of the Merck material. For the Nu Plasma results, the improved isotope ratio measurement precision again led directly to a smaller total uncertainty. In this case, the total uncertainty was a factor of 6 smaller than for the ELAN 6000 results, whereas the measurement precision of the Nu Plasma was around 16-times lower than the ELAN 6000. The relative contribution of the mass discrimination factor was similar for both instruments, so it can be concluded that as observed for the Pb measurements, isotope ratio measurement precision was not the limiting factor for reducing the uncertainty. A review of the remaining uncertainty contributors shows that uncertainty in the certificate value of the spike ^{203}Tl/^{205}Tl ratio is having a large effect. In addition, the spike concentration uncertainty and even the small spike weighing uncertainty (around 0.04% for each weighing) are significant contributors to the total uncertainty. Evidently, improvements in weighing and spike certification are also required to further drive down measurement uncertainties in isotope dilution ICP-MS studies.

The isotopic composition results obtained for Pb using each instrument, in terms of percentage molar abundance of each isotope, are presented below, together with the corresponding calculated molar mass (Table 3).

Table 3 *Merck ^{206}Pb-enriched isotopic composition and molar mass results for both instruments*

Instrument	^{204}Pb (%)	^{206}Pb (%)	^{207}Pb (%)	^{208}Pb (%)	Molar mass (g mol^{-1})
ELAN 6000	0.000250(42)	99.9765(22)	0.00955(96)	0.0137(13)	205.974813(36)
Nu Plasma	< 0.0003	99.965(28)	0.0097(24)	0.0152(24)	205.97464(58)

The material supplied by Merck was found to very highly enriched in ^{206}Pb (>99.95%). The ^{206}Pb/^{204}Pb, ^{206}Pb/^{207}Pb and ^{206}Pb/^{208}Pb ratios were found to be of the order of 4×10^5, 1×10^4 and 7×10^3 respectively. For the ELAN 6000, the latter ratio could be measured with confidence above the blank ^{208}Pb signal for the 5 ng g^{-1} Pb samples. However, due to the low signal intensities involved, to make meaningful measurements of the other ratios it was necessary to analyse the 200 ng g^{-1} solutions and then normalise the instrument response via the ^{208}Pb signal to correct for any instrument sensitivity change between the analyses. The normalisation calculation was included in the uncertainty budget and was found to not contribute significantly to the final uncertainty. The Faraday cup detectors of the Nu Plasma were not sufficiently sensitive to measure the ^{204}Pb signal above the blank in the 200 ng g^{-1} Pb solutions, so the ^{206}Pb/^{204}Pb could not be measured. However, values for the ^{206}Pb/^{207}Pb and ^{206}Pb/^{208}Pb ratios, which agreed within the measurement uncertainty with those measured on the ELAN 6000 were obtained. None of the parent 10 μg g^{-1} sample remained after the above samples had been prepared, so it was not possible to attempt to derive a better value of the ^{206}Pb/^{204}Pb ratio using this more concentrated solution, without preparing a fresh set of solutions. However, it can be expected that had this been possible, meaningful measurements of the ^{206}Pb/^{204}Pb ratio may have just been achievable on the parent 10 μg g^{-1} sample. Finally, Table 3 shows that both instruments gave abundance values for the four Pb isotopes that agreed with each other within the measurement uncertainties.

The isotopic composition results obtained for Tl using each instrument, in terms of percentage molar abundance of each isotope, are presented below, along with the corresponding calculated molar mass (Table 4).

Table 4 *Merck ^{203}Tl-enriched isotopic composition and molar mass results for both instruments*

Instrument	^{203}Tl (%)	^{205}Tl (%)	Molar mass (g mol^{-1})
ELAN 6000	98.131(16)	1.869(16)	203.00975(34)
Nu Plasma	98.1456(19)	1.8543(19)	203.009455(40)

The molar amount fraction results for ^{203}Tl and ^{205}Tl, and as a result the molar mass result also, for the Nu Plasma are around 10 times more precise than the results obtained using the ELAN 6000. Since the ^{203}Tl material was less enriched (around 98%) than the ^{206}Pb sample and the instrumental Tl blank signal was much lower (by up to a factor of 10) than for Pb, this improvement was entirely due to the higher isotope ratio measurement precision achieved with the multi-collector instrument. The ^{203}Tl/^{205}Tl ratios and consequently the abundance values for the two Tl isotopes measured with each instrument agreed within the measurement uncertainty.

5 REFERENCES

1   Quétel, C. R., Prohaska, T., Vogl, J., Nelms, S.M. and Taylor, P. D. P., *Fres. J. Anal. Chem.*, 2000, **368**, 148.

2   Heumann, K., Isotope Dilution Mass Spectrometry in *Inorganic Mass Spectrometry*, 1988, J. Wiley and Sons Inc., ed. F. Adams *et al.*

3   Rosman, K.J.R. and Taylor, P.D.P., *Pure and Appl. Chem.*, 1998, **70**, 217

4   Nelms, S.M., Quétel, C. R. and Dyckmans, B., IRMM, Geel, Belgium, internal report number GE/R/IM/53/99.

5   ISO/GUM, *Guide to the expression of uncertainty in measurement*, International Organisation for Standardisation, Geneva, Switzerland, 1995, ISBN 92-67-10188-9.

6   Kragten, J., *Analyst*, 1994, **119**, 2161.

7   GUM Workbench®, Metrodata GmbH, D-79639 Grenzach- Wyhlen, Germany.

8   Miller, J.C. and Miller, J.M., in *Statistics for Analytical Chemistry*, Ellis Horwood, 3rd Edition, 1993, p. 55.

9   Quétel, C. R., Prohaska, T., Nelms, S.M., Diemer, J. and Taylor, P. D. P., *7th International Conference on Plasma Source Mass Spectrometry*, Durham, 14th October 2000, Lecture 6, afternoon session.

10   Held, A. and Taylor, P., *J. Anal. At. Spectrom.*, 1999, **14**, 1075.

11   Nelms, S.M., Quétel, C. R., Prohaska, T., Vogl, J. and Taylor, P. D. P., *J. Anal At. Spectrom.*, submitted for publication.

WHICH WAY FOR OS?
A COMPARISON OF PLASMA-SOURCE VERSUS NEGATIVE THERMAL IONISATION MASS SPECTROMETRY FOR OS ISOTOPE MEASUREMENT

D.G. Pearson and C.J. Ottley

Department of Geological Sciences, Durham University, South Road, Durham City, DH1 3LE, UK

1 INTRODUCTION

The huge resurgence of interest in the Re-Os isotope system by the geological community in the 1990's was largely driven by the possibility of analysing pg levels of both Re and Os by negative-ion thermal ionisation mass spectrometry [1,2]. In contrast, despite early demonstration of the sensitivity of ICP-MS for Re and Os analysis [3-5], little work has been performed by quadrupole ICP-MS (Q-ICP-MS). The advent of plasma ionisation multi-collector mass spectrometers (PIMMS) provides the means for making highly precise isotopic measurements for systems such as Re-Os and there has been much recent activity aimed at applying this methodology to Os isotope measurements. This contribution evaluates the relative merits of plasma source mass spectrometry and N-TIMS for geological applications of the Re-Os isotope system.

2 ANALYTICAL REQUIREMENTS

Two Os isotopes are the stable daughter products of long-lived radioactive decays[6];

$$^{187}Re - {}^{187}Os \quad \lambda = 1.666 * 10^{-11} \text{ yr}^{-1}$$
$$^{190}Pt - {}^{186}Os \quad \lambda = 1.54 * 10^{-12} \text{ yr}^{-1}$$

and hence ratio measurement involving both these decay products is desirable in Earth Sciences. Re and Os are some of the least abundant elements in the Earth's crust, with average crustal values of <1 ng/g and < 0.05 ng/g respectively [7]. Even in the mantle, average Re levels are c. 0.3 ng/g and Os approximately 3 μg/g [8]. Many geologically important samples can have less than 100 pg total Os and numerous recent studies into dating diamonds [9] or Os isotope variations in seawater [10,11] have used total analytes in the sub pg to 10's of fg range.

These low abundances dictate that any technique for isotope ratio measurement must have good sensitivity. Counterbalancing this need for sensitivity is the large variation in the commonly used Os isotope ratio, $^{187}Os/^{188}Os$, observed in natural systems compared to commonly used radiogenic isotope systems. For instance, the crust-mantle system shows a total variation in $^{187}Os/^{188}Os$ of between 0.09 and > 100, i.e., a relative variation of approximately 10^5 % compared to the mantle values of 0.13 [6]. Even in more limited Earth reservoirs, such as the ocean island basalt mantle source, there is over 20 % relative variation in $^{187}Os/^{188}Os$ compared to approximately 0.1 % shown by variations in $^{143}Nd/^{144}Nd$ resulting from the ^{147}Sm - ^{143}Nd decay system.

Precisions of 0.1 % are sufficient for resolutions of +/- 150 Ma in Re-Os model ages, adequate for many studies. Ultimately, all absolute isochron ages are limited by the +/- 1 % error in the decay constant of ^{187}Re, however, relative isochron ages can be potentially resolved to differences of 0.5 % in geologically undisturbed systems (e.g., [12]).

Requirements of the ^{190}Pt - ^{186}Os system are more stringent due to the much longer half-life of ^{190}Pt and the low natural abundance of ^{190}Pt, restricting parent-daughter isotope variability and limiting radiogenic in-growth. These factors result in $^{186}Os/^{188}Os$ variations in many systems of interest of only 30 ppm total, requiring external precisions exceeding this.[13,14]

3   N-TIMS

The high first ionisation potentials of both Re (7.9 eV) and Os (8.7 eV) make them unsuited to positive TIMS. The possibility of negative thermal ionisation of Re as an oxide species has long been known.[15,16] The group of K. Heumann at the University of Regensburg, Germany, first developed N-TIMS analytical techniques for Os, in 1989[2,17] and its use in Earth Sciences was popularised by Creaser et al (1991). The technique relies upon use of Pt filaments and the action of an "emitter" such as $Ba(NO_3)_2$ [1], or $Ba(OH)_2$ [17] to lower the work function of the filament material, promoting the production of negative thermal ions. It was apparent that Re, Os and Ir could be ionised efficiently as negative-ion oxide species, each combining with different numbers of oxygen atoms to produce, respectively, ReO_4^-, OsO_3^-, and IrO_2^-. This difference in oxide species raised the possibility that all 3 elements could be loaded on the same filament [1] because potential isobaric overlaps between ^{187}Re and ^{187}Os might be avoided. In practice, small but significant amounts of other oxide species are formed, such as ReO_3^-, which are sufficient to compromise precise Os isotope ratio measurement.

A major attraction of N-TIMS is its high ionisation efficiency, permitting isotopic analysis of sub-pg quantities of Os [9,11,18]. Ionisation efficiency varies depending on sample cleanliness, which is ultimately a measure of sample preparation procedure quality. Estimates for the ionisation efficiency of Os standard solutions are up to c. 30 % [17] and can be > 5 % for samples. Such efficiencies are generally achieved by bleeding oxygen or freons into the ion-source[17] and permit isotopic analysis of as little as 10 fg at the 1 % (2SE) internal precision level for $^{187}Os/^{188}Os$ (Figure 1). Internal precisions of 0.1 % (2SE) and better can be achieved for 1 pg standards[19]. Short-term external precisions on larger runs can be as good as 0.03 % [19] (Figure 2) but longer-term (6 months to > 1 year) external reproducibility is usually considerably worse than this and varies from 0.18 to 0.5 % between different laboratories over periods of a year or more. [20-24] The poor long-term external precision for sub-ng sized Os samples, compared to within-run precisions, is probably attributable to non-systematic interferences problems encountered in N-TIMS.

For larger sample sizes (> 5ng) analysed by multiple faraday collector arrays, external precision approaches internal precision much more closely, e.g., Brandon et al., [14] report internal precisions for $^{187}Os/^{188}Os$ on >5ng Os of 0.003 % and external precisions as good as 0.005 %.

Other advantages to N-TIMS Os analysis are that sample to sample memory is very low, provided that the ion-source is regularly cleaned, and that the main instrumental blank is from the Pt filament, which can be reduced to c. 5 fg or less [25]. Recent advances incorporating multiple secondary electron multipliers (SEM) onto TIMS instruments, and fast switching magnets will act to increase internal precision obtainable from very small samples. Whether external precision is improved at the fg level is largely dependant on the long-term gain stability between multipliers and on interference-free runs at very small signal sizes.

Fundamental drawbacks with the N-TIMS technique include the production of strong thermal electron emission which may lead to arcing in the ion-source, or de-focussing of the ion-extraction lenses, and the dramatic lowering of ionisation efficiency due to surface impurities on the filament, which alter the filament work function. Impurities on the filament surface (especially Re) or in the sample solution loaded onto the filament can also cause isobaric interferences. Occasional interferences on mass 235 ($^{187}Os^{16}O_3^-$) that are not related to Re, can occur, particularly at high thermal electron emission currents. Such interferences are very difficult to correct for. Very clean sample chemistries are required to minimise these effects.

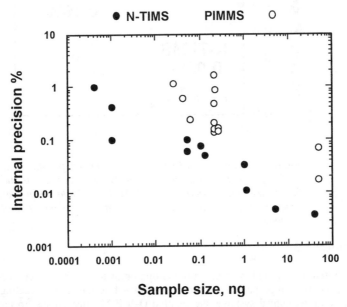

Figure 1 *Internal precision of analysis (% 2 standard deviation) vs. sample size (in ng) for Os standards run by N-TIMS, PIMMS.* [2, 14, 17, 19, 21, 20, 32, 34, 35, 36, 37]. *Data for sector-field ICP-MS with single collector included with PIMMS data*

Highly variable ambient Re levels in commercially produced Pt filament materials have led to much work in maintaining consistent filament Re-blanks, either by chemical cleaning with an agent such as $KHSO_4$ [23] or, elaborate alternatives such as V_2O_5 coated Ni filaments [26]. For Os isotope analysis, the need for clean sample chemistry combined with instrumental and filament requirements needed to avoid Re interferences leads to relatively low sample throughputs, with several hours of instrument time required per sample. Re analysis is more rapid but can still require one hour of instrument time per sample plus significant effort to obtain clean filaments.

External precisions of 0.025 % or better, required for $^{186}Os/^{188}Os$ measurements to be geologically useful are possible by N-TIMS using 20 to 50 ng samples (Figure 3) which commonly yield ^{186}Os signals of 100mV on Faraday cups (10^{-11} ohm resistor), of sufficient duration for 100 ratios to be collected by multi-dynamic measurement routines.

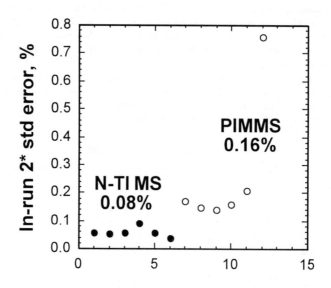

Figure 2 *Comparison of short-term external reproducibility of N-TIMS and PIMMS for Os solutions of between 100 and 250 pg (N-TIMS[39]) and 125 pg (PIMMS[35]). Data taken over a period of 1 to 2 days. Errors are 2 standard deviations of the mean*

4 Q-ICP-MS

Although limited in their ability to measure isotope ratios to <<0.3 % internal precision (2* standard error), the large parent-daughter elemental fractionations affecting the Re-Os decay scheme in geological systems has enabled Q-ICP-MS instruments to produce useful data [27]. Our experience of the Perkin Elmer Sciex Elan 6000 instrument using direct injection nebulisation for 0.5 ng of Os standard solution is that it is difficult to achieve external short term precisions of better than 3 %. While this is sufficient for some geological problems it is inadequate for many. Recently, Bandura et al., [28] have reported greatly improved isotope ratio precisions from a Q-ICP-MS system equipped with a dynamic reaction cell. The dynamic reaction cell creates collisional damping of ion density fluctuations in the ion beam which increase internal and external precisions of isotope ratio measurements. These authors report external precision for ^{207}Pb/^{206}Pb and ^{208}Pb/^{206}Pb on a 10 ppb solution of 0.022 to 0.034 % and 0.014 to 0.028 % RSD respectively. No experiments have been reported so far for Os isotope measurements using this type of system.

The major obstacle that has prevented wide-spread use of Q-ICP-MS for Os isotope studies is the sample to sample memory problem in the introduction system of the mass spectrometer [29-31]. As more analyte is introduced to the plasma, to increase signal and improve counting statistics, the sample to sample memory deteriorates such that any gain in internal precision are completely negated by compromised accuracy and poor external precision. Similarly, the signal intensity gains from introducing oxidised Os into the plasma [29] are countered by the increased memory of these species in the mass spectrometer system, requiring elaborate washing/cleaning. Pearson et al.[30] found that of current nebuliser designs, the direct injection nebuliser system offered by Cetac was optimal in producing minimal sample to sample memory, even for oxidised Os species but that this system, linked to standard Q-ICP-MS is limited to internal precisions of a few % on sub-ng sized Os samples.

5 PIMMS

The advent of magnetic sector ICP-MS instruments such as the Finnigan Element and multi-collector magnetic sector plasma mass spectrometers has increased the potential for more precise Os isotope ratio measurement using plasma sources. Peak switching tests using a secondary electron multiplier (SEM) on a variety of PIMMS instruments running in solution mode with an unradiogenic standard using micro-concentric nebulisers and total Os of between 250 to 500 pg of Os yield internal precisions of *c.* 0.5 to 0.8 % (2* standard error) for $^{187}Os/^{188}Os$. Using triple SEM measurement on c. 360 pg Os internal precisions of c. 0.5 % (2* standard error) were achieved and 0.3 % for c. 800 pg Os. Short-term external precision for runs of c. 100 pg of Os were 0.72 % (2*standard deviation). Carlson[32] used a VG Axiom instrument, in rapid-scan mode, with a single SEM to achieve 0.25% (2*standard deviation) for 6 runs of an Os standard solution, consuming 60 pg total Os per run and using a "standard" i.e., non-desolvating nebuliser and spray chamber. Such performance, at least for standard solutions, approaches that available by N-TIMS at this sample level (Figure. 2); however, long-term external reproducibility on a system routinely running samples and standards of varying isotopic compositions has not yet been reported.

The increased ionisation of Os when introduced in an oxidised form, i.e., OsO_4 in ICP sources has been known for some time [4,33]. Recently, Hassler et al., [34] and Schoenberg et al., [35] have used sparging and direct distillation respectively to introduce OsO_4 into plasma sources for isotope ratio measurement. Hassler et al.,[34] sparged Os into a Finnigan Element instrument equipped with single SEM. They obtained internal precisions of c 0.2% (2*standard errors; 0.91 % 2*standard deviation) for samples as small as 200 pg. External reproducibility of 80 pg to 1.2 ng standards over 6 months were 1.6 % (2*standard deviation). Using only 1.2 ng standards, external precision was improved to 0.46 % (2* standard deviation). In contrast, the same authors report long-term reproducibility, over a wide range of sample sizes, for N-TIMS of 0.28 % (2*standard deviation). Data acquisition time for the plasma measurements was 7 minutes and the ion yields are estimated to be c. 0.005 %.

Schoenberg et al [35] report in-run precisions of 0.21 to 0.15% (2*standard error) for standards, directly distilled into the plasma as OsO_4, of 200 to 250 pg in size, using a Nu instruments PIMMS equipped with triple ion counting and using a 2-step peak jump. Using 200 ng unradiogenic standards total beam intensities of $8*10^{-14}$ A over a 17 minute period are obtained, equating to total ion efficiencies of c. 0.8 %. Efficiency appears to be independent of sample size. A 25 pg standard gave an internal precision of 1.2%. Short-term external precision of standards in the range 25 to 250 pg Os was 0.16 % (2*standard deviation) which is comparable to the long-term external precision usually quoted for N-TIMS. External precision rose to 0.32 % (2*standard deviation) for standards that were spiked with ^{190}Os, similar to totally spiked geological samples. The same authors measured $^{186}Os/^{188}Os$ values on 50 ng standards achieving a value within error of the N-TIMS measurement and an external precision of 0.017 % (5 runs; 2*standard deviation) compared to an external precision of 0.0037 % reported for the same standard by N-TIMS. In both sparging and direct distillation methods, cleaning of the sample introduction system procedures and the use of new inlet tubing for each sample greatly reduce memory effects compared to solution analysis but lead to increased analysis time and greater operator demands. The direct distillation approach, distilling OsO_4 from a purified sample, is the more promising approach in that far less impurities are being transported into the mass spectrometer than with the sparging approach.

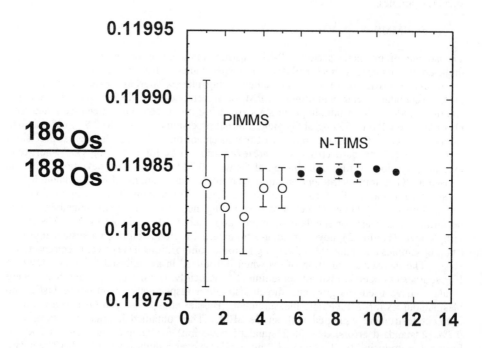

Figure 3 Comparative *precision of $^{186}Os/^{188}Os$ measurements for Os standard solutions by PIMMS[35] and N-TIMS[14] for similar total Os levels analysed (20-50 ng). External re-producibility (2*standard deviation of the mean) is 2.8 ppm for N-TIMS data and 21 ppm for PIMMS data*

Schoenberg et al [35] also report $^{186}Os/^{188}Os$ for 20 to 50 ng samples. At the moment external precisions for these measurements using PIMMS (2.8 ppm; 2*standard deviation), for similar sized Os total analyte levels, are a factor of 8-10 worse then for N-TIMS (21 ppm ; 2*standard deviation) and internal precision is approximately a factor of 10 worse (Figure 3). This is largely due to the much higher ionisation efficiencies of N-TIMS, providing much larger beam intensities. The c. 21 ppm precision obtained by PIMMS is adequate for $^{186}Os/^{188}Os$ studies of mantle derived samples, however, the much better external and internal reproducibility of N-TIMS means that sample sizes can be reduced by a factor of 10 compared to PIMMS, increasing the applicability of the technique to smaller sample sizes.

Gronvold et al [36] have recently reported high precision $^{187}Os/^{188}Os$ and $^{186}Os/^{188}Os$ measurements using triple ion counting PIMMS. These authors release Os from large weights (20g) of powdered rocks or minerals via furnace heating and trap OsO_4 into 1% nitric acid and then nebulise this solution, thereby effectively introducing Os as the tetraoxide and taking advantage of the high ionisation efficiency. There are no details of how memory problems are addressed with this approach. Quoted precisions are 2 ppm for both isotope ratios; however, given the problems of cross-calibrating electron multipliers this level of precision seems unlikely. In addition, from consideration of counting statistics alone, using 20 to 40 ng total Os, this procedure would require on the order of 100 % ionisation and instrument transmission to obtain the levels of precision quoted.

The use of direct injection nebulisation on PIMMS for Os isotope measurements is likely to enable low cross-contamination analysis of Os and may be particularly effective for oxidised Os [30] but this has yet to be fully evaluated or reported by multi-collector instrument users.

Hemming et al [37] increased internal and external precisions for PIMMS analysis of Os by a factor of c. 3 by internally spiking solutions with 100 ppb Ir. Internal normalisation of Os signals measured on a single SEM to the ^{191}Ir signal measured on faraday buckets gave short-term external precision of 0.36 % (2* standard deviation; n=7) for samples of 150 to <1000 pg Os. This compared with an external precision of 1.1 % for the same solutions when Ir normalisation is not used. A similar approach has been used for laser-ablation PIMMS analysis of Os in sulfides by Pearson et al [38] who introduce Ir as a dry aerosol into the ICP during ablation of sulfide minerals and used the $^{191}Ir/^{193}Ir$ ratio for mass bias correction. Introduction of Ir as a vapour, via a desolvating nebuliser produced more precise data than using indigenous Ir within the sulfide, probably because of the enhanced signal sizes using aspirated Ir. Precisions of c. 0.5 % (2* standard error) are obtained for sulfides with over 50 μg/g Os, giving total Os signals of > 0.01 V. These precisions are obtainable despite having to make substantial ^{187}Re corrections. Precise solution data has been obtained for these authors with $^{187}Re/^{188}Os$ ratios of up to 0.26, underscoring the value of PIMMS for this type of analysis.

The internal precision using laser ablation and ion-counting is c. a factor of 5 worse than that routinely available via N-TIMS with similar sized samples but is sufficiently precise to produce geologically useful data. The data is particularly useful in that it is spatially resolved and obtained without resorting to chemistry. Although internal precision is not quite as good as N-TIMS, external precision may be comparable because of the lower levels of molecular interferences.

6 SUMMARY

Recent advances in PIMMS combined with high sensitivity direct introduction methods for OsO_4 permit Os isotope analysis of many geological samples. The ability to directly measure some minerals, e.g., sulfides, using laser ablation adds an extra dimension to PIMMS, allowing detailed, spatially resolved studies of Os isotope variations within minerals that will greatly enhance our understanding of bulk-rock geochemistry and allow new problems to be addressed[38]. Internal and external precision via solution and even laser can approach or, in some cases equal conventional N-TIMS for measurement of $^{187}Os/^{188}Os$ on >100 pg samples, although long-term external precision estimates from instruments running samples on a daily basis are not yet available. Many recent geochemical studies have involved the measurement of << 10 pg of Os; such levels are still only routinely accessible to N-TIMS, and the option on multiple ion-counting, now available on some instruments, will increase sensitivity still further. Although external reproducibility for

^{186}Os/^{188}Os measurements at 50 ng levels, adequate for geological applications, has been recently demonstrated by PIMMS[35], the much greater sensitivity of N-TIMS for these measurements allows much broader applicability of the Pt-Os isotope system at present.

The growing distribution of PIMMS instruments around the world will make Os isotope analysis available to a much larger user-base and should diversify the application of this isotope system. Developments in PIMMS techniques and instrumentation over the next few years may allow comparable performance at all levels with N-TIMS, but for the present, both analytical approaches have their place for the geochemist interested in using Os isotopes as a tracer/chronometer in as many applications as possible.

7 REFERENCES

1   Creaser, R.A., Papanastassiou, D.A. and Wasserburg, G.J., *Geochimica et Cosmochimica Acta*, 1991, **55**, 397.

2   J. Volkening, T. Walczyk and K.G. Heumann *International Journal of Mass Spectrometry and Ion Processes* , 1991, **105**, 147.

3   Masuda, A., Hirata, T. and Shimizu, H., *Geochemical Journal*, 1986, **20**, 233.

4   Russ, G.P., Bazan, J.M. and Date, A.R., *Anal. Chem.*, 1987, **59**, 984.

5   Dickin, A.P., McNutt, R.H. and Andrew, J.L., *J. Anal. At. Spectrom.*, 1988, **3**, 337.

6   Shirey, S.B. and Walker, R.J., *Annu. Rev. Earth Planet. Sci.*, 1998, **26**, 423.

7   Esser, B.K., and Turekian, K.K., *Geochim. Cosmochim. Acta*, 1993, **57**, 3093.

8   Morgan, J.W., *J. Geophys. Res.*, 1986, **91**, 12375.

9 .   Pearson, D.G., Shirey, S.B., Harris, J.W. and Carlson, R.W., *Earth Planet. Sci. Lett.*, 1998, **160**, 311.

10   Sharma, M., Papanastassiou, D. and Wasserburg, G.J., *Geochim. Cosmochim. Acta*, 1997, **61**, 3287.

11   Levasseur, S., Birk, J.L., and Allegre, C.J., *Science*, 1998, **282**, 272.

12   Smoliar, M, Walker, R.J., Morgan, J.W., *Science*, 1996, **271**, 1099.

13   Walker, R.J. et al., *Geochimica et Cosmochimica Acta*, 1998, **61**, 4799.

14   Brandon, A.D., Norman, M.D., Walker, R.J., and Morgan, J.W., *Earth Planet. Sci. Lett.* 1999, **174**, 25.

15   Daley, N.R., and Rudley, R.G., *Nature*, 1964, **202**, 896.

16   Kawano, H. and Page, F.M., *Int. J. Mass Spectrom. Ion Physics*, 1983, **50**, 1.

17   Walczyk, T., Hebeda, E.H. and Heumann, K.G., *Fresenius J Anal Chem*, 1991, **341**, 537.

18   Birck, J.L., Roy-Barman, M. and Capmas, F., *Geostand. Newslett*, 1997, **20**, 19.

19   D. Tuttas, *Finnigan MAT Application News*, 1992, **No. 1**, 1.

20   Hauri, E.H. and Hart, S.R., *Earth Planet. Sci. Lett.*, 1993, **114**, 353.

21   Burnham, O.M., Rogers, N.W., Pearson, D.G., van Calsteren, P.W. and Hawkesworth, C.J., *Geochim. Cosmochim. Acta*, 1998, **62**, 2293.

22   Lambert, D.D., Foster, J.G., Frick, L.R., C.Li and Naldrett, A.J., *Lithos*, 1999, **47**, 69.

23   Shirey, S.B. *Pers. comm.*, 2000

24   Walker, R.J., Storey, M., Kerr, A.C., Tarney, J. and Arndt, N.T., *Geochim. Cosmochim. Acta*, 1999, **63**, 713.

25   Pearson, D. G., Shirey, S. B., Bulanova, G. P., Carlson, R. W., and Milledge, H. J., *Proc. 7th Int. Kimb. Conf.*, 1999, Eds J.J. Gurney, J.L. Gurney, M.D. Pascoe & S.H. Richardson, National Book Printers, Goodwood, S. Africa, 637.

26   Walczyk, T., Hebeda, E.H. and Heumann, K.G., *Int. J. Mass Spectrom Ion Proc.* 1994, **130**, 237.

27   Hulbert, L.J. and Gregoire, D.C., *Canadian Mineralogist*, 1993, **31**, 861.
28   Bandura, D.R., Baranov, V.I. and Tanner, S.C., *J. Anal. At. Spectrom.* 2000, **15**, 921.
29   Gregoire, D.C., *Anal. Chem.*, 1990, **62**, 141.
30   Pearson, D.G., Ottley, C.J. and Woodland, S.J., *In*: J.G. Holland and S.D. Tanner (Eds), Plasma mass spectrometry - New developments and applications. Royal Society of Chemistry, London, 1999, 2
31   Pearson, D.G. and Woodland, S.J., *Chem. Geol.*, 2000, **165**, 87.
32   Carlson, R.W., *Pers. comm.*, 2000.
33   Hirata, T., Akagi, T, Shimizu, H., and Masuda, A., *Anal. Chem.*, 1989, **61**, 2263.
34   Hassler, D.R., Peucker-Ehrenbrink, B. and Ravizza, G.E., *Chem. Geol.*, 2000, **166**, 1.
35   Schoenberg, R., Nagler, T.F. and Kramers, J.D., *Int. J. Mass Spectrom.*, in press.
36   Gronvold, K., Oskarsson, N, Sigurdsson, G. and Sverrisdottir, G., *J. Conf. Abstracts*, 2000, **5**, 462.
37   Hemming, N.G., Goldstein, S.L. and Fairbanks, R.G., *Ninth Ann. V.M. Goldschmidt Conference (Abst)*, 1999, LPI Contribution no **971**, Lunar & Planetary Institute, Houston, 1999, 120.
38   Pearson. N.J., Alard, O., Griffin, W.L. Graham, S. and Jackon, S.E., *J. Conference Abst.*, 2000, **5**, 777.
39   Burnham, O.M., PhD Thesis, Open University, 1995.

MATRIX-INDUCED ISOTOPIC MASS FRACTIONATION IN THE ICP-MS

Richard W. Carlson, Erik H. Hauri and Conel M.O'D. Alexander

Department of Terrestrial Magnetism, Carnegie Institution of Washington, 5241 Broad Branch Road, N.W., Washington, DC, 20015 USA

1 INTRODUCTION

Isotopic variation in natural materials, particularly that induced by mass fractionation during low-temperature chemical reactions, tends to be small, usually on the order of a few parts per thousand or less. In order to resolve natural isotopic variations a measurement technique must be capable of distinguishing isotopic variations in the sample from those that are introduced by the measurement procedure. Many aspects of laboratory analysis are capable of modifying the isotopic composition of natural samples. These include mass fractionation caused by precipitation, evaporation, or interaction with the sample container during sample collection and transport to the laboratory, followed by additional mass fractionation that can occur during chemical separation, particularly with the use of ion-exchange chromatography. A final stage of isotopic fractionation is introduced by the instrumentation employed for isotopic analysis. For example, thermal ionization mass spectrometry (TIMS) causes mass dependent fractionation ranging from parts per thousand to percent per AMU depending on the element, whereas ICP-MS can cause between circa 1% per AMU fractionation of heavy elements and 15% per AMU at light elements such as Li.

Several avenues are available to correct for instrument-induced mass fractionation. The one capable of highest precision in isotope ratio determination is applied when the element of interest has more than two stable isotopes. Using an estimate of the "true" isotope ratio of one stable isotope pair, the difference between the measured and expected values of this normalizing isotope ratio is used to determine the instrumental mass fractionation factor. This factor can then be applied to correct the other measured isotope ratios of the element assuming some mass dependency law for the fractionation, usually either power or exponential in mass [1]. This "internal" mass fractionation correction provides isotope ratio measurement precision of up to 0.001% and is primarily responsible for allowing application of a number of radioactive decay schemes to geochronology and isotope geology (e.g. Rb-Sr, Sm-Nd, Lu-Hf). The drawbacks to this approach are two-fold. First, not all elements of interest have more than two stable isotopes. This is particularly true at light mass (e.g. Li, B, and C), but also true of one of the most widely used radiometric geochronometers, the U-Th-Pb isotopic system. Of the four stable isotopes of Pb, three are the product of radioactive decay and vary greatly in abundance in nature. Secondly, using the internal method to correct for instrumental mass fractionation also removes natural mass dependent isotope fractionation that may be present in the sample. In many cases, such natural mass fractionation is the "signal" needed to trace a wide variety of geologic processes.

Where internal fractionation correction cannot be done, the traditional approach, as applied for decades in gas-source mass spectrometry, is to compare the sample isotopic compositions to bracketing measurements of a laboratory standard. In ICP-MS applications, this approach has proven particularly suitable for the light elements (e.g. Li, B, Mg; [2-4]) where relative mass differences are sufficiently large to press the accuracy of the fractionation law used for correction.

Another approach, unique to ICP-MS applications, involves spiking the sample solution with an element of similar mass [5]. The expectation of a simple mass dependent, element invariant mass fractionation in the ICP ion source then allows the measured instrumental fractionation of the spike element to be applied to the element of interest in much the same way as the internal correction procedure. Current examples of this approach include: Tl to correct for Pb (e.g. [5]; [6]), Cu for Zn (and visa-versa - [7]), and Pd to correct for Ag [8]. These procedures provide isotope ratio precision on the order of 0.01%, roughly a factor of 10-20 improvement compared to previous TIMS techniques for these elements. However, using either standard-bracketing or external fractionation correction, the precision obtained on individual measurements, primarily representing counting statistics, often are a factor of 5 to 10 better than the reproducibility of repeat measurements. This suggests that some additional error source is present in these approaches.

We discuss here the evidence that the instrumental mass fractionation present in the ICP ion source indeed contains an element dependent term leading to matrix dependence for the accuracy of isotope ratios measured either by standard-sample bracketing or external normalization. Furthermore, the element dependence of the matrix effect is not purely related to the amount of matrix elements present, but also to which elements are present. Different matrix elements are shown to enhance or reduce "normal" instrument mass fractionation leading to lessened precision in isotope ratio measurements.

2 "TUNING" INDUCED MASS FRACTIONATION

Mass fractionation in the ICP ion source primarily is due to space charge induced scattering when the plasma passes through the cones and electrons are separated from ions in the ion-extraction/acceleration optics of the mass spectrometer [9, 10]. However, changes in instrument tuning parameters, such as torch position, cone cleanliness, gas flows, and extraction lens voltages, introduce second-order changes in mass fractionation that are not purely mass dependent, and hence, not clearly related to space charge alone. An example of this effect in Cu and Zn isotopic measurement was described in detail by [7].

As a preface to our discussion of matrix-induced fractionation, we offer a similar example to that discussed by [7], but in this case for analyses when Pd is used to correct for the instrumental mass fractionation experienced by Ag. For all the Pd-Ag results discussed in this paper, a 1N HNO_3 solution containing approximately 100 ppb Pd and 50 ppb Ag was introduced into a VG-Elemental P54 multicollector magnetic sector ICP-MS using a CETAC MCN6000 desolvating nebulizer and microconcentric nebulizer with an uptake rate of approximately 70 µl per minute. This solution produced ^{106}Pd and ^{107}Ag beams of between 3-6 x 10^{-11} amps. The faraday cups were set to simultaneously monitor the 104, 105, 106, 108 and 110 masses of Pd, along with ^{107}Ag, ^{109}Ag and the potential isobaric interferences caused by Cd and Ru (monitored at ^{111}Cd and ^{102}Ru). No detectable Cd or Ru signals were present for the measurements discussed here. Isotopic ratios were determined by simultaneous detection, with analyses consisting of 90 ratio measurements of 5 seconds integration each. Correction for mass fractionation in all cases assumed an exponential mass dependence to the fractionation [1].

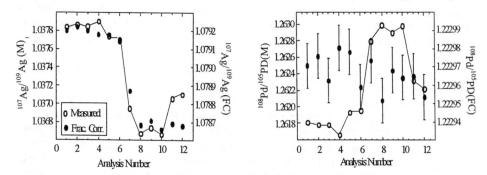

Figure 1 *Results for consecutive analyses of a solution containing 100 ppb NIST SRM 3138 Pd and 50 ppb NIST 978a Ag standards. The left panel shows the measured $^{107}Ag/^{109}Ag$ (open symbols - scale on left) and the Ag isotopic composition after correction for fractionation (closed symbols - scale on right) using Pd. The right panel shows both the measured $^{108}Pd/^{105}Pd$ during this analysis (open symbols - scale on left) and the fractionation corrected $^{106}Pd/^{105}Pd$ (closed symbols - scale on right). Note the varying scale ranges for the different isotope ratios, and particularly that the total range shown for $^{106}Pd/^{105}Pd$ approaches the precision of the individual runs, as shown by the error bars on the points*

    The fact that the instrumental mass fractionation is not purely mass dependent is shown clearly during a "typical" daylong analytical session measuring a single mixed Pd-Ag standard solution (Figure 1). In this particular experiment, the measured $^{107}Ag/^{109}Ag$ drifted by 0.12% (0.06% per AMU) and the simultaneously measured $^{108}Pd/^{105}Pd$ drifted sympathetically, albeit only over a range of 0.04% per AMU. Using the measured $^{108}Pd/^{105}Pd$ to correct for fractionation, the total range in fractionation corrected $^{106}Pd/^{105}Pd$ is approximately 0.002%. Clearly, the mass dependency of the Pd fractionation is being reasonably well accounted for since the precision obtained on $^{106}Pd/^{105}Pd$ approaches that determined by counting statistics alone. However, while the Pd-based correction for Ag mass fractionation removes about half of the drift in measured $^{107}Ag/^{109}Ag$, the fractionation corrected $^{107}Ag/^{109}Ag$ still shows a residual variation of 0.05%, roughly 25 times that displayed by the fractionation corrected $^{106}Pd/^{105}Pd$. The shape of the variation in fractionation corrected $^{107}Ag/^{109}Ag$ parallels that of the uncorrected Ag isotopic composition, indicating that the Ag is experiencing a slightly different isotopic fractionation than is Pd.

    Over longer time periods, these differential changes in fractionation can be more severe. Figure 2 shows various aspects of the results of over 1-year of sporadic measurements of the same mixed Pd-Ag solution. Correcting for mass fractionation using $^{108}Pd/^{105}Pd$ and a purely mass-dependent exponential law [1] results in a 2σ precision for $^{107}Ag/^{109}Ag$ of +/- 0.075%. Surprisingly, a similar reproducibility ($2\sigma = 0.082\%$) is achieved if the Ag is not corrected using Pd, but is instead normalized by the average measured value of bracketing standards. In other words, the sample to sample changes in fractionation that occur over short (i.e. < 30 minute) analytical intervals are approximately of the same size as the changes that occur between the fractionation experienced by Ag and by Pd over longer intervals. On the short time scale, however, changes in Pd fractionation do track those experienced by Ag. Consequently, by first using Pd to correct the mass fractionation of Ag, and then normalizing to bracketing standards also corrected in this manner, the long-term precision of the fractionation corrected $^{107}Ag/^{109}Ag$ can be reduced to 0.013%. This result indicates that external correction using Pd does track the fractionation experienced by Ag during short periods, i.e. the time needed to run a sample with bracketing standards. On the longer time scale, fractionation of Pd differs from that experienced by Ag, but most of this differential fractionation can be corrected for by normalizing sample measurements to bracketing measurements of the standard.

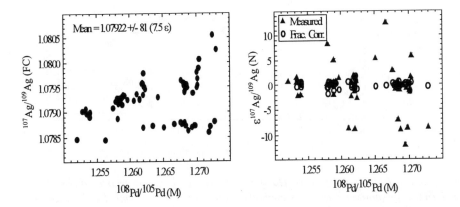

Figure 2 *Results from over 1 year of sporadic measurements of the NIST Pd-Ag standard solution. Left panel shows the range in Ag isotopic composition after correction for fractionation using Pd, only. Triangular symbols in the right panel show the measured Ag isotopic compositions (not corrected using Pd) normalized to bracketing measurements of the same Ag standard (difference in $^{107}Ag/^{109}Ag$ expressed in ε-units – parts in 10,000). These data have a 2σ precision of +/- 0.082%. The open circles in the right panel represent Ag data that were first corrected using Pd, then normalized to bracketing measurements (also corrected using Pd) of the Ag standard. The 2σ precision of this data set is +/- 0.013%*

3 MATRIX INDUCED FRACTIONATION

Since the measurements shown in Figure 2 were done with clean Pd-Ag solutions, the results do not reflect the added uncertainty that may be caused by matrix-induced, element-specific, effects. The significance of matrix effects for differential fractionation of Pd and Ag can be clearly demonstrated. Figure 3 shows the results for measurements of the same Pd-Ag solution to which varying amounts of Ir were added. The overall drift in fractionation corrected Ag isotopic composition with time reflects the instrument-induced differential Pd-Ag fractionation discussed in the preceding section. Of more concern is the fact that addition of Ir to the Ag solution causes roughly a 0.01% increase in the fractionation corrected $^{107}Ag/^{109}Ag$. The fact that the offset in Ag isotopic composition is independent of the amount of Ir added shows that the affect is not caused by an isobaric interference proportional to the amount of Ir in solution. The constant offset in Ag isotopic composition in spite of the widely varying Ir concentrations investigated also suggests that the effect is not caused by space-charge scattering, since this process should vary in proportion to the intensity of the overall ion beam entering the mass spectrometer. Because the intensity of the ion beam entering the mass spectrometer is dominated by Ar^+ from the plasma, it is likely that the analyte ions do not contribute significantly to space-charge effects at the concentration levels of these experiments.

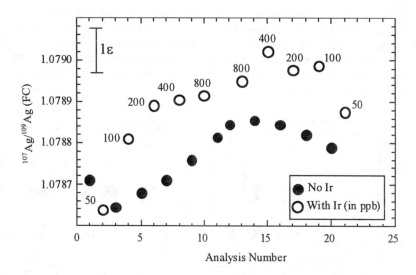

Figure 3 *Pd fractionation corrected Ag isotopic composition measured for the NIST standard Pd-Ag solution with (open symbols) and without (closed symbols) the addition of Ir. Amount of Ir added is listed beside each point in ppb in the analyzed solution*

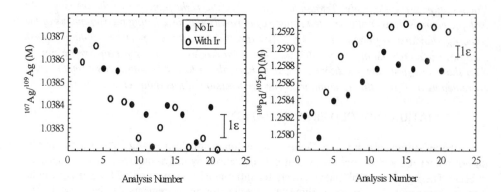

Figure 4 *Measured Ag and Pd isotopic composition, without fractionation correction, for the analyses shown in figure 3. Individual points have a measurement precision of approximately 0.0012%*

Figure 4 explores how the Ir addition causes the offset in Pd-corrected Ag isotopic composition. Addition of Ir causes no resolvable variation in measured Ag isotopic composition, but does change the fractionation experienced by Pd as shown by the offset in measured $^{108}Pd/^{105}Pd$ with Ir addition. This "extra" fractionation induced in Pd is then translated into the Ag isotopic composition through the mass fractionation correction. Surprisingly, although the Ir addition is most clearly expressed in the isotopic composition of Pd and not at all in Ag, Ir addition at all concentration levels studied suppresses the Ag ion current by approximately 25% while only slightly (6%) enhancing that of Pd. If this matrix effect were the result of charge exchange between Ir and Pd-Ag in the plasma, one might expect that the magnitude of change of mass fractionation would track that of signal suppression/enhancement, yet this is not the case in the Ir-Pd-Ag experiment.

That the matrix effect is dependent on the element that constitutes the matrix can be demonstrated through the results of two additional experiments. Figure 5 shows that the addition of Sm or Rb causes the Pd-corrected Ag isotopic composition to increase by approximately 0.01%, similar to that observed when Ir was added. This offset in Ag isotopic composition, however, is the result of different phenomena depending on whether Sm or Rb constitutes the added matrix element. Opposite to the effect observed with Ir addition, addition of Rb has essentially no effect on the measured Pd isotopic composition, but causes the measured $^{107}Ag/^{109}Ag$ to increase by approximately 0.01%. In contrast, Sm added to the solution causes the measured $^{107}Ag/^{109}Ag$ to decrease, but leads to an increase in the measured $^{108}Pd/^{105}Pd$ of sufficient magnitude to compensate for the decrease in measured $^{107}Ag/^{109}Ag$, causing the Pd-corrected Ag isotopic composition to increase.

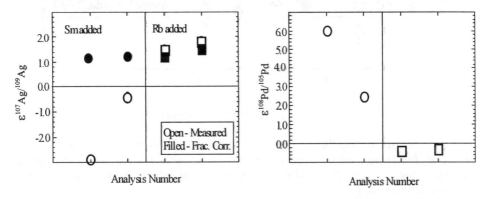

Figure 5 *NIST Pd-Ag solution to which Sm or Rb were added to produce a ^{152}Sm signal of 6 x 10^{-11} amps or $^{87}Rb = 9$ x 10^{-11} amps compared to ^{105}Pd and ^{107}Ag signal sizes of approximately 4 x 10^{-11} amps. Left panel shows Ag isotopic composition, normalized to bracketing standards not containing Sm or Rb, both before (open symbols) and after (filled symbols) fractionation correction using Pd. Right panel shows the measured Pd isotopic composition normalized only to bracketing standards. All isotopic compositions are expressed in ε-unit differences (parts in 10,000) compared to the average of bracketing standard measurements*

A similar element dependency to the matrix effect can be demonstrated with addition of various matrix elements to a standard solution of Mg. In this experiment, a solution (1 N HNO_3) containing Mg extracted from olivine crystals removed from a mantle xenolith from San Carlos, Arizona, was introduced into the TJA Solutions Axiom multicollector ICP-MS using either a CETAC Aridus desolvating nebulizer or a Glass Expansion Meinhard-style nebulizer and combination cyclonic-impact spray chamber. The concentration of Mg in the solution was adjusted (approximately 1 ppm with desolvating nebulizer, 5 ppm for Meinhard nebulizer) to obtain 4-6 x 10^{-11} amps of ^{24}Mg. Signals for the 3 isotopes of Mg were collected simultaneously in faraday cups with individual measurements consisting of 60 signal integrations of 5 seconds each. Using this technique, repeat measurements of the standard solution alone, without matrix element added, show a reproducibility of +/- 0.025% in

$^{26}Mg/^{24}Mg$ when the data are corrected by dividing by the mean of bracketing measurements of the same solution.

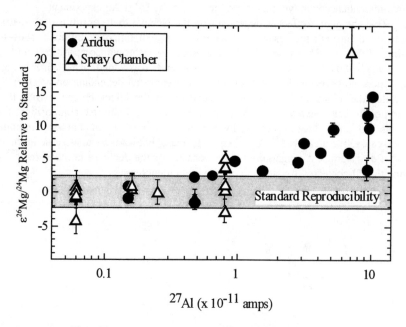

Figure 6 *Measured $^{26}Mg/^{24}Mg$ expressed in ε-unit differences (parts in 10,000) from the mean of bracketing standard measurements. The dotted band shows a standard reproducibility of +/- 0.025%. Al was added to the Mg solution to produce the ^{27}Al ion currents shown on the x-axis*

When Al is added to the Mg solution, there is an increase in the measured $^{26}Mg/^{24}Mg$ (Figure 6). This increase in $^{26}Mg/^{24}Mg$ is correlated with the amount of Al added so that by the time Al reaches approximately 2 ppm in solution, the $^{26}Mg/^{24}Mg$ is increased by approximately 0.1 to 0.2%. A similar offset can be seen in the data taken with both desolvating and Meinhard nebulizers indicating that the effect does not originate in the desolvating membrane. That the increase in $^{26}Mg/^{24}Mg$ reflects mass fractionation of the Mg and not an isobaric interference is shown by the fact that the $^{25}Mg/^{24}Mg$ increases by half that of $^{26}Mg/^{24}Mg$.

To investigate how this matrix effect depends on the matrix element, the monoisotopic (or nearly so in the case of V) transition metals were added to the solution, in part because the transition metals are potential contaminants in natural Mg-rich materials. The monoisotopic transition metals were used to avoid possible isobaric interferences from doubly charged species. Figure 7 shows that addition of Sc and Mn to the Mg solution causes $^{26}Mg/^{24}Mg$ to increase at the same rate as does Al, so that by the time the matrix element is present at signal sizes approaching 10^{-10} amps, $^{26}Mg/^{24}Mg$ is between 0.1% and 0.2% higher than in the matrix-free analyses. In contrast, V and Co have minimal affect on Mg fractionation below signal sizes of about 6×10^{-11} amp, but then cause $^{26}Mg/^{24}Mg$ to deviate to lower values at higher matrix element concentration. The changes in mass fractionation are accompanied by changes in the Mg signal intensities upon addition of the matrix elements.

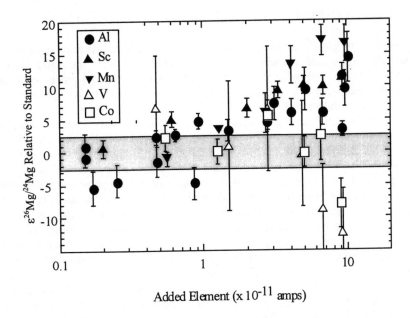

Figure 7 *Measured $^{26}Mg/^{24}Mg$ relative to bracketing standards with addition of the various matrix elements listed on the legend. The dotted band shows standard reproducibility for a clean Mg standard solution. ^{26}Mg signals for these measurements ranged from 3-6 x 10^{-11} amps.*

The explanation for the differing matrix effects is unclear. The phenomena is not due solely to mass, though heavy elements are known to suppress signals from lighter ions due to space charge repulsion [10]. No connection exists between magnitude or direction of the effect and the ionization potential and electron-shell configuration of the different matrix elements, suggesting that the phenomena is not caused by ion-molecule reactions in the plasma. No clear connection exists with the vaporization temperature or vaporization enthalpy of the various matrix elements, but this possibility depends strongly on the chemical species that is being evaporated. For example, Co and V both have substantially higher enthalpies of vaporization, and slightly higher vapor temperatures than do Sc, Mn and Al as metals, but as oxides, Sc, Al and V have much higher enthalpies of vaporization than does Mn. Similarly, Ir and Sm_2O_3 have higher, and Rb lower, enthalpies of evaporation than does Ag, whereas as metals, both Sm and Rb have lower evaporation enthalpies compared to either Ag or Pd. Differences in vapor point between the matrix and analyte elements potentially could change the position of the initial vaporization of the sample in the plasma, as suggested by [10] for more general matrix effects. If due to differences in the vapor temperature of the matrix and the analyte elements, then the observed change in mass fractionation might reflect a positional sensitivity of the ion extraction from the plasma with the fractionation caused either by incomplete evaporation or by a space charge induced expansion of the ion cloud in the plasma [10].

4 CONCLUSIONS

Isotope ratio determinations using the ICP-MS are affected by matrix elements in the analyte. Besides affecting signal intensities, matrix elements also alter the mass fractionation experienced by the analyte element. When internal correction for mass fractionation is possible, the consequences of the matrix effects can be eliminated and isotope ratio precisions approaching those determined purely by counting statistics and other machine parameters, for example amplifier and faraday gain stability, can be achieved. When internal correction is not possible, matrix effects can limit isotope ratio precision. Matrix elements introduce element dependent (i.e. not purely mass dependent) mass fractionation of as much as 0.01 to 0.03% per AMU for heavy elements such as Ag and Pb [11] and as much as 0.05 to 0.1% at Mg. Consequently, the matrix effects cannot be completely eliminated by external fractionation correction since the element added to determine the mass fractionation factor may not experience the same degree of mass fractionation experienced by the analyte element. Nevertheless, where Pd is used to correct for the mass fractionation experienced by Ag, the combination of external fractionation correction and normalization to bracketing standard analyses can improve the precision of Ag isotopic determination by approximately a factor of 6 over that obtained without using the external normalization.

The cause of the matrix effects is not obvious. For Pd-Ag, the demonstrated matrix effects are dependent on the matrix element, but are not dependent on the concentration of the matrix element. This suggests that space charge dispersion of the ions is not responsible for the element-dependent fractionation. At Mg, the matrix effects again are only mildly dependent on the concentration of the matrix element, but are apparently dependent on the vaporization point of the matrix element. This observation might suggest that the matrix elements change the position of vaporization of the sample element in the plasma [10] and that ion extraction into the mass spectrometer is sensitive to this phenomenon. Identification of the ultimate causes of the matrix effects we have observed will require a substantial effort. Until such matrix effects can be quantified in a predictive model, this study makes clear that high-precision isotope ratio determinations by ICP-MS require the measurement of clean analyte solutions with a minimum of contaminating elements.

5 REFERENCES

1.   W.A. Russell, D.A. Papanastassiou and T.A. Tombrello, Ca isotope fractionation on the Earth and other solar system materials, *Geochim. Cosmochim. Acta* **42**, 1075-1090, 1978.

2.   P.B. Tomascak, R.W. Carlson and S.B. Shirey, Accurate and precise determination of Li isotopic compositions by multi-collector sector ICP-MS, *Chem. Geol.* **158**, 145-154, 1999.

3.   H.-E. Gabler and A. Bahr, Boron isotope ratio measurements with a double-focusing magnetic sector ICP mass spectrometer for tracing anthropogenic input into surface and ground water, *Chem. Geol.* **156**, 323-330, 1999.

4.   L.-A. Nguyen, C.M.O.D. Alexander and R.W. Carlson, Mg isotope variation in bulk meteorites and chondrules, *Lunar Planet. Sci.* **31**, 2000.

5.   A.J. Walder, I. Platzner and P.A. Freedman, Isotope ratio measurement of lead, neodymium and neodymium-samarium mixtures, hafnium and hafnium-lutetium mixtures with a double focusing multiple collector inductively coupled plasma mass spectrometer, *J. Anal. At. Spectrom.* **8**, 19-23, 1993.

6.   M. Rehkamper and A.N. Halliday, Accuracy and long-term reproducibility of lead isotopic measurements by MC-ICPMS using an external method for correction of mass discrimination, *Int. J. Mass Spectrom. Ion Proc.* **181**, 123, 1998.

7.   C.N. Marechal, P. Telouk and F. Albarede, Precise analysis of copper and zinc isotopic compositions by plasma-source mass spectrometry, *Chem. Geol.* **156**, 251-273, 1999.

8.   R.W. Carlson and E.H. Hauri, Extending the 107Pd-107Ag Chronometer to Low Pd/Ag Meteorites with the MC-ICPMS, *Geochim. Cosmochim. Acta*, in review, 2000.

9.   S.D. Tanner, in: Plasma Source Mass Spectrometry: Developments and Applications, G.P. Holland and S.D. Tanner, eds., pp. 13-27, Royal Society of Chemistry, Cambridge, 1997.

10. J.W. Olesik, I.I. Stewart, J.A. Hartshome and C.E. Hensman, Sensitivity and matrix effects in ICP-MS: Aerosol processing, ion production and ion transport, in: Plasma Source Mass Spectrometry: Developments and Applications, G.P. Holland and S.D. Tanner, eds., pp. 3-19, Royal Society of Chemistry, Cambridge, 1999.

11. M.F. Thirwall, Precise Pb isotope analysis of standards and samples using an Isoprobe multicollector ICP-MS: Comparisons with doublespike thermal ionization data, in: 7th International Conference on Plasma Source Mass Spectrometry, abstract, Durham, 2000.

6 ACKNOWLEDGEMENTS

The multiple collector ICP-MS's at DTM were purchased with funds provided by National Science Foundation grant EAR-9724409 and the Carnegie Institution of Washington. We are greatly appreciative of the installation, training, and continued technical support provided for the ICPs by Ian Bowen and Steve Guilfoyle. On a day to day basis, Tim Mock ensures that the instruments are running in peak condition and Mary Horan assists in many aspects of the accompanying work in the chemistry laboratory. The work described in this paper could not have been done without their assistance.

SIMULTANEOUS ACQUISITION OF ISOTOPE COMPOSITIONS AND PARENT/DAUGHTER RATIOS BY NON-ISOTOPE DILUTION SOLUTION-MODE PLASMA IONISATION MULTI-COLLECTOR MASS SPECTROMETRY (PIMMS)

G.Nowell[1,2] and R.R.Parrish[1]

NERC Isotope Geosciences Laboratory[1], Wills Memorial Building, Nicker Hill, Keyworth, NG12 5GG, UK.

Dept. Earth Sciences[2] , University of Bristol, Bristol, BS8 1RJ, uk

1 INTRODUCTION

Precise and accurate daughter isotope compositions (< a few 10's of ppm) and parent/daughter ratio (1-5‰) measurements are prerequisite for high precision geochronology and also those isotope tracer studies where a correction for *in situ* radiogenic ingrowth of the daughter isotope needs to be applied.

The conventional method for obtaining such high precision daughter isotope compositions and parent/daughter ratios has been isotope dilution - thermal ionisation mass spectrometry (ID-TIMS). Although very precise, ID-TIMS is time consuming, both in terms of sample preparation and mass spectrometric analysis and the precision of the measurements is dependant upon optimising spike addition in order to minimise the error amplification factor and ensuring complete spike-sample isotopic equilibrium. Nevertheless, if these conditions can be satisfied then isotope dilution almost certainly provides the highest precision on both isotope compositions and parent/daughter ratios.

The advent of Plasma Ionisation Multi-Collector Mass Spectrometry (PIMMS) in the early-mid 1990's[1-2] has allowed for the development of alternative and more rapid methods for the acquisition of daughter isotope compositions and parent/daughter ratios. As the plasma source operates at steady-state, with the instrumental mass fractionation factor for neighbouring elements varying in a consistent way and mass fractionation being constant for a given instrument set-up, it has become possible to run mixed parent-daughter solutions and measure isotopic compositions and parent/daughter ratios simultaneously to a precision similar to ID-TIMS[3]. The main advantage of the simultaneous ID-PIMMS of Luais et al[3] over the more conventional ID-TIMS measurements is the reduction of sample preparation to a single digestion and column separation and the need for only a single PIMMS analysis. However, although applicable for Sm-Nd measurements, this simultaneous ID-PIMMS technique is less appropriate for beta-decay radiogenic parent-daughter isotope pairs, such as Lu-Hf or Rb-Sr, where there are only two parent isotopes, one of which isobarically overlaps the radiogenic daughter isotope.

Furthermore, by coupling a laser ablation system to a PIMMS (LA-PIMMS) it has also become feasible to acquire isotope compositions and parent/daughter ratios in situ.[4-5] Although such an in-situ method requires no sample preparation and is extremely rapid it is not an all encompassing technique as its application is restricted to mineral phases with sufficient concentrations of the elements of interest, generally >100's ppm. Furthermore, whereas it is possible to obtain high precision isotope compositions by laser ablation there is usually a sacrifice in the precision and reproducibility of the parent/daughter ratio. For example, in a study of Hf in kimberlitic zircons, Griffin et al[5] reported a typical analysis precision of 1-2% (2SE) and obtained a reproducibility on the $^{176}Lu/^{177}Hf$ ratio for the Nancy 91500 standard zircon of only 16% (2SD, n=60). This is despite the fact that the 91500 standard 'appears to be a reliable standard material; it is quite homogenous in terms of Yb and Lu contents, as well as in isotopic composition'.[5] The best reproducibility that has been obtained on parent/daughter ratios by laser ablation is for $^{238}U/^{206}Pb$ at ~1.5 - 2% (2SD)[6-7] and it is likely that this is close to the best that will be possible by laser ablation.

Here we describe a new solution-based PIMMS procedure developed at NIGL for rapid and simultaneous acquisition of isotope compositions and parent/daughter ratios which does not require the utilisation of mixed isotopic spikes, as in the Luais et al[3] technique. Its main advantages are that sample preparation is much reduced, precision on the isotope and parent/daughter ratio measurements can approach that obtainable by ID-TIMS, yet the precision is not dependant on optimising spike addition or obtaining spike-sample isotopic equilibrium. Furthermore, chemistry permitting it should be applicable to most parent-daughter isotope pairs. This paper represents the first real alternative to isotope dilution methods for obtaining such high precision measurements.

2 ANALYTICAL PROCEDURES

The new analytical procedure described below was initially developed for the simultaneous measurement of $Hf_{IC}+Lu/Hf_R$ and $Nd_{IC}+Sm/Nd_R$ ratios (where the subscripts IC and R refer to isotopic composition and isotope ratio respectively) for zircon and monazite samples respectively as these are routinely processed at NIGL for U-Th-Pb dating. The main emphasis of this work is the mass spectrometric analysis procedure and so the chemical separation is described only briefly.

2.1 Chemistry

As the use of mixed parent-daughter isotopic spikes is avoided in this new analytical procedure it is essential not to induce any fractionation of the parent/daughter ratio prior to analysis. It is therefore necessary to ensure quantitative recovery of Lu & Hf and/or Sm & Nd from the sample dissolution and column chemistry.

Fractionation of Lu/Hf during sample dissolution and chemical separation is easily induced[8] so we employed measures to minimise/eliminate any fractionation. After hydrothermal dissolution of zircon in a bomb at 240C for 36h in 29N HF the sample is loaded from the Teflon dissolution capsule onto the U-Pb anion exchange column in 3.1N HCl. As Hf is prone to 'sticking' to any surfaces of the capsule and/or column without the presence of trace amounts of hydrofluoric acid it is a distinct possibility that fractionation may occur at this stage with some Hf remaining in the capsule as no HF is used in the column-loading, having been previously evaporated. When the eluant containing the Lu-Hf is removed from the column it is done so into the original dissolution capsule. The U-Pb separation is then performed before the dissolution capsule is replaced beneath the column and a 29M HF wash (which will remove any residual Hf from the column) is collected. This is then dried and converted to a 2% HNO3-0.1M HF solution prior to analysis. We are currently in the process of verifying that this procedure eliminates Lu-Hf fractionation.

The collection of Sm-Nd from monazite U-Pb chemistry is straightforward as the bulk rare earth elements (REE) are collected in the first column wash, as a group, with no retention whatever in anion resin in 3.1 N HCl. After collection the bulk REE wash is dried down and taken up in 2% HNO3 for analysis.

Although we have not tried processing whole-rock samples we anticipate that this might be a relatively straightforward single cation exchange column separation for Sm-Nd. However, it will doubtless prove difficult for Lu-Hf because of the ease in inducing fractionation both during the dissolution and column separation stages.

2.2 Mass spectrometry

All the analytical development work described below was carried out on the NIGL VG Elemental Plasma 54 PIMMS instrument. A CETAC MCN6000 was used for sample/standard introduction with a sensitivity of approximately 100V/ppm for Hf and Nd at an uptake rate of approximately 0.05ml/min. A typical analysis of ca. 15mins consumed between 75 and 150ng of Hf or Nd and gave a total signal intensity of 10 - 20V.

Data were collected in the static multi-collection mode i.e. the isotopes of interest (up to a maximum of 9) were assigned a unique faraday cup and were collected simultaneously in a single acquisition sequence. The collector configurations used for $Hf_{IC}+Lu/Hf_R$ and $Nd_{IC}+Sm/Nd_R$ analyses are shown in Tables 1 and 2 respectively. Each analysis consisted of five blocks of twenty cycles, each cycle having a five second integration time. Baseline measurements and peak centring routines were performed at

Table 1 *Faraday cup configuration used for Lu-Hf analyses. Monitor peaks used for making interference corrections are shown in bold*

Cup No:	Low 4	Low 3	Low 2	Low 1	Axial	High 1	High 2	High 3	High 4
Isotope	^{172}Yb	^{173}Yb	**^{175}Lu**	^{176}Hf	^{177}Hf	^{178}Hf	^{179}Hf	^{180}Hf	-
Overlap 1				^{176}Lu					
Overlap 2				^{176}Yb					

Table 2 *Faraday cup configuration used for Sm-Nd analyses. Monitor peaks used for making interference corrections are shown in bold*

Cup No:	Low 4	Low 3	Low 2	Low 1	Axial	High 1	High 2	High 3	High 4
Isotope	^{142}Nd	^{143}Nd	^{144}Nd	^{145}Nd	^{146}Nd	**^{147}Sm**	^{148}Nd	**^{149}Sm**	^{150}Nd
Overlap	^{142}Ce		^{144}Sm				^{144}Sm		^{150}Sm

the start and between each block. After peak-stripping (see below) measured Hf and Nd isotopic ratios were exponentially corrected for instrumental mass bias using $^{179}Hf/^{177}Hf = 0.7325^9$ and $^{146}Nd/^{145}Nd = 2.071943$ (equivalent to a $^{146}Nd/^{144}Nd$ ratio of 0.7219)[10] respectively.

The two most critical aspects to the mass spectrometric analysis which must be satisfied in order to make precise and accurate Hf_{IC+Lu}/Hf_R and Nd_{IC+Sm}/Nd_R measurements are described in detail.

2.2.1. Corrections for isobaric overlaps   As Lu (and Yb) and Sm (and Ce) are not separated from Hf and Nd respectively, it is necessary to accurately correct for isobaric overlap between ^{176}Lu, ^{176}Yb and ^{176}Hf and between ^{142}Ce, $^{144,148,150}Sm$ and $^{142, 144,148,150}Nd$ respectively.

For simultaneous Lu-Hf measurements the Lu and Yb overlaps must be stripped away from the Hf, both to obtain the correct $^{176}Hf/^{177}Hf$ and $^{176}Lu/^{177}Hf$. The isobaric overlap of Lu and Yb on Hf is on the ^{176}Hf daughter isotope peak. Therefore any ratio between ^{176}Hf and a stable Hf isotope is unknown preventing the stripping of ^{176}Hf away from the ^{176}Lu and ^{176}Yb peaks. Corrections for isobaric interferences from ^{176}Lu and ^{176}Yb on ^{176}Hf are hence made by monitoring the intensities of the interference-free stable ^{175}Lu and ^{173}Yb isotopes. By knowing the natural $^{176}Lu/^{175}Lu$ and $^{176}Yb/^{173}Yb$ ratios appropriate ^{176}Lu and ^{176}Yb intensities can be subtracted from the $^{176}total$ beam to yield the ^{176}Hf and ^{176}Lu intensities. True $^{176}Lu/^{175}Lu$ and $^{176}Yb/^{173}Yb$ ratios were determined on the P54 by analysing two sets of the international Hf standard JMC 475; one set with varying Lu/Hf, the second with varying Yb/Hf. $^{176}Lu/^{175}Lu$ and $^{176}Yb/^{173}Yb$ ratios were then calculated from the inverse of the slope of a regression through $^{175}Lu/^{177}Hf$ and $^{173}Yb/^{177}Hf$ versus $^{176}Total/^{177}Hf$ (all ratios corrected for mass bias using $^{179}Hf/^{177}Hf$). The 'natural' ratios so derived are given in Table 3 and are either identical to or within error of published values ($^{176}Lu/^{175}Lu = 0.02669$[10] and 0.02652[9]; $^{176}Yb/^{173}Yb = 0.788 \pm 0.015$[10] and 0.7938[4]).

Table 3 *Natural $^{176}Lu/^{175}Lu$ and $^{176}Yb/^{173}Yb$ isotope ratios used for isobaric overlap corrections from Lu and Yb on Hf*

$^{176}Lu/^{175}Lu$	$^{176}Yb/^{173}Yb$
0.02653	0.794753

Prior to peak-stripping it is necessary to convert these true ratios to 'measured' ratios as the Lu and Yb in the sample/standard are, like any element, subjected to instrumental mass fractionation such that at the detectors the $^{176}Lu/^{175}Lu$ and $^{176}Yb/^{173}Yb$ ratios deviate from the natural values. For Lu, which has only one interference-free isotope, calculation of 'measured' $^{176}Lu/^{175}Lu$ relies on the mass bias determined from a second element such as Hf. For Yb, however, calculation of the 'measured' $^{176}Yb/^{173}Yb$ could be based on either the mass bias determined from the $^{172}Yb/^{173}Yb$ ratio, as both isotopes are present in the collector configuration

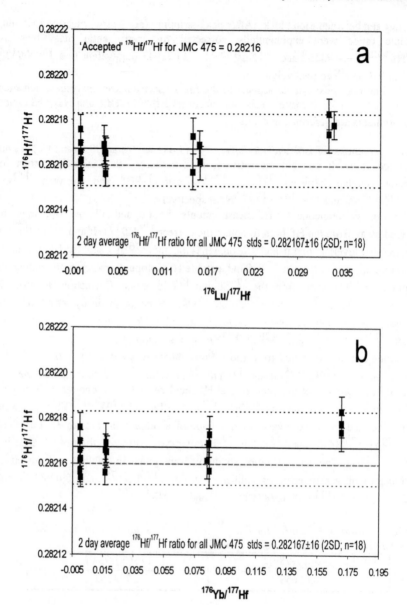

Figure 1 *Accuracy and reproducibility of the $^{176}Hf/^{177}Hf$ ratio for JMC 475 after correction for isobaric overlaps from Lu and Yb with a) varying $^{176}Lu/^{177}Hf$ and b) varying $^{176}Yb/^{177}Hf$ ratios*

(Table 1), or from the Hf-derived mass bias. In this work 'measured' ratios were calculated using the mass bias derived from the $^{179}Hf/^{177}Hf$ ratio by applying the exponential law in reverse to the natural $^{176}Lu/^{175}Lu$ and $^{176}Yb/^{173}Yb$ ratios.

The precision, accuracy and reproducibility of the isobaric corrections for Lu and Yb on the ^{176}Hf/^{177}Hf ratio is illustrated in Figure 1 for a set of 200ppb JMC 475 standards doped with varying amounts of both Lu and Yb. The Lu and Yb doping of JMC 475 was intended to extend to higher Lu/Hf and Yb/Hf ratios than would be expected to be encountered in routine analysis of geological materials. The average ^{176}Hf/^{177}Hf obtained for all JMC 475 standards, including those doped with Lu and Yb, over the two days of the development work was 0.282167±16 (2SD, n=18). This is within error of that recommended for the pure JMC 475 standard (0.282160)[12] and the reproducibility is comparable to the best obtainable by PIMMS on pure JMC 475 standard (0.282163±9; 2SD)[8].

In simultaneous Sm-Nd measurements two methods are available for the correction of isobaric overlap between Sm, Ce and Nd. One is to strip the ^{142,144,148,150}Nd away from the Sm and Ce (the reverse to the peak-stripping routine applied for the Lu-Hf measurements). This is possible as the Sm and Ce isobaric overlaps are on stable Nd isotopes and the ratio of these Nd isotopes with the interference-free ^{145}Nd or ^{146}Nd isotopes are fixed and well established[6]. The other method, which was utilised in this study, is similar to the Lu-Yb peak-stripping and involves stripping the Sm away from the Nd. The only disadvantage of this latter peak-stripping method is that, given the chosen collector configuration it is not possible to strip ^{142}Ce away from ^{142}Nd as there is no interference-free Ce isotope monitor peak. However, there are nevertheless two stable Nd isotope ratio measurements made in each analysis to ensure that a correct Nd measurement is obtained.

Corrections for isobaric overlaps from ^{144,148,150}Sm on ^{144,148,150}Nd are made by monitoring the intensity of the interference-free ^{147}Sm isotope. The natural ^{144}Sm/^{147}Sm, ^{148}Sm/^{147}Sm and ^{150}Sm/^{147}Sm ratios used for peak stripping (Table 4) were taken directly from Wasserburg et al.[10] as the composition of Sm is well established. 'Measured' ^{144}Sm/^{147}Sm, ^{148}Sm/^{147}Sm and ^{150}Sm/^{147}Sm ratios were derived from the natural ratios in Table 4 using the mass bias derived from the measured ^{149}Sm/^{147}Sm ratio and taking the accepted ^{149}Sm/^{147}Sm value of 0.921600.[10]

Table 4 *'True' isotope ratios used for isobaric interference corrections from Sm on Nd*

^{144}Sm/^{147}Sm	^{148}Sm/^{147}Sm	^{150}Sm/^{147}Sm
0.20504	0.74970	0.49213

The precision, accuracy and reproducibility of the isobaric corrections for Sm on the ^{143}Nd/^{144}Nd, ^{148}Nd/^{144}Nd and ^{150}Nd/^{144}Nd ratios is illustrated in Figure 2 for our pure in-house J&M Nd standard and J&M doped with Sm to give a Sm/Nd ratio of 0.25, typical for the type of geological materials processed at NIGL. The average ^{143}Nd/^{144}Nd ratios obtained for the pure J&M and Sm-doped J&M agree with one another to within 10ppm and the overall PIMMS average (0.511161±17 2SD) to within 18ppm of the long-term average NIGL TIMS value. The agreement

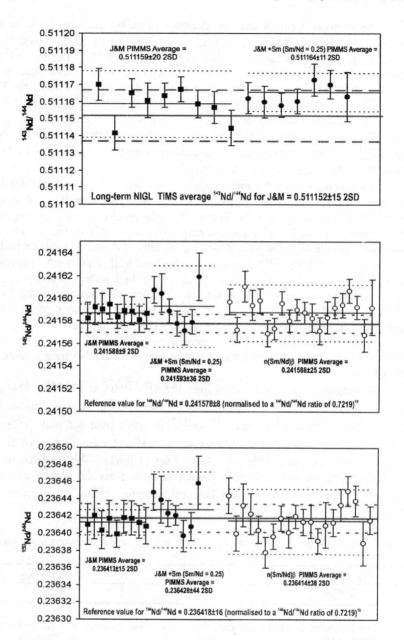

Figure 2 *Accuracy and reproducibility obtained on a) $^{143}Nd/^{144}Nd$, b) $^{148}Nd/^{144}Nd$ and c) $^{150}Nd/^{144}Nd$ for the pure NIGL internal J&M Nd isotopic standard and for Sm-doped J&M (Sm/Nd ratio = 0.25) and the n(Sm/Nd)β shelf solution, after correction for the isobaric overlap from Sm. Also shown, as heavy grey lines in b) and c) are the reference values of Wasserburg et al.[10]*

between the average ^{148}Nd/^{144}Nd and^{150}Nd/^{144}Nd ratios for the pure J&M ND standard and both the Sm-doped J&M and CIT n(Sm/Nd)β shelf solution[10] is excellent (24 and 63ppm respectively) despite a Sm correction being applied on each of the ^{144,148,150}Nd peaks. The increased error on the reproducibility of the stable Nd isotope ratios for Sm-doped J&M and CIT n(Sm/Nd)β shelf solution is due to the incorporation of the error on the Sm corrections. The overall average ^{148}Nd/^{144}Nd and^{150}Nd/^{144}Nd ratios agree to within 63 and 42ppm respectively of the values of Wasserburg et al.[10]

Although there are potential interferences from barium oxides in the Sm-Nd mass range (mass 146, 148 and 150) no attempt has been made to correct for these as the oxide generation levels are minimised by the use of the MCN6000 (CeO/Ce<0.3%). Furthermore, the excellent agreement between our measured Nd ratios, after correction for Sm overlaps, and the accepted TIMS ratios (Figure 2a-c) suggests that any residual oxide or hydride interferences are negligible. Only once the corrections for isobaric overlaps from Sm can be demonstrated to be accurate and reproducible can parent/daughter ratio measurements be made successfully.

2.2.2. Corrections for element fractionation of the parent/daughter ratio   To test the accuracy of the analytical method in measuring parent/daughter ratios two gravimetric shelf solutions with precisely known parent/daughter (ROM Lu-Hf shelf[13] with ^{176}Lu/^{177}Hf = 0.13267±0.5‰ 2SE and CIT n(Sm/Nd)β shelf[10] with ^{147}Sm/^{144}Nd = 0.19655±0.5‰ 2SE) ratios were analysed. Even after correction of the measured ^{176}Lu/^{177}Hf and ^{147}Sm/^{144}Nd ratios for instrumental mass bias the corrected ratios are consistently inaccurate, although very precise (<0.01% 2RSE). The deviation of the mass bias-corrected^{176}Lu/^{177}Hf or ^{147}Sm/^{144}Nd ratios from the accepted values for the shelf solutions is typically within 3 to 15%. This inaccuracy appears to be due to an elemental, as opposed to mass, fractionation that occurs within the plasma and/or interface region. Furthermore, the degree of deviation of the parent/daughter ratio from the accepted shelf solution value appears to be related to torch position and gas flow conditions and can to some extent be reduced in magnitude by changing optimising torch position and refocusing. The degree of elemental fractionation, however, appears to remain relatively constant on any one day and is relatively stable for prolonged periods of 5-6hours so long as *no* instrumental parameters (lens settings, ion energy, extraction voltage, cool gas, auxiliary gas, and nebuliser gas flows) are adjusted. If using the MCN 6000 desolvating nebuliser introduction system it is also necessary to keep the sweep gas flow rate constant.

Because the parent/daughter element fractionation in the PIMMS is stable it is possible to apply a correction to the measured parent/daughter ratios on unknowns. This is achieved by running the appropriate gravimetric shelf solution and deriving a correction factor which can be made to account for either the elemental fractionation alone (by dividing the mass bias corrected parent/daughter ratio by the shelf ratio) or for both the elemental fractionation and mass bias concurrently (by dividing the measured parent/daughter ratio by the shelf ratio). Once the correction factor is derived the PIMMS parameters are left unchanged and the factor entered into the analysis program as a constant and used on-line to correct the parent/daughter ratio of unknowns. The shelf solution is run repeatedly throughout the sample analyses as an unknown to provide an estimate of the accuracy and reproducibility of parent/daughter ratio measurements. As instrumental settings may drift slightly throughout the day it is necessary to run sufficient shelf solutions as unknowns in order to monitor stability and to be able to take remedial action as soon as any drift occurs. If and when drift does occur the instrument is refocused and the shelf once again run as a known in order to derive a new correction factor before any further unknowns are analysed.

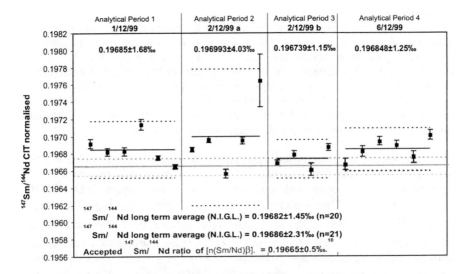

Figure 3 *Accuracy and reproducibility of the $^{147}Sm/^{144}Nd$ ratio after correction for elemental fractionation using the $n(Sm/Nd)\beta$ shelf solution[10] for 4 analytical sessions over three days. Internal errors are 2SE, while reproducibility for each analytical period is 2SD. Dark grey shading represents the accepted $^{147}Sm/^{144}Nd$ ratio of the $n(Sm/Nd)\beta$ with 2SD error [10]*

Figure 3 shows an example of the effect of instrumental drift on the corrected $^{147}Sm/^{144}Nd$ ratio of the shelf. In the second analytical period (2/12/99 a) the corrected $^{147}Sm/^{144}Nd$ ratio of the shelf solution, run as an unknown, had drifted towards a markedly higher value than previous analyses of the shelf solution, resulting in a reproducibility of only 4.03‰ 2SD. The shift in the corrected $^{147}Sm/^{144}Nd$ ratio was associated with a concomitant increase in internal errors suggesting a change in instrument parameters. This analytical session was terminated and a new correction factor derived before continuing. The subsequent analytical session (2/12/99 b) yielded a better reproducibility (1.15‰ 2SD) on the corrected $^{147}Sm/^{144}Nd$ ratio of the shelf solution. A similar instrument drift can be seen for the corrected $^{176}Lu/^{177}Hf$ ratio in Figure 4, though in this case the drift is considerably greater. We therefore adopted a protocol in which unknowns, both shelf solutions and samples, were run in short analytical periods interspersed with refocusing and derivation of new correction factors in order to maximise the reproducibility of the parent/daughter ratios.

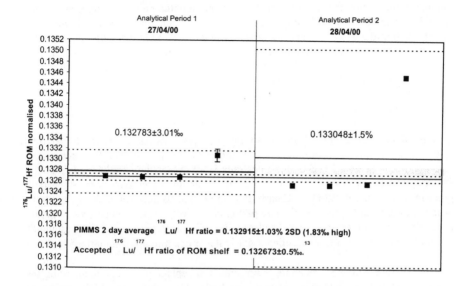

Figure 4 *Accuracy and reproducibility of the $^{176}Lu/^{177}Hf$ ratio after correction for elemental fractionation using the ROM Lu-Hf shelf solution[13] for 2 analytical sessions over two days. Internal errors are 2SE, while reproducibility for the different analytical periods is 2SD. Dark grey shading represents the accepted $^{176}Lu/^{177}Hf$ ratio of the ROM Lu-Hf shelf with 2SD error [13]*

Figures 3 and 4 illustrate the stability of the mass bias and element fractionation-corrected $^{147}Sm/^{144}Nd$ and $^{176}Lu/^{177}Hf$ ratios on the CIT $n(Sm/Nd)\beta$[10] and ROM Lu-Hf[13] shelf solutions measured as unknowns over four and two analytical sessions respectively. The reproducibility of the corrected parent/daughter ratios is clearly variable between analytical periods reflecting variable PIMMS stability. However, the overall reproducibility obtained on the $^{176}Lu/^{177}Hf$ and $^{147}Sm/^{144}Nd$ ratios for the shelf solutions, run as unknowns after derivation of the appropriate element fractionation correction factors, are 1.0% and 0.15-0.25% (2SD), respectively. This reproducibility compares favourably with that obtained on $^{176}Lu/^{177}Hf$ and $^{147}Sm/^{144}Nd$ parent/daughter ratios by conventional ID-PIMMS or ID-TIMS measurements of <1%[8] and 0.5%[14] respectively.

It is also clear from Figures 3 and 4 that the reproducibility of the $^{176}Lu/^{177}Hf$ measurements is less than that obtained for $^{147}Sm/^{144}Nd$ measurements. This is attributed to two factors. First, the overall stability of the P54 was simply not as good as during the execution of the Sm-Nd experiments. Second, the Lu/Hf element ratio (~1) of the ROM shelf solution used to derive the correction factor is unrealistically high for common geological materials, and as such the corrections for both isobaric overlap between ^{176}Lu and ^{176}Hf and element fractionation are extremely sensitive to instrument stability. There is no reason why, in theory, the reproducibility on Lu/Hf ratio measurements should not be equally as good as that obtained during the Sm/Nd experiments.

Just as reproducibility varies between analytical periods, it is also clear from Figures 3 and 4 that the accuracy of the parent/daughter ratio measurement is also variable. Therefore, in addition to the online correction for element fractionation, sample parent/daughter ratios measured during each analytical period are also corrected for the offset between the average parent/daughter ratio obtained for the shelf and the accepted ratio. This eliminates any systematic bias in corrected sample parent/daughter ratios between analytical periods.

3 DISCUSSION

With the development of PIMMS the isotope geologist now has a variety of analytical methods available for obtaining isotope compositions and parent/daughter ratios; conventional ID-TIMS/PIMMS, simultaneous ID-PIMMS, the simultaneous non ID-PIMMS procedure described above and, at generally lower parent/daughter ratio precisions, in-situ LA-PIMMS. Clearly it is important to decide which of the above procedures is most appropriate for any particular project. A non-exhaustive summary of some of the advantages and disadvantages of the respective procedures which should be considered before selecting a particular procedure are summarised in Table 5.

The new PIMMS procedure is not aimed at replacing but complementing the more conventional isotope dilution methods for obtaining high precision parent/daughter ratios. The major advantages of this new procedure are that it provides measurement precisions approaching that of the best isotope dilution-methods yet without the associated difficulties of optimising spike and achieving complete spike-sample equilibrium. Furthermore, it is relatively rapid and is less costly in terms of consumables and both staff and facility time. It is anticipated that the new PIMMS procedure will find its real niche between ID-TIMS and LA-PIMMS methods, which currently occupy opposite ends of the spectrum in terms of precision, reproducibility and speed/ease of analysis. The new procedure described here would be ideal for those projects which require high precision isotope compositions and parent/daughter ratios, thereby essentially eliminating LA-PIMMS, on large sample sets which would be prohibitively time consuming and/or costly by conventional ID-methods.

NIGL processes significant numbers of zircons and monazites for TIMS U-Th-Pb dating and the Lu-Hf and Sm-Nd, respectively, come as free by-products of the chemistry and can therefore potentially provide important additional isotopic information. However, the number of samples processed at NIGL precludes the use of ID methods for the analysis of the Lu-Hf and Sm-Nd fractions. However, with the new rapid non-ID PIMMS procedure described above we can now routinely collect all zircon and monazite U-Th-Pb column washes for subsequent Lu-Hf and Sm-Nd isotopic analysis.

4 CONCLUSIONS

A new solution-based PIMMS procedure has been developed at NIGL that allows for very rapid and simultaneous measurements of isotope compositions and parent/daughter ratios without the use of enriched isotopic spikes. The reproducibility that can be obtained for $^{176}Hf/^{177}Hf + ^{176}Lu/^{177}Hf$ and $^{143}Nd/^{144}Nd +$

$^{147}Sm/^{144}Nd$ is similar to that obtained by more conventional isotope dilution methods and is independent of optimising spike addition and ensuring complete spike-sample mixing during dissolution. The key to obtaining precise and accurate isotope compositions and parent/daughter ratios is the use of appropriate shelf solutions to derive a correction for elemental fractionation in the plasma. Reproducibility on the other hand is ultimately dependent upon instrument stability, as the instrument parameters must remain unchanged throughout the period of analysis.

Table 5 Summary of some of the main advantages and disadvantages of the new PIMMS and conventional ID procedures

Technique	Advantages	Disadvantages
New PIMMS D$_{IC}$+P/D$_R$ Procedure	1] No spiking required. 2] Chemical separation procedure is simple and reduced to a single column. 3] Isotope composition and parent/daughter ratios obtained simultaneously. 4] Analysis very rapid: One analysis of ~15mins. 5] Maximum number of samples per day ≥15 (plus 20-25 standards and reference solutions).	1] Quantitative recovery of parent and daughter isotopes essential. 2] Isobaric overlaps must be corrected for. 3] Elemental fractionation in PIMMS plasma/interface requires correction. 4] Static analysis is necessary because of signal fluctuation. Ultimate precision on isotope ratio is therefore less than TIMS. 5] Sample memory requires wash out times of ~10mins plus. 6] Difficult to apply to whole-rock Lu-Hf studies. 7] Precision of Parent/daughter ratio depends on PIMMS stability
ID-TIMS	1] Use of double spike negates the need for quantitative recovery of elements. 2] Little or no isobaric interferences to correct for during analysis. 3] Only concentrations are determined therefore no 'elemental fractionation' during analysis. 4] Mulit-dynamic analysis improves precision on isotope ratio (but increases analysis time). 5] TIMMS capable of analysing smaller quantities of Nd or Sr than laser ablation or solution-mode PIMMS.	1] Spike addition needs to be optimised. 2] Spike-sample isotopic equilibrium must be achieved. 3] Separation of parent and daughter isotopes necessary: longer chemical separation procedures. 4] Parent and daughter isotopes must be analysed separately. 5] For ID-TIMS the source chamber must be evacuated: requires ~2 hours. 6] ID-TIMS analysis time slow: ~1 to 2 hours for each element. 7] Maximum number of samples per day by TIMS limited. 6-7 samples about maximum (including 2 parent and daughter isotopic standards).

Simultaneous D$_{IC}$+P/D$_R$ ID-PIMMS	1] Use of double spike at dissolution stage negates the need for quantitative recovery of elements. 2] Chemical separation procedure is simple and reduced to a single column. 3] Isotope composition and parent/daughter ratios obtained simultaneously. 4] Analysis very rapid: One analysis of ~15mins. 5] Maximum number of samples per day ≥15 (plus 20-25 standards and reference solutions).	1] Spike addition needs to be optimised. 2] Spike-sample isotopic equilibrium must be achieved. 3] Sample memory requires wash out times of ~10mins plus. 4] More difficult to apply to beta-decay radiogenic isotope systems.
Laser-ablation	1] Minimal sample preparation. 2] Very rapid. Many 10's per day. 3] High spatial resolution.	1] Element fractionation may occur at ablation site: Difficult to correct for. 2] Elemental fractionation in PIMMS plasma/interface requires correction. 3] Lack of ablation standards for element fractionation correction. Precision on parent/daughter ratio probably limited to several % level. 4] Static analysis essential. 5] Requires sufficient element concentrations.

5 REFERENCES

1. A.J. Walder and P.A. Freedman, *J. Anal. Atom. Spectrom.* 1992, **7**, 571.
2. A.J. Walder, I. Platzner and P.A. Freedman, *J. Anal. Atom. Spectrom.* 1992, **8**, 19.
3. B. Luais, P. Telouk and F. Albarede, *Geochim et Cosmochim Acta.* 1999, **61**, 4847.
4. M.F. Thirlwall and A.J. Walder, *Chem. Geol. (Isotope Geoscience Section).* 1995, **122**, 241.
5. W.L. Griffin, N.J. Pearson, E. Belousova, S.E. Jackson, E. van Achterbergh, S.Y. O'Reilly and S.R. Shee, *Geochim et Cosmochim Acta.* 2000, **64**, 133.
6. M.S.A. Horstwood, R.R. Parrish, G.M. Nowell and S.R. Noble, 7th Int. Conf. On Plasma Source Mass Spectrometry, Durham, 2000.
7. M. Horstwood, *pers comm.*
8. J. Blichert-Toft, C. Chauvel, F. Albarede, *Contrib. Mineral. Petrol.* 1997, **127**, 248.
9. P.J. Patchett and M. Tatsumoto, Contrib. Mineral. Petrol. 1980, **75**, 263.
10. G.J. Wasserburg, S.B. Jacobsen, D.J. DePaolo, M.T. McCulloch and T. Wen, *Geochim et Cosmochim Acta.* 1981, **45**, 2311.
11. P.J. De Bievre, M. Gallet, N.E. Holden and I.L. Barnes, *J.Phys. Chem. ref. data.* 1984, **13**, 809.
12. G. Nowell, P.D. Kempton, S.R. Noble, J.G. Fitton, A.D. Saunders, J.J. Mahoney and R.N. Taylor, *Chem. Geol. (Isotope Geoscience Section).* 1998, **149**, 211.
13. S.R.Noble *pers.comm.*
14. J. Blichert-Toft, F. Albarede, M. Rosing, R. Frei and D. Bridgwater, *Geochim et Cosmochim Acta.* 1999, **63**, 3901.

COMBINED Pb-, Sr- AND O-ISOTOPE ANALYSIS OF HUMAN DENTAL TISSUE FOR THE RECONSTRUCTION OF ARCHAEOLOGICAL RESIDENTIAL MOBILITY

P. Budd[1], J. Montgomery[2], J. Evans[3] and C. Chenery[3]

[1]Department of Archaeology, University of Durham, South Road, Durham, DH1 3LE, UK.
[2]Department of Archaeological Sciences, University of Bradford, Bradford, BD7 1DP, UK.
[3]NERC Isotope Geosciences Laboratory, British Geological Survey, Keyworth, Nottingham, NG12 5GG, UK.

1 INTRODUCTION

Characterising immigration, invasion and settlement is a recurrent theme in archaeology. In Britain, episodes such as the Roman 'invasion' of the 1st century AD and early medieval Anglo-Saxon immigration are prime examples. Over the years, the archaeological interpretation of such events has swung from wholesale invasions and large scale population movements to the arrival and rapid mingling of small elite bands. Traditionally, archaeologists have concentrated on grave goods, burial practices and other aspects of the material record to assess migration, but such factors may be more revealing of cultural affiliations than place of origin. Independent scientific methods to identify immigrants, have included the analysis of skeletal (and particularly cranial) morphology, but this approach is limited by intra-population variation. DNA analysis offers new possibilities, but cannot be used to distinguish original settlers from their descendants. Here, we present a new approach to the identification of first generation immigrants among burial populations from the combined isotopic analysis of human dental tissue.

The possibility of reconstructing patterns of residency and mobility among prehistoric populations from the analysis of dental remains arises from the systematic variation within nature, and between localities, of stable and radiogenic isotopes which become incorporated in teeth during life. Strontium isotope analysis has been used in this way for several years[1-6]. Prior to this work, lead isotope measurement of archaeological tissues for similar purposes had rarely been attempted, but lead isotope analysis of modern dental tissues had been shown to have applications in source tracing ingested pollutants and in forensic science to identify place of origin[7-9]. The oxygen isotope composition of archaeological human dental enamel has previously been measured to obtain palaeoclimate data[10], but it's potential as an independent parameter for life history reconstruction had only hitherto been exploited in a single pioneering study[11]. We have applied all three techniques to the analysis of archaeological human dental enamel for the purpose of reconstructing residential mobility. We have also shown elsewhere that this approach can be used to identify first generation immigrants among burial groups[12-15] and even comment directly upon their place of origin[16,17].

1.1 Strontium and lead isotope variation and incorporation in human dental tissues

Strontium and lead may be ingested *in vivo* via foodstuffs and water. The isotopic composition of both elements varies in a systematic manner throughout the geosphere as a result of radiogenic isotope evolution so that the strontium and lead isotope composition of the diet can depend on the geology of the region in which foods were obtained. There is no significant fractionation of strontium or lead in soil formation and biological processes. Recent experimental studies confirm this for strontium showing that it's isotopic composition is not significantly altered as rock-derived strontium enters soil, becomes bioavailable and moves up the food chain[18]. For prehistoric, and certainly pre-metallurgical, societies dietary lead and strontium are highly likely to have been derived from locally sourced foodstuffs and hence related to the geology of the place of residence. By Classical and early medieval times however, human exposure to lead is generally dominated by manufactured products[19]. Nevertheless, the isotopic composition of such lead may be highly revealing as there is evidence that the exploitation of different ore deposits with restricted networks of supply and circulation gave rise to regional patterns of metal circulation with distinctive isotopic characteristics[20].

Some strontium from the diet substitutes for calcium in the inorganic (hydroxyapatite) mineral lattice of bones and teeth, typically resulting in hard tissue strontium contents of a few hundred parts per million[21]. Skeletal tissues may also accumulate lead from the blood supply, although the mechanism is not properly understood[19]. As the isotopic composition of the dietary strontium and lead incorporated is unaltered by the process of transport and bioaccumulation, the isotope ratios found in hard tissues after death reflect a time averaged signal representative of diet over some period of life, depending on the tissue under consideration. Teeth are particularly advantageous as different dental tissues preserve metals ingested at particular stages of life. Deciduous tooth formation is instigated in the developing foetus within 14-19 weeks of fertilisation and enamel mineralisation is complete within a year of birth. Once formed, the enamel is not remodelled so that its metal content is considered a reliable indicator of *in utero* or neonatal exposure[8]. In contrast, secondary dentine, laid down within the pulp cavity after tooth formation, is thought to accumulate lead and strontium from the blood supply[22] and has been considered representative of post-natal exposure[9]. Although development of the permanent first molar is initiated *in utero*, most of the permanent dentition is formed in childhood from 3-4 months after birth until about 12 or more years of age. As with deciduous teeth, permanent enamel is thought to preserve metals ingested only during the period of formation whereas dentine is known to accumulate them from the blood and, in adults, can be considered to represent a time-averaged signal accumulated prior to tooth loss or death. These differences between the retention characteristics of different human dental tissues make teeth potential archives of both the recent and more remote exposure of single individuals.

1.2 Oxygen isotope variation and incorporation in human dental tissues

In contrast to strontium and lead, the high relative mass difference between ^{16}O and ^{18}O are such that isotopic fractionation readily occurs in nature resulting in considerable variation among meteoric waters depending on climate and geography. Such differences are expressed as relative differences in the heavier isotope relative to Standard Mean Ocean Water and reported as $\delta^{18}O_{SMOW}$. In the body, the oxygen isotope composition of skeletal tissue is directly related to that of the oxygen consumed and this has been shown to be dominated by drinking water[23]. Metabolic processes elevate the level of ^{18}O in skeletal tissues relative to drinking water increasing the value of $\delta^{18}O$. However, among mammals, skeletal tissues form at a relatively constant body temperature so that the oxygen isotope fractionation that does take place is very similar both within and between species[23-29]. Although there is some inter-species variation due to body mass, diet and metabolism, a number of researchers have developed calibrations to relate skeletal $\delta^{18}O$ to that of drinking water. Those calculated by Longinelli, Luz, Levinson and co-workers[23,24,30] are perhaps best known.

Calibrations accounting for the fractionation between skeletal tissue $\delta^{18}O$ and drinking water $\delta^{18}O$ have been used for a number of years in studies of the climatological and hydrological environment of various species including man[10,26,31-34]. For much of the archaeological past, during which climate parameters have been similar to those prevailing today, the measurement of dental enamel offers a potential insight into place of origin independent from, but complementary to, that from strontium and lead[12,35].

1.3 Post mortem diagenesis of archaeological bone and teeth

The success of the isotopic approach to the reconstruction of residency and migration rests on the presumption that the elements of interest preserved within archaeological tissue are those incorporated *in vivo*, and that these are, at least substantially, unaffected by subsequent contamination. Processes of Strontium diagenesis has been examined by Sillen[36,37] and Sealy[1] and both have developed a decontamination procedure for bone based on a suggested difference in solubility between biogenic and diagenetic apatite. Recently however, doubt has been cast as to the reliability of such pre-treatments for the recovery of biogenic strontium from archaeological bone and dentine, although enamel appears to be a reliable reservoir of *in-vivo* strontium, whether treated or not[38].

Previous studies comparing modern and pre-industrial bone lead burdens[39-41] appeared to show that modern bones have considerably elevated lead compared with their archaeological correlates, arguing for a lack of serious contamination of the latter. However, similar studies of Roman individuals initially appeared to indicate very high lead burdens[42,43], and it is now understood that measured levels in bone were considerably elevated by diagenetic accumulation[44]. As with strontium, it appears that dental enamel is the only tissue normally resistant to post-mortem lead diagenesis and this receives confirmation from recent pilot studies comparing the lead content of archaeological and modern human teeth[12,13,15,45]. Preliminary results suggest that ancient and modern tooth enamel are comparable, but that dentine analysis remains problematic[38].

Although the biological apatite that makes up tooth enamel is predominantly hydroxyapatite of the approximate formula $Ca_{10}(PO_4)_6(OH)_2$, in humans about 2-3% of the mineral is carbonate of which about 90% substitutes for the phosphate group and the remainder for the hydroxyl group. There are therefore three inorganic phases in which oxygen can be found, phosphate, carbonate and hydroxyl. These also have varied oxygen contents, containing 35, 3.3 and <1.6 wt.% oxygen respectively. It has already been noted that the highly mineralised nature of enamel makes this tissue uniquely resistant to diagenesis and this stability extends to the oxygen-bearing mineral phases. Both carbonate and phosphate phases of biological apatites have been shown to preserve biogenic oxygen under most circumstances, although of the two the carbonate phase is generally considered more susceptible to diagenesis[46]. Here, we have used a laser fluorination technique to extract total oxygen from the inorganic phase of the tissue which is dominated by and closely approximates phosphate oxygen (O_p).

2 SAMPLES AND METHODS

2.1 Sites and samples

Archaeological human tooth samples were obtained from three sites on the Cretaceous Chalk of southern England. The chalk geology minimises some of the potential complications for studies of residential mobility in providing for excellent skeletal preservation, a naturally low-lead environment and one that is relatively homogenous with respect to the isotopic composition of bioavailable strontium and lead. The uniformity of the local (and diagenetic) isotope signal maximises the probability of detecting immigrants from elsewhere. The earliest samples were obtained from four Neolithic burials from Monkton-up-Wimbourne, Dorset (MUW). Permanent teeth were obtained for one adult female (C) and samples of both permanent and deciduous teeth were obtained from a further three juveniles (A, B & D) from the same burial group. Five permanent teeth of Romano-British date were analysed from the site of a 4th century AD cemetery at the Eagle Hotel site, Winchester (EHW), including two teeth from one individual (G339) interred within a sealed lead coffin. A permanent tooth from a 7th century AD adult male burial from Stonehenge (STH) was also analysed.

2.2 Sample preparation

Each tooth was cleaned ultrasonically for 5 minutes in high-purity water and then acetone. Each was then divided longitudinally using a flexible diamond edged rotary dental saw to produce half-tooth samples for mass spectrometry. Enamel and dentine samples were separated using a tungsten carbide dental bur. Enamel was cleaned of all adhering dentine and abraded from the surface to a depth of >100µm using the bur. The surface material was discarded to produce a sample of core enamel tissue, some of which was set aside for O-isotope analysis. Samples of primary, crown dentine were obtained by removing all secondary dentine from the pulp cavity. Clean core enamel and primary dentine samples were then transferred to a clean (class 100, laminar flow) working area at NIGL for further preparation.

Soil (and, for MUW, underlying chalk) samples were collected for lead and strontium isotope analysis to assess the isotopic composition of metals available to aqueous phases in the burial environment. For each sample, about 2g of finely divided soil (or powdered rock) was placed in each of two acid-leached teflon beakers. For each soil, one sample was leached overnight at room temperature in 5ml of high purity water and one in 5ml of 10% acetic acid. The leachates were then removed for analysis. Aqueous leaches failed to produce sufficient lead for reliable analysis.

At NIGL, tissue samples for strontium and lead analysis were first cleaned ultrasonically in high purity water to remove dust, rinsed twice, dried down in high purity acetone and then weighed into pre-cleaned Teflon beakers. Solid samples were dissolved in Teflon distilled 16M HNO_3 and an aliquot (10% by volume) was transferred to a second pre-cleaned Teflon beaker and spiked with ^{208}Pb tracer solution for lead concentration determination using the isotope dilution method. Lead was collected from all samples using conventional anion exchange methods and the washes from the larger aliquot of each sample were spiked with ^{84}Sr tracer solution. Strontium was collected from this fraction using conventional ion exchange methods.

Dental enamel sub-samples for oxygen isotope analysis were cleaned in dichloromethane, soaked overnight in concentrated (100 volume) H_2O_2, and rinsed in double deionised water (<0.5mS cm^{-1}). Samples were allowed to dry and then heated for

1hr at 600°C to remove all interstitial water and reduce remaining organic material to carbon.

2.3 Analytical

2.3.1. Lead and strontium. The lead isotope compositions of the MUW samples were determined by Plasma Ionisation Multi-collector Mass Spectrometry (PIMMS) using a VG Elemental P54 mass spectrometer. Data were corrected for mass discrimination using the Tl spiking technique[47,48]. The lead isotope compositions of the other samples, all lead and strontium concentrations and the strontium isotope composition were determined by Thermal Ionization Mass Spectrometry (TIMS) using a Finnigan Mat 262 multi-collector mass spectrometer. Errors, determined from replicate analysis of the NBS 981 standard, were 0.05% for $^{207}Pb/^{206}Pb$ and 0.10% for the $^{208}Pb/^{206}Pb$ (n=60, 2σ). The international standard for strontium gave a value of $^{87}Sr/^{86}Sr$ = 0.710200 ± 32 (n=10, 2σ) during the period of analysis. All strontium ratios have been corrected to the accepted value for NBS 987 of 0.710235. The mean strontium blank was 300pg (range 33-400pg).

Isotope dilution elemental concentration accuracy can exceed 1% (2σ) for homogenous samples in favourable circumstances[49] and this value is widely quoted for strontium concentration determinations on rock powders[50]. However, dental tissues have not been well characterised with respect to compositional heterogeneity and preliminary LA-ICP-MS studies suggest a degree of lead and strontium variation within tooth enamel[15]. Estimates of reproducibility for the current study are based on a preliminary study of three replicate tooth enamel measurements undertaken for both a permanent and deciduous tooth. These suggest errors of ±14% for lead and ± 10% for strontium (n=3, 2σ).

2.3.2. Oxygen. Oxygen extraction was undertaken by laser fluorination to maximise efficiency and minimise sample size. Approximately 2mg of dental enamel was weighed and placed into one of 16 wells of a nickel sample holder. The holder was placed into a stainless steel sample chamber fitted with a barium fluoride window. The chamber was evacuated overnight and pre-fluorinated with ClF_3 for one hour at room temperature. Oxygen released during pre-fluorination was below detection limits. The samples were reacted with 1.5×10^{-4} moles of ClF_3 using a 10 watt CO_2 laser beam (10.6mm wavelength) until a clear melt bead was formed and ceased to react. Cryogenic, and KBr traps were used to purify the sample oxygen prior to conversion to CO_2 over hot platinized carbon. The yield was then measured and the CO_2 transferred cryogenically to sample collection vessels to be analysed off line. A VG Isotech Optima dual inlet isotope ratio mass spectrometer operating Micromass DI2.47 software, was used to determine the enamel oxygen isotope composition.

Apatite reference material certified with respect to O-isotope composition is not available, but a number of researchers have made use of two NBS phosphate minerals, 120b and 120c, as laboratory standards. Although the laser fluorination procedure is new and still under development for this application, our initial assessments of external reproducibility are 21.54 ± 0.21‰ (2σ, n=6) for NBS120b and 20.70 ± 0.23‰ (2σ, n=2) for NBS120c. These values are comparable with previously reported data. Initial assessments of reproducibility based on replicate analyses of single tooth enamel specimens gave $\delta^{18}O_p$ values of 15.75 ± 0.14‰ (2σ, n=3) and 18.69 ± 0.15‰ (2σ, n=3) for a modern and an archaeological tooth respectively.

Table 1   *Strontium concentration and isotope ratio composition data for the archaeological teeth. EHW - Eagle Hotel, Winchester; STH - Stonehenge; MUW - Monkton-up-Wimbourne. na - not analysed*

Site	Sample	Tooth	Enamel Sr (ppm)	$^{87}Sr/^{86}Sr$	Dentine Sr (ppm)	$^{87}Sr/^{86}Sr$
EHW	G318	P^1right	56	0.70828	59	0.70834
EHW	G319	P_1left	91	0.70848	na	0.70823
EHW	G326	P^1right	80	0.70863	na	0.70838
EHW	G339a	M^3right	93	0.70915	111	0.70862
EHW	G339b	P_2right	79	0.70931	127	0.70863
STH	STH1	P^1right	55	0.70832	112	0.70794
MUW	A_{decid}	c^1right	75	0.70955	230	0.70788
MUW	A_{perm}	M^1right	71	0.70878	207	0.70782
MUW	B_{decid}	c^1left	103	0.70844	248	0.70790
MUW	B_{perm}	P^1left	57	0.70928	184	0.70789
MUW	C_{perm}	P^2right	55	0.71007	162	0.70866
MUW	D_{decid}	c^1left	72	0.70849	282	0.70809
MUW	D_{perm}	C^1left	54	0.70897	250	0.70792

3 RESULTS

3.1 Strontium isotope data

Sr concentration and isotope compositional data are given in Table 1 and plotted in Figure 1. Strontium isotope data for the associated soil samples are given in Table 2

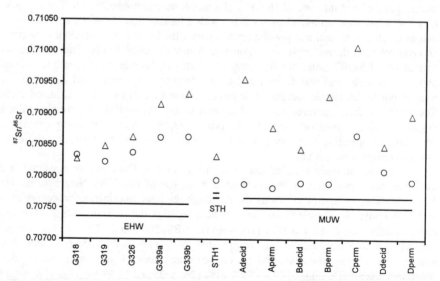

Figure 1   *Enamel (triangles) and dentine (circles) Sr isotope ratios from each tooth. Solid lines indicate the range of Sr isotope ratios obtained from aqueous and acetic acid leachates of soil samples from each of the three sites. Measurement errors (2σ) are smaller than the symbols*

The enamel strontium concentrations (Table 1) are similar to those of modern people[21] suggesting that the enamel is representative of the biogenic signal and this is confirmed by more detailed analysis of the data[38]. It is immediately apparent that there are very significant differences between enamel strontium isotope values for each of the people examined and those of the soil-derived strontium from the respective burial sites. In one case, EHW-G318, enamel and dentine strontium concentrations and isotope ratios are closely similar and the former comparable with *in-vivo* levels in similar tissues of modern people. For this tooth it appears that both enamel and dentine preserve biogenic strontium largely unaffected by diagenesis. For the other samples however, dentine strontium isotope values are intermediate between those of the enamel and the soil. In these samples, dentine also appears significantly elevated in strontium concentration compared with the enamel (Table 1). The results suggest diagenetic addition of soil strontium to the dentine in most cases. The evidence for diagenesis of the dentine is discussed in detail elsewhere[38]. For the great majority of the samples examined however, enamel strontium isotope composition would appear to be an accurate reflection of the biogenic signal.

Table 2   *Strontium isotope data for aqueous and acetic acid leachates from soils taken from the three sites. EHW - Eagle Hotel, Winchester; STH - Stonehenge; MUW - Monkton-up-Wimbourne. For MUW: a - burial soil, b - soil from elsewhere in the excavated area, c - underlying chalk rock*

Site	Aqueous $^{87}Sr/^{86}Sr$	Acetic acid $^{87}Sr/^{86}Sr$
EHW	0.70756	0.70736
STH	0.70766	0.70774
MUWa	0.70754	0.70765
MUWb	0.70766	0.70753
MUWc	0.70754	0.70750

3.2 Lead isotope data

Lead concentrations and isotope compositional data for tooth enamel samples from the archaeological teeth are given in Table 3 and plotted in Figure 2. Lead isotope data for the associated soil samples are given in Table 4.

Lead isotope ratios are reported here as the non-standard $^{207}Pb/^{206}Pb$ and $^{208}Pb/^{206}Pb$ ratios which are widely used in archaeological literature. Although it is desirable to facilitate the wider comparison of data by use of the standard ^{204}Pb ratios, the very low lead concentrations of the teeth in this case resulted in ^{204}Pb ratios of relatively poor precision. Since the main purpose of the study involves the inter-comparison of the teeth and local soil, ^{206}Pb ratios were adopted.

Significant differences in the lead concentrations and isotope composition of tooth enamel between the groups examined are apparent (Table 3, Figure 2). As might be expected the Romano-British EHW group appear to have much higher lead concentrations than the prehistoric group with closely similar isotope compositions typical of UK lead ores. It is notable that G339b has essentially the same isotopic composition as the rest of the group, but that concentration data clearly show that this sample was contaminated with lead derived from the coffin in which the tooth was buried. A second tooth from the same burial, G339a, also has an enamel lead concentration high enough to suggest contamination from the coffin. In general, the lead isotope ratios of tooth enamel from the EHW site are radically different to those of soil derived lead from the burial environment. Similar conclusions are likely for the STH individual. Although soil leaches from this specific burial environment did not yield sufficient lead for accurate isotope ratio analysis, it is highly likely that the soil lead isotope composition would be similar to those of the EHW and MUW sites given the similarity of the underlying geology. Tooth enamel from the STH burial however, again has a very different isotope composition, similar to that of UK lead ores.

Table 3 *Lead concentration and isotope ratio composition data for enamel samples from the archaeological teeth. EHW - Eagle Hotel, Winchester; STH - Stonehenge; MUW - Monkton-up-Wimbourne. *contamination from lead coffin*

Site	Sample	Tooth	Pb (ppm)	$^{207}Pb/^{206}Pb$	$^{208}Pb/^{206}Pb$
EHW	G318	P^1right	2.07	0.8479	2.0830
EHW	G319	P_1left	1.81	0.8476	2.0840
EHW	G326	P^1right	8.56	0.8474	2.0833
EHW	G339a	M^3right	41.8	0.8475	2.0810
EHW	G339b	P_2right	~1500*	0.8472	2.0810
STH	STH1	P^1right	2.24	0.8496	2.0978
MUW	A$_{decid.}$	c^1right	0.26	0.8378	2.0484
MUW	A$_{perm.}$	M^1right	0.33	0.8321	2.0501
MUW	B$_{decid.}$	c^1left	0.29	0.8324	2.0502
MUW	B$_{perm.}$	P^1left	0.15	0.8318	2.0499
MUW	C$_{perm.}$	P^2right	0.23	0.8498	2.0786
MUW	D$_{decid.}$	c^1left	0.68	0.8356	2.0532
MUW	D$_{perm.}$	C^1left	0.25	0.8279	2.0391

Table 4 *Lead isotope data for acid leachates from soils taken from two of the sites. EHW - Eagle Hotel, Winchester; MUW - Monkton-up-Wimbourne. The Stonehenge soil produced too little Pb for reliable analysis. For MUW: a - burial soil, b - soil from elsewhere in the excavated area, c - underlying chalk rock*

Site	$^{207}Pb/^{206}Pb$	$^{208}Pb/^{206}Pb$
EHW	0.8271	2.0476
MUWa	0.8256	2.0414
MUWb	0.8245	2.0391
MUWc	0.8275	2.0449

The prehistoric MUW burials show a very different pattern of lead concentrations and isotope ratios. All have very considerably lower lead concentrations and, although the adult female (C) has a tooth enamel lead isotope composition close to that expected for UK lead ores, all of the juveniles have quite different lead isotope ratios plotting either close to the field established for soils overlaying the chalk or intermediate between these values and that for the ores.

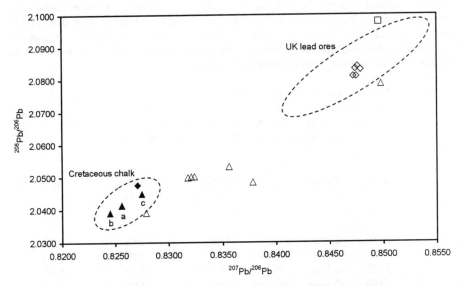

Figure 2 *Lead isotope data for tooth enamel samples (open symbols) from each of the three sites: EHW - diamond symbols, STH - square symbols, MUW - triangular symbols. Shown with their respective soil samples (filled symbols). Ellipses show, for comparison, the approximate extent of UK ore lead values and the range of values so far obtained for soils overlaying Cretaceous chalk in southern central England. Analytical measurement errors (2σ) are smaller than the symbols for the MUW data and approximately the same size as the symbols for all other data*

3.3 Oxygen isotope data

The measured oxygen isotope composition of the tooth enamel samples ($\delta^{18}O_p$) is given in Table 5. Data are calibrated, after Levinson[30], to give the isotopic composition of water ingested *in vivo* which relates to meteoric water at the place of childhood residence. For the juveniles considered here the situation is complicated by oxygen isotope fractionation in antenatal and postnatal environment. Deciduous tooth mineralisation takes place *in utero*. We have shown elsewhere[19] that tooth enamel phosphate oxygen in the developing foetus is enriched in ^{18}O by a further ~2‰ compared with the mothers drinking water as a result of additional metabolic processing. The effect continues after birth with breast feeding and only ceases at weaning. This effect is expected to result in estimates of $\delta^{18}O_w$ from deciduous teeth which are approximately 2‰ less negative than the true values. Considering the MUW individuals, the permanent tooth from individual D, a canine, is also highly likely to have formed prior to weaning which explains it's apparently high value for $\delta^{18}O_p$.

4 CONCLUSIONS AND DISCUSSION

For all of the individuals studied tooth enamel strontium isotope ratios would appear to be substantially unaffected by diagenesis and offer the possibility to be used for residency reconstruction. On the basis of the strontium data, none of the individuals examined would appear to have had a childhood diet restricted to the chalk geology alone, but one that involved foodstuffs drawn also from a catchment involving a more radiogenic environment. Such a conclusion is perhaps unsurprising for the Romano-British and early medieval period, but quite revealing of the Neolithic, supporting progressive archaeological thinking that the period may have been characterised by a more diverse economy and less sedentary existence than was once thought.

Table 5 *Mean oxygen isotope ratio composition data for enamel samples from the archaeological teeth and calibrated, after Levinson[30], to give the isotopic composition of drinking (meteoric) water with standard deviation. EHW - Eagle Hotel, Winchester; STH - Stonehenge; MUW - Monkton-up-Wimbourne*

Site	Sample	Tooth	Enamel $\delta^{18}O_p$ SMOW ‰	Water $\delta^{18}O_w$ SMOW ‰	σ_w	n
EHW	G318	P^1right	17.63	-3.86	0.05	2
EHW	G319	P_1left	18.00	-3.04	0.23	4
EHW	G326	P^1right	17.24	-4.70	0.69	3
EHW	G339a	M^3right	17.42	-4.30	0.52	3
EHW	G339b	P_2right	17.74	-3.60	0.49	3
STH	STH1	P^1right	15.97	-7.46	0.18	2
MUW	$A_{decid.}$	c^1right	18.62	-1.69	0.38	3
MUW	$A_{perm.}$	M^1right	17.54	-4.05	0.26	4
MUW	$B_{decid.}$	c^1left	17.94	-3.17	0.36	3
MUW	$B_{perm.}$	P^1left	17.40	-4.34	0.62	2
MUW	$C_{perm.}$	P^2right	17.83	-3.40	0.22	3
MUW	$D_{decid.}$	c^1left	18.45	-2.06	1.83	2
MUW	$D_{perm.}$	C^1left	18.85	-1.19	0.16	2

Significant differences between the lead isotope composition of tooth enamel and associated soil in almost all of the cases considered suggest that for lead too the tissue is not likely to have been significantly affected by diagenesis. Although, for post-metallurgical burials in sites for which manufactured lead products are a potentially significant source of post-mortem contamination, the situation remains inherently ambiguous. There can be little doubt that individual G339, buried within a lead coffin, has tissues now swamped by diagenetic lead from the burial environment. For the other EHW individuals however, the situation is less clear. Their tooth enamel lead concentrations, whilst significantly higher than those of their pre-metallurgical correlates, were not so excessive as to rule out ingestion during life as the source. Indeed, the STH individual, for which there was no evidence for contamination of the burial environment with lead or lead products, has a similar tooth enamel lead concentration and isotope ratios to the Romano-British group. At both EHW and STH, for the majority of individuals lead concentrations and isotope ratios could be explained by *in vivo* exposure to lead-based products derived from UK sources, but, for EHW at least, lead diagenesis remains a possibility which cannot be excluded on the evidence presented here.

For the pre-metallurgical, Neolithic, group dietary lead, like strontium, can be linked directly to the geology of place of childhood residence. In this case, the three juveniles examined offered the opportunity to analyse both deciduous and permanent tooth enamel and comment on residential mobility between birth (or in utero), early childhood and death. The data are discussed in detail elsewhere[16], but can be used to chart a pattern of movement between the chalk and a more radiogenic strontium environment and back to the chalk within the lifetimes of the two eldest children. The distinctive strontium and lead isotope composition of the adult female's (C) tooth enamel indicates that her place of childhood residence was probably 80km north east of her burial in the Mendips and may hint at the location of the other region involved in this pattern of movement.

Preliminary oxygen isotope data reported here (Table 5), broadly confirm the suggestion of a local UK origin for the individuals investigated. Although current estimates of $\delta^{18}O_p$ reproducibility based on replicate measurements of NBS120b and tooth enamel are approximately 0.2‰ (2σ), the currently available calibration and limitations on the number of replicate measurements possible for each individual limit the precision with which we can estimate $\delta^{18}O_w$ (σ_w, Table 5). Nevertheless, preliminary conclusions are revealing. The oxygen isotope composition of meteoric water from southern central England today is in the range -7.5 to -5.5‰. The 7th century AD individual from Stonehenge is at the more negative end of this distribution, reflecting climatic conditions similar, or perhaps slightly cooler, than those prevailing in the area today. Taking account of the anticipated additional fractionation in deciduous tooth formation, the Neolithic MUW individuals suggest a drinking water oxygen isotope composition around -4 to -3‰, significantly less negative than those of today. Less negative values of $\delta^{18}O$ in the Neolithic in this area can probably be explained by the warmer conditions of the later Climatic Optimum, which continued to influence southern England in the third millennium BC. The EHW group also appear to have been drinking water with a rather less negative value for $\delta^{18}O$ than would be expected today (around -5 to -3‰). There is no strong evidence for a significantly warmer climate prevailing in southern England during the Roman period, but the results may be explained in part by water resource supplementation with groundwater drawn from the chalk aquifer, which is not yet characterised.

The case studies discussed here illustrate both the great potential of the combined isotopic analysis approach to archaeological life history reconstruction and some of the difficulties. The successful isolation of biogenic strontium, lead and oxygen from preserved human tooth enamel are now allowing us an unprecedented insight into the residential mobility of individuals long after death and burial, but care must be exercised in the interpretation of data. Different considerations apply to pre- and post-metallurgical populations, but the possibility exists to extract detailed and highly complementary data from all three isotopic systems for archaeological burials of any date.

5 REFERENCES

1.   J. C. Sealy, N. J. van der Merwe, A. Sillen, F. J. Kruger, and H. W. Krueger, *J. Arch. Sci.*, 1991, **18**, 399.
2.   J. C. Sealy, R. Armstrong, and C. Schrire, *Antiquity*, 1995, **69**, 290.
3.   G. Grupe, T. D. Price, P. Schröter, F. Söllner, C. M. Johnson, and B. L. Beard, *App. Geochem.*, 1997, **12**, 517.
4.   G. Grupe, T. D. Price, and F. Söllner, *App. Geochem.*, 1999, **14**, 271.
5.   T. D. Price, G. Grupe, and P. Schröter, *App. Geochem.*, 1994, **9**, 413.
6.   T. D. Price, C. M. Johnson, J. A. Ezzo, J. E. Ericson, and J. H. Burton, *J. Arch. Sci.*, 1994, **21**, 315.
7.   B. L. Gulson, C. W. Jameson, and B. R. Gillings, *J. Forensic Sci.*, 1997, **42**, 787.
8.   B. L. Gulson, *Environ. Health Persp.*, 1996, **104**, 306.

9.   B. L. Gulson and D. Wilson, *Arch. Environ. Health*, 1994, **49**, 279.
10.   H. C. Fricke, J. R. O'Neil, and N. Lynnerup, *Geology*, 1995, **23**, 869.
11.   H. P. Schwarcz, L. Gibbs, and M. Knyf, in *Snake Hill: An Investigation of a Military Cemetery from the War of 1812*, ed. S. Pfeiffer and R. F. Williamson, Dundurn Press, Toronto, 1991, p. 263.
12.   B. Barreiro, P. Budd, C. Chenery, and J. Montgomery, in Applied Isotope Geochemistry Conference, Lake Louise, Canada, September 1997.
13.   P. Budd, J. Christensen, R. Haggerty, A. N. Halliday, J. Montgomery, and S. M. M. Young, in 213th ACS National Meeting, San Francisco April 13-17 1997, p. 35-GEOC.
14.   P. Budd, J. Montgomery, P. Rainbird, R. Thomas, and S. Young, in *The Pacific from 5000 to 2000 BP: colonisation and transformations*, ed. J.-C. Galipaud and I. Lilley, Institut de Recherche pour le Développement, Paris, 1999, p. 301.
15.   J. Montgomery, P. Budd, A. Cox, P. Krause, and R. G. Thomas, in *Metals in Antiquity, BAR Int. Series 792*, ed. S. M. M. Young, A. M. Pollard, P. Budd, and R. A. Ixer, Archaeopress, Oxford, 1999, p. 290.
16.   J. Montgomery, P. Budd, and J. Evans, *Euro. J. Arch.*, 2000, **3(3)**, 407.
17.   J. Montgomery, PhD Thesis, University of Bradford, in prep.
18.   J. D. Blum, E. H. Taliaferro, M. T. Weisse, and R. T. Holmes, *Biogeochem.*, 2000, **49**, 87.
19.   P. Budd, J. Montgomery, J. Evans, and B. Barreiro, *Sci. Tot. Environ*, in press.
20.   P. Budd, R. Haggerty, A. M. Pollard, B. Scaife, and R. G. Thomas, *Antiquity*, 1996, **70**, 168.
21.   E. J. Underwood, *Trace Elements in Human and Animal Nutrition*, Academic Press, London, 1977.
22.   I. M. Shapiro, H. L. Needleman, and O. C. Tuncay, *Environ. Resch.*, 1972, **5**, 467.
23.   A. Longinelli, *Geochim. Cosmochim. Acta*, 1984, **48**, 385.
24.   B. Luz, Y. Kolodny, and M. Horowitz, *Geochim. Cosmochim. Acta*, 1984, **48**, 1689.
25.   B. Luz and Y. Kolodny, *Earth and Planetary Sci. Let.*, 1985, **75**, 29.
26.   B. Luz, A. B. Cormie, and H. P. Schwarcz, *Geochim. Cosmochim. Acta*, 1990, **54**, 1723.
27.   D. D'Angela and A. Longinelli, *Chem. Geol. (Iso. Geosci. Sectn.)*, 1990, **86**, 75.
28.   D. D'Angela and A. Longinelli, *Chem. Geol. (Iso. Geosci. Sectn.)*, 1993, **103**, 171.
29.   J. D. Bryant and P. N. Froelich, *Geochim. Cosmochim. Acta*, 1995, **59**, 4523.
30.   A. A. Levinson, B. Luz, and Y. Kolodny, *App. Geochem.*, 1987, **2**, 367.
31.   J. D. Bryant, B. Luz, and P. N. Froelich, *Palaeogeo. Palaeoclim. Palaeoecol.*, 1994, **107**, 303.
32.   J. D. Bryant and P. N. Froelich, *Geology*, 1996, **24**, 477.
33.   P. Iacumin, H. Bocherens, A. Mariotti, and A. Longinelli, *Palaeogeo. Palaeoclim. Palaeoecol*, 1996, **126**, 15.
34.   C. Lecuyer, P. Grandjean, and C. C. Emig, *Palaeogeo. Palaeoclim. Palaeoecol*, 1996, **126**, 101.
35.   C. D. White, M. W. Spence, H. L. Q. Stuart-Williams, and H. P. Schwarcz, *J. Arch. Sci.*, 1998, **25**, 643.
36.   A. Sillen, *Palaeobiol.*, 1986, 12, 311.
37.   A. Sillen and R. LeGeros, *J. Arch. Sci.*, 1991, **18**, 385.
38.   P. Budd, J. Montgomery, B. Barreiro, and R. G. Thomas, *App. Geochem.*, 2000, **15**, 687.

39. P. Grandjean, *CRC Crit. Rev. Toxicol.*, 1988, **19**, 11.
40. A. Hisanaga, Y. Eguchi, M. Hirata, and N. Ishinishi, *Biol. Trace Ele. Res.*, 1988, **16**, 77.
41. C. C. Patterson, J. E. Ericson, M. Manea-Krichten, and H. Shirahata, *Sci. Tot. Environ.*, 1991, **107**, 205.
42. A. Mackie, A. Townshend, and H. A. Waldron, *J. Arch. Sci.*, 1975, **2**, 235.
43. H. A. Waldron, A. Mackie, and A. Townshend, *Archaeometry*, 1976, **18**, 221.
44. H. A. Waldron, *Am. J. Phys. Anthrop.*, 1981, 55, 395.
45. P. Budd, J. Montgomery, A. Cox, P. Krause, B. Barreiro, and R. G. Thomas, *Sci. Tot. Environ.*, 1998, **220**, 121.
46. P. Iacumin, H. Bocherens, A. Mariotti, and A. Longinelli, *Earth and Planetary Sci. Let.*, 1996, **142**, 1.
47. M. E. Ketterer, M. J. Peters, and P. J. Tisdale, *J. Anal. Atom. Spec.*, 1991, **6**, 439.
48. A. J. Walder, I. Platzner, and P. A. Freedman, *J. Anal. Atom. Spec*, 1993, **8**, 19.
49. A. P. Dickin, *Radiogenic Isotopes*, Cambridge University Press, Cambridge, 1995.
50. J. A. Evans, *J. Geol. Soc. London*, 1996, **153**, 101.

6 ACKNOWLEDGEMENTS

The authors would like to thank Drs Jackie McKinley and Charlotte Roberts for their help and Martin Green, Julian Richards, Phil Greatorex, Andy Young, Geraldine Barber and Mike Pitts for providing the site information and samples for analysis. Thanks also to Drs Barbara Barreiro and Geoff Nowell for analytical support at NIGL. Sampling and analyses were carried out whilst PB and JM were supported by the NERC: NIGL Publication No. 425.

Section 6

Speciation

ISOTOPE DILUTION ANALYSIS FOR TRACE METAL SPECIATION

J. Ignacio García Alonso, Jorge Ruiz Encinar, Cristina Sariego Muñiz, J. Manuel Marchante Gayón and Alfredo Sanz Medel.

Department of Physical and Analytical Chemistry. Faculty of Chemistry. University of Oviedo. Julián Clavería 8, 33006 Oviedo, Spain.

1 INTRODUCTION

The speciation of trace elements in clinical, biological and environmental samples can give important information on the availability, cycling, bioaccumulation and fate of those elements. That information could be used to asses the environmental impact of a certain compound or to establish bioaccumulation pathways or detoxification procedures. It is clear that the quality of the information obtained will be directly related to the accuracy and precision of the analytical data obtained.

Among the different approaches used to undertake elemental speciation in biological, environmental and clinical samples hyphenated techniques are preferred[1]. In this way, a separation technique (chromatography or electrophoresis) is coupled on-line to an element-specific atomic detector. High Performance Liquid Chromatography (HPLC) is the most popular technique for the separation of proteins in human serum prior to atomic detection while Gas Chromatography is usually selected for the analysis of organometallic compounds in environmental samples. For the separation of proteins different mechanisms has been used for this purpose, including size-exclusion[2,3,4] and ion-exchange[5,6,7,8]. For detection both Inductively Coupled Plasma Atomic Emission Spectrometry (ICP-AES)[9,10] and, more recently, Inductively Coupled Plasma Mass Spectrometry (ICP-MS)[11], both with quadrupole[2,8] and sector field mass analysers[4,6], are the main atomic techniques used for the specific on-line elemental detection. In contrast with ICP-AES, ICP-MS exhibit better detection limits and the possibility of performing isotopic measurements. A further step are sector field ICP-MS which offer the advantage of having sufficient resolution to overcome polyatomic ions with the same nominal m/z as the element of interest[12,13].

For the determination of butyltin compounds in environmental samples the preferred techniques are GC-AAS, GC-MIP-AES, GC-FPD, GC-MS[14] and, more recently, GC-ICP-MS[15]. The extremely low concentrations of organotin compounds in environmental and biological matrices demand powerful detection techniques[16]. The coupling of gas chromatography to ICP-MS appears to be one of the techniques of choice to perform this type of speciation analysis due to its extremely high sensitivity[17,18] and multi-isotopic capabilities. The measurement of isotope ratios would open the way for isotope dilution analysis procedures to be performed for this type of compounds.

Despite the excellent advantages of coupled techniques for elemental speciation studies, accurate results are not easily achievable. For the speciation of trace elements in blood serum, one of the main reasons for that is the lack of commercially available standards for all the metal-species under study in this material. On the other hand, for the analysis of organotin compounds in sediments, it is necessary to improve the quality of the methods used[19-21]. It is well known that organotin speciation involves a number of discrete analytical steps comprising extraction, clean up, preconcentration, derivatisation, separation and specific detection which all could be a source of problems[22,23]. All these critical steps make speciation analysis of organotin compounds a difficult task.

An attractive approach to get reliable determinations in speciation analysis consists on the application of Isotope Dilution (ID) methodologies on-line both with the couplings HPLC-ICP-MS or GC-ICP-MS. All the inherent advantages of ID (high accuracy and precision) are implemented to the separation/determination procedure and it is likely to have a wider use in the future[24]. This concept was first presented by Rottmann and Heumann[25] for the determination of heavy metal interactions with dissolved organic materials in natural aquatic systems. Two different approaches of the method were described corresponding to the spiking method used[26,27,28]: a) the species-specific spiking mode[27,28] and b) the species-unspecific spiking mode[25,26]. From the two different spiking methods described, the species-unspecific spiking method can be considered suitable for protein speciation in body fluids. This mode is applied when the structure of the compounds is unknown or difficult to determine and standards are not available. The separation is performed first and the enriched isotope is added post-column. On the other hand, the species-specific spiking mode can be applied when the structure of the compound is well known and adequate spikes are available or can be synthesised. In this latter case the spike is added at the beginning of the analytical procedure as in any other isotope dilution experiment. Examples of this approach have been published for the speciation of chromium[29], iodine[27], lead[30], selenium[31], mercury[32] and tin[33].

In this paper we discuss recent advances in our laboratory on isotope dilution analysis for trace metal speciation using both species-specific and species-unspecific spiking. First, we will report on the application of isotope dilution analysis on-line with Fast Protein Liquid Chromatography (FPLC) coupled to a double focusing ICP-MS (ELEMENT, Finnigan MAT) for the speciation of Fe, Cu and Zn in human serum. For this analysis the species-unspecific spike mode is suitable due to the complexity of the structure of the main proteins binding Fe (transferrin[7]), Cu (ceruloplasmin[34]) and Zn (albumin[35]) in human serum. Secondly, the use of a [119]Sn enriched mono-, di- and tributyltin spike for the accurate determination of all three tin species in sediments will be described.

2 EXPERIMENTAL

2.1 Instrumentation

2.1.1 Speciation of trace metals in human serum. The chromatographic equipment used for the separation of serum proteins consisted on a Shimadzu LC-10 AD HPLC pump (Kyoto, Japan) with a Rheodyne Model 7125 (Cotati, CA, USA) sample injection valve fitted with a 100 μl loop. The chromatographic column was a Mono-Q HR 5/5 FPLC analytical column (50x5 mm id) (Pharmacia LKB Biotech, Uppsala, Sweden), a strong anion-exchanger based on a beaded hydrophilic resin with a particle size of 10 μm. The UV measurements were performed with a Waters 484 UV/VIS absorption detector (Waters Corporation, Massachusetts, USA) and a Shimadzu C-R6A recording integrator. A scavenger column (25x0.5 mm id) was placed between the pump and the injection valve to decontaminate the mobile phases. The column was packed with Kelex-100 (Schering, Germany) impregnated silica C_{18} material (20 µm particle size) (Bondapack, Waters Corporation, Massachusetts, USA). The double focusing ICP mass spectrometer used was the ELEMENT from Finnigan-MAT (Bremen, Germany), which was operated at medium resolution ($m/\Delta m=3000$) through all the experiments. All measurements were made using a Scott type spray chamber working at room temperature, a Meinhard concentric nebuliser and a demountable torch, in which the injector can be easily extracted for cleaning. The instrument was placed in a clean room. Two peristaltic pumps were used, a Minipuls 2 peristaltic pump (Gilson, France) for the waste from the spray chamber and a Spectec peristaltic pump (Erding, Germany) to insert the enriched isotope at the end of the chromatographic column through a T piece. The experimental set-up is shown in Figure 1.

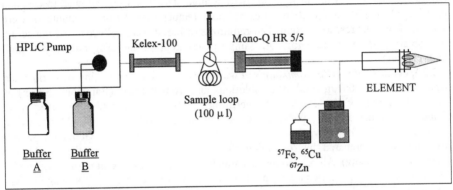

Figure 1 *Experimental set-up*

2.1.2 Speciation of butyltin compounds in sediments. A Hewlett Packard (Palo Alto, CA. USA) gas chromatograph model 6890, fitted with a split/splitless injector and a HP-5 capillary column (crosslinked 5% phenyl-methyl siloxane, 30 m x 0.32 mm i.d. x 0.25 µm coating), was used for the separation of the organometallic compounds. The gas chromatograph was coupled to a Hewlett Packard model HP-4500 Inductively Coupled Plasma Mass Spectrometer (Yokogawa Analytical Systems, Tokyo, Japan), with the transfer line described in detail in reference 15. Basically, the last 10 cm of the separation column is inserted into a copper tube which is maintained at 250 °C with the help of a metallic block equipped with an electric heater and a temperature sensor and connected to the GC instrument. The temperature of the copper tube is controlled directly by the gas chromatograph (back injector controller). The top of the copper tube is connected to a Swagelok 1/4" T piece and inserted inside the T so the copper tube terminates beyond the side arm of the T piece. A flow of argon carrier gas of ca. 1 L min^{-1} is introduced through the side arm of the T piece to transport the analytes to the plasma. The copper tube is narrower inside the T piece and the argon carrier flow goes externally to this copper tube providing a sheathing flow which prevents condensation of the analytes inside the T piece and on the PTFE tube which connects the exit of the T piece to the plasma. Performance characteristics of this interface for the speciation of butyltin compounds can be found in reference 33.

2.2 Reagents and materials

2.2.1 Speciation of trace metals in human serum.   The Tris (hidroxymethyl) ammino-methane and the ammonium acetate used on the preparation of the mobile phases and also the HCl used for pH adjusting of the mobile phases were purchased from Merck (Darmstadt, Germany). Human protein standards (albumin, apotransferrin and inmunoglobulin G) were obtain from Sigma (St. Louis, MO, USA). Stable isotope solutions enriched in ^{57}Fe, ^{65}Cu and ^{67}Zn with certified abundances (92.45% of ^{57}Fe, 99.61% of ^{65}Cu and 94.60% of ^{67}Zn), were purchased from Spectrascan (Teknolab A.S. Dröbak, Norway) and were diluted appropriately for the preparation of the aqueous solution for post-column addition. They were kept refrigerated at 4°C. Serum samples from patients on hemodialysis were obtained through the Research Unit from the Hospital Central de Asturias. Blood samples were extracted from the arterial line at the end of the dialysis session using a standard syringe. The blood was allowed to clot and centrifuged at 800 xg for 20 minutes at room temperature. Healthy volunteers were selected from our research group. Ten millilitres of blood were collected and the serum was obtained as described before. All samples were kept on the refrigerator at 4°C until measurement.

2.2.2 Speciation of butyltin compounds in sediments.   Inorganic tin was obtained from Merck as 1000 µg mL^{-1} stock solutions and from Panreac (Barcelona, Spain) as metal. Dilutions of the stock were performed with 1% v/v subboiled nitric acid. Tributyltin chloride (TBT, 96%), Dibutyltin chloride (DBT, 97%) and Monobutyltin chloride (MBT, 95%) were obtained from Aldrich (Steinheim, Germany). Stock solutions were prepared by dissolving the corresponding salt in methanol (Merck, Darmstadt, Germany). All organometallic standard solutions were kept in the dark at 4 °C and diluted working solutions were prepared daily before the analysis. Tetraethyltin was obtained by ethylation of Sn(IV) (Merck) with sodium tetraethylborate (Strem Chemicals, Bisheheim, France) at pH of 5.4 and extraction into hexane. Sediment reference materials tested were PACS-2 and CRM-646 purchased from NRCC (Ottawa, Ontario, Canada) and BCR (Retieseweg, Geel, Belgium) respectively. Tin metal, enriched in ^{119}Sn, was obtained from Cambridge Isotope Laboratories (Andover, MA, USA) and the different butylhalides were purchased both from Aldrich.

All other reagents were of analytical reagent grade. Ultrapure water was obtained from a Milli-Q 185 system (Millipore, Molsheim, France).

2.3 Procedures

2.3.1 Speciation of trace metals in human serum.   Serum samples were injected directly into the column with no dilution. After equilibration of the column with buffer A for ten minutes, a linear gradient of 20 min of ammonium acetate (0-0.5 mol l^{-1}) (buffer B) was used for the separation of the protein species. The chromatographic conditions are indicated in Table 1. The mobile phases were previously degassed with Helium for 10 minutes. The flow coming from the column was mixed with the enriched isotopes (^{57}Fe, ^{65}Cu and ^{67}Zn) and the mixture was nebulised into the plasma. The signals for ^{56}Fe, ^{57}Fe, ^{63}Cu, ^{65}Cu, ^{64}Zn and ^{67}Zn were monitored with time. The intensity (counts/s) chromatogram was converted to a mass flow chromatogram (ng/min) and the amount of metal in each peak was obtained directly by integration of the chromatographic peaks. Protein detection was made on the UV detector working at 295 nm using the same chromatographic conditions described before.

2.3.2 Speciation of butyltin compounds in sediments.   <u>Ethylation of Sn compounds</u>:
Mixed standard solutions of different organotin compounds were adjusted to pH 5.4 with
3 mL of a 1 M acetic acid/sodium acetate buffer and 1 mL of 1% w/v sodium
tetraethylborate in 0.1 M NaOH and 1 mL of hexane were added both to derivatise and
extract the tetraalkyl compounds formed. After five minutes manual shaking the organic
layer was transferred to a glass vial and stored at -18 °C until measurement.

Table 1 *Chromatographic conditions used for the separation of proteins in human
serum*

Chromatographic column:	Mono Q HR 5/5
Injection volume:	100 μl
Flow rate:	0.9 ml min^{-1}
Mobile Phases:	
Buffer A:	Tris-HCl 0.05 mol l^{-1} (pH=7.4)
Buffer B:	(A)+ammonium acetate 0.5 mol l^{-1}
Gradient:	Time (minutes) B (%)
	0 0
	20 100
Flow rate of the spike solution:	0.1 ml min^{-1}

2.3.3 Synthesis and analysis of ^{119}Sn enriched MBT, DBT and TB.   A mixture of
90mg of ^{119}Sn enriched tin metal was mixed with 241 μL of n-butyl chloride (molar ratio
1:3) and 32 μL of triethylamine, used as transfer phase catalyst, and 11.7 mg of iodine,
used as reaction catalyst, were added and heated in a sealed tube at 160^0 C for 14 h with
magnetic stirring in a sand bath. After cooling, the brown suspension was washed with 2
mL of MeOH and 1 mL of CH$_2$Cl$_2$ and the final reaction mixture was stored at -18 °C in
the dark. The purity and isotopic composition of the final material was checked by
dilution of the stock solution with a 1+3 methanol-acetic acid mixture and ethylation as
described before. The recovery was measured by reverse isotope dilution analysis using
natural MBT,DBT and TBT chloride standards after ethylation. All measurements were
performed by GC-ICP-MS.

2.3.4   Extraction and derivatisation of organotin compounds from sediments.
Approximately 0.25 g of sediment were spiked with a diluted solution of the ^{119}Sn
enriched mixture of MBT, DBT and TBT and mixed with 1 mL of methanol and 3 mL of
acetic acid. The resultant slurry was shaken mechanically for 12 hours in stoppered glass
test tubes. After centrifugation, 200 μL of the extract were ethylated as described before.

2.3.5 Measurement of isotope ratios using GC-ICP-MS. Typical operating conditions
used for the gas chromatograph separation and the ICP-MS detection are illustrated in
Table 2. Daily optimisation of the ICP-MS conditions was performed after connection of
the GC to the ICP-MS using m/z = 80 (^{40}Ar$_2$$^+$). Integration of the chromatographic peaks
was performed using the software of the ICP-MS instrument. Isotope ratios were
measured always as peak area ratios. Mass bias was corrected using ethylated natural
organotin standards. No dead time correction was necessary on the HP-4500[36,33]

Table 2 *Chromatographic and ICP-MS conditions used for the separation and detection of butyltin compounds*

Injector parameters -	
- Injection mode	Split/splitless
- Split time	0.5 min
- Injection volume	1 µL
- Splitting ratio	1:20
- Injection temperature	250 ^{0}C
GC parameters -	
- Column	HP-5 (30 m x 0.25 mm x 0.25 µm)
- Carrier gas/inlet pressure	He/ 15 psi
- GC program	50 ^{0}C (0.5 min) to 250^{0}C (2 min) at 30^{0}C min^{-1}
- Transfer line PFA tube	80 cm length, 1.5 mm i.d.
- Heating block temperature	250^{0}C
ICP-MS parameters -	
- RF power	1300 W
- Carrier gas flow rate	1 L min^{-1}
- Intermediate gas flow rate	1 L min^{-1}
- Outer gas flow rate	15 L min^{-1}
Data acquisition parameters -	
- Points per peak	1
- Integration time per point	0.066
- Isotopes selected	3 (118, 119 and 120)

3 RESULTS AND DISCUSSION

3.1 Speciation of trace elements in serum

3.1.1 On-line isotope dilution equation.   After the chromatographic separation, the flow containing the sample (s) is pumped at a flow rate f_s (ml min^{-1}) and mixed with the spike solution (sp) pumped at a flow rate f_{sp} (ml min^{-1}). The total flow rate (atom min^{-1}) for isotopes a and b of the sample is:

$$\text{a) } N_s^a \cdot f_s \cdot d_s + N_{sp}^a \cdot d_{sp} \cdot f_{sp} = N_m^a \cdot (f_s + f_{sp})$$

$$\text{b) } N_s^b \cdot f_s \cdot d_s + N_{sp}^b \cdot d_{sp} \cdot f_{sp} = N_m^b \cdot (f_s + f_{sp})$$

(1)

with N (atom/ml) and d_s and d_{sp} the density of the sample and the spike respectively.

Dividing (1a) by (1b) and rearranging the expression:

$$N_s^a \cdot d_s \cdot f_s + N_{sp}^a \cdot d_{sp} \cdot f_{sp} = R_m \cdot N_{sb} \cdot d_s \cdot f_s + N_{sp}^b \cdot d_{sp} \cdot f_{sp} \cdot R_m \tag{2}$$

where

$$R_m = \frac{N_m^a}{N_m^b}$$

Dividing (2) by $N_s^a \cdot d_s \cdot f_s$ and N_{sp}^b and rearranging we obtain:

$$1 - R_m \cdot R_s = \frac{d_{sp} \cdot f_{sp}}{d_s \cdot f_s} \cdot \frac{R_m - R_{sp}}{N_s^a / N_{sp}^b} \tag{3}$$

where Rs and Rsp are the isotope ratios in the sample (b/a) and the spike (a/b) respectively. Taking into account the abundances in the sample and the spike:

a)

$$N_{sp}^b = \frac{A_{sp}^b \cdot N_{sp}}{100}$$

b)

$$N_s^a = \frac{A_s^a \cdot N_s}{100}$$

(3) becomes:

$$N_s = N_{sp} \cdot \frac{d_{sp} \cdot f_{sp}}{d_s \cdot f_s} \cdot \frac{A_{sp}^b}{A_s^a} \cdot \frac{R_m - R_{sp}}{1 - R_m \cdot R_s} \tag{4}$$

or expressed in concentration:

$$M_s = C_{sp} \cdot f_{sp} \cdot d_{sp} \frac{AW_s}{AW_{sp}} \cdot \frac{A_{sp}^b}{A_s^a} \cdot \frac{R_m - R_{sp}}{1 - R_m \cdot R_s} \tag{5}$$

Where M_s is the mass flow in ng min^{-1}, C_{sp} is the concentration of the spike in ng g^{-1}, f_{sp} is the flow rate of the spike in ml min^{-1}, d_{sp} is the density of the spike in g ml^{-1}. AW_s and AW_{sp} are the atomic weights in the sample and the spike respectively. A_{sp}^b is the abundance of the isotope b (^{57}Fe, ^{65}Cu and ^{67}Zn) in the spike and A_s^a is the abundance (At%) of the isotope a (^{56}Fe, ^{63}Cu and ^{64}Zn) in the sample. Finally R_m, R_{sp} and R_s are the isotope ratios in the mixture ($^{56}Fe/^{57}Fe$, $^{63}Cu/^{65}Cu$ and $^{64}Zn/^{67}Zn$), the spike ($^{57}Fe/^{56}Fe$, $^{65}Cu/^{63}Cu$ and $^{67}Zn/^{64}Zn$) and the sample ($^{57}Fe/^{56}Fe$, $^{65}Cu/^{63}Cu$ and $^{67}Zn/^{64}Zn$) respectively. If the concentration changes with time M_s will also change with time. The integration of the chromatographic peak in the mass flow chromatogram will give the amount of metal in that fraction. The concentration is easily calculated knowing the sample volume injected. Concentration can be expressed in weight units by using the density value of the sample. Equation (5) was used in the present work.

3.1.2 Chromatographic separation. Following previous studies for the speciation of trace and ultratrace elements in human serum[5,7] using the Mono-Q HR 5/5, the chromatographic separation of Fe, Cu and Zn was optimised. The chromatographic separation was optimised with standards of proteins, 1 g l^{-1} of inmunoglobulin G, 0.5 g l^{-1} apotransferrin and 5 g l^{-1} albumin prepared in 0.01 mol l^{-1} Tris-HCl. The pH of the mobile phases was kept to the physilogical value (pH=7.4) as suggested by recent experiments[3]. The concentration of buffer B was increased to 0.5 mol l^{-1}. This allowed the elution of the serum proteins in about twenty minutes compared to the 0.25 mol l^{-1} value previously used[7] maintaining the resolution of the peaks. The flow rate of the mobile phases was also modified. To keep the total flow rate (buffer and spiked solution) to 1 ml min^{-1} different flow rates on the HPLC pump were tested. The final flow rate used was 0.9 ml min^{-1} which allowed the elution of proteins in an acceptable time. The gradient was modified using a lineal gradient of 20 minutes from 0 to 100% of buffer B. The chromatogram for the separation of proteins in undiluted human serum with UV detection (λ=295 nm) is shown in Figure 2. As can be seen from the retention times, inmunoglobulin eluted first at ca. 1 minute, followed by transferrin at 11.5 minutes and albumin at 14.5 minutes. The shape of this last peak suggests that this peak could be in reality a mixture of different proteins eluting together.

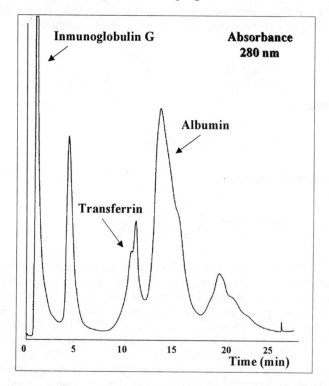

Figure 2 *Chromatogram obtained for undiluted human serum by UV detection.*

3.1.3 On-line isotope dilution analysis for Fe, Cu and Zn in human serum.   The calculation of the amount of metal bound to each protein fraction will be indicated taking Fe as an example. Copper and Zn were calculated in a similar manner. After the separation of Fe bound to different proteins in human serum, the intensity signals of ^{56}Fe and ^{57}Fe were measured on the ELEMENT. The signal of ^{56}Fe corresponded to the Fe present in human serum eluting from the column with a certain elution pattern. The intensity of ^{57}Fe corresponded to the post-column enriched isotope added and therefore was a continuous flat signal except for a small dip at the column dead time. The typical iron chromatogram obtained is shown in Figure 3.

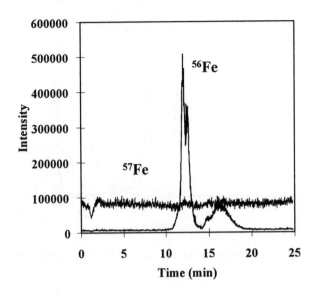

Figure 3 *Chromatogram obtained for human serum of a healthy volunteer at masses 56 and 57 (3000 resolution) after post-column spiking with enriched ^{57}Fe*

First, every single intensity value in the chromatogram had to be corrected for the dead time of the detector using the well known equation:

$$I_{real} = \frac{I_{meas}}{(1 - \tau \cdot I_{meas})}$$ (6)

where τ is the dead time which has been calculated previously as 50 ns[37]. Then, the isotope ratio ($^{56}Fe/^{57}Fe$) can be calculated for each point obtaining a chromatogram of isotope ratio vs. time. Of course, isotope ratios will need to be corrected for mass bias. We have applied the linear model previously found to be adequate[37] and used the expression:

$$R_{real} = \frac{R_{meas}}{(1 + K \cdot \Delta M)}$$ (7)

where K is the mass bias factor (in this case -0.02) and ΔM the mass difference between the isotopes. Moving average smoothing was applied next (n=5) and finally the mass flow was computed from each point in the chromatogram using the measured ratio as R_m in equation (5). Figure 4 shows a typical chromatogram obtained for Fe, Cu and Zn from a healthy volunteer.

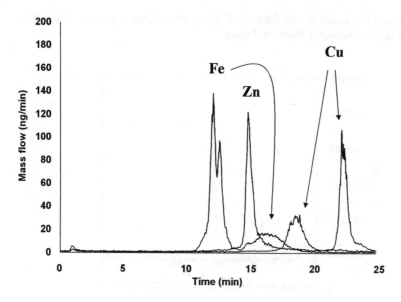

Figure 4 *Mass flow chromatogram for Fe, Cu and Zn in human serum from a healthy volunteer*

As can be observed, Fe is eluted mainly at the retention time of transferrin with a second broad peak eluting at the retention time of albumin. Zinc elutes as a single unidentified peak and copper shows two distinct peaks not corresponding to the retention times of any of the protein standards tested. The integration of the total chromatogram will provide the total concentration of Fe, Cu and Zn in the sample. These data can be compared to the total elemental composition determined separately by direct isotope dilution analysis as described previously[37] for mass balance calculations. The integration of each peak will provide the amount of the corresponding element in each protein fraction. The next section will provide examples of those determinations.

3.1.4 Analysis of human serum samples.   After the chromatographic optimisation, the analysis of human serum samples was undertaken. To check the speciation procedure, the sum of the content of metal in each of the fractions was compared to the total content of the metal present in the serum sample. The total content of Fe, Cu and Zn present in the samples was determined by total ID analysis following the method developed previously in our laboratory[37]. The recoveries were checked by the injection of three serum samples from healthy volunteers and three serum samples from patients on hemodialysis and the results obtained are shown in Table 3. As can be observed, good agreement was obtained for Fe and Zn for the six samples under study. However the copper results obtained with the chromatographic method were elevated. The difference was about the same order in every single sample and around 35%. The reason for this difference has not been identified yet.

The elution profile for the three metals under study is shown on Figure 5 for two healthy volunteers and two patients on hemodialysis. There are some significative differences which deserve some comments. As mentioned previously, Fe elutes in two fractions in healthy subjects. The first could be identified as Fe bound to transferrin and the second Fe bound to albumin. In healthy volunteers Fe bound to transferrin was about 72% on the three samples while in uremic patients was about 91%. The peak identified as albumin, seems to be formed by two different species. A first small peak overlapped by a second big Fe-albumin peak. Comparing the elution profile of the healthy subjects to the patients on hemodialysis it can be seen how the albumin peak disappears, making the small peak visible which was not possible to identify.

Table 3 *Mass balance: concentrations found by speciated ID analysis vs the total content obtained by direct ID analysis*

Subjects*	H1	H2	H3	HD1	HD2	HD3
Fe Speciated content ($\mu g\ g^{-1}$)	1.59	1.83	1.13	0.70	1.04	1.11
Fe Total content found ($\mu g\ g^{-1}$)	1.54	1.74	1.03	0.64	0.97	1.00
Zn Speciated content ($\mu g\ g^{-1}$)	0.86	0.84	1.09	0.84	0.70	0.70
Zn Total content found ($\mu g\ g^{-1}$)	0.87	0.80	0.92	0.60	0.55	0.71

*H: Healthy volunteers, HD: patients on hemodialisys.

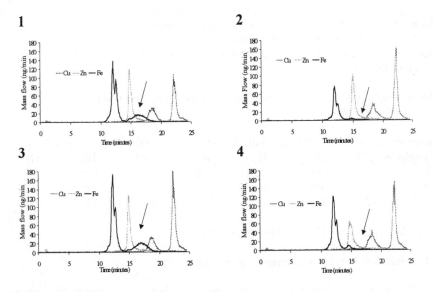

Figure 5 *Comparison of the elution profiles of healthy volunteers (1,3) and uraemic patients undergoing hemodialysis (2,4)*

Copper elutes in four different peaks. The first one a small peak at dead volume. The second coincident with the small unidentified peak of Fe, the third one at 18 minutes and the last at 22 minutes which contains a big fraction of the total copper present. This peaks could not be identified because a protein standard was not available. Previous studies with this column were protein identification was performed suggest that the fraction eluting at 22 minutes is Cu bound to ceruloplasmin[5]. Copper is known to be bound to ceruloplasmin in about 95% in human serum[38], being the rest bound to albumin and about 1% ionic or bound to low molecular weight fractions[34]. A recent study performed over 25 healthy subjects using an off-line separation of the loosely and firmly metal bound to proteins, showed that between 90.6-99.7% of Cu was bound to ceruloplasmin while 0.3-9.4% was bound to albumin[35]. It is important to stress that similar results were obtained for the injection of several serum samples from different individuals. The only differences obtained between healthy volunteers and uremic patients were observed on the total content of Cu which was higher for the last group under study.

Finally Zn elutes in a big fraction between albumin and the second Cu peak. This peak is coincident with the small peak observed for Fe and for its elution could be attributed to Zn bound to α2-macroglobulin[5]. Again, Zn is known to be bound to albumin (60-70%) and α2-macroglobulin (30-40%)[38]. Inagaki el al.[35] found that between 12.4-31.3% was bound to α2-macroglobulin (firmly bound specie) and 68.7-87.6% was bound to albumin (loosely bound specie). It is obvious that for the speciation of Cu and Zn some interactions are altering the speciation of these metals.

Due to the differences found for the speciation of Fe (Figure 5) in the two groups under study the quantification of the fractions is presented in Table 4 (expressed in ng g^{-1}). It can be seen the differences on the contents of Fe on the two species. The content of Fe decreases in hemodialysis patients. This is in agreement with the general knowledge that this patients have lower Fe contents in contrast to healthy individuals as the result of anaemia. However, the ratio of the two fractions showed to be lower in hemodialysis patients indicating that Fe bound to albumin decreases in this group. Whether this difference is atributed to the ilness itself or it is a consecuence of anaemia it can not be said at the moment.

Table 4 *Contents of Fe bound to proteins in human serum from healthy volunteers and patients on hemodialysis*

Concentration in the peaks (ng g^{-1})	Healthy volunteers*			Patients on hemodialysis*		
	H1	H2	H3	HD1	HD2	HD3
Fe-Transferrin	1170	1350	780	625	880	1000
Fe-Albumin	430	500	360	70	60	91
Ratio Fe-A/Fe-T	0.37	0.37	0.46	0.11	0.07	0.09

*H: Healthy volunteers, HD: patients on hemodialysis.

3.2 Speciation of butyltin compounds in sediments

3.2.1 Optimisation of the measurement conditions. Separation conditions described previously on the HP-5 capillary column were used[15,33] and are indicated in Table 2. Ion lens conditions were optimised daily using mass 80 (Ar_2^+) as it was found that optimum lens settings varied significantly between wet and dry plasma conditions[33]. Compromise conditions for the data acquisition parameters had to be found due to the opposite influence of integration time on sensitivity and chromatographic peak profiles. The chromatographic peak was perfectly defined for very short integration times but an increase in the noise was detected due to poorer counting statistics. On the other hand, using very long integration times, sensitivity improved but the peak definition was dramatically worsened and peak areas exhibited poor reproducibility. A maximum of 200 ms total integration time was selected and this provided adequate peak profiles. So, for isotope dilution analysis the tin isotopes measured were 118, 119 and 120 using a 66 ms integration time per isotope.

Tin isotope ratios were computed as peak area ratios between the tin isotopes measured. Under the optimised measurement conditions typical precision obtained for these ratios were between 0.1 to 2% depending on the peak size. Average precisions were around 0.8% throughout all the measurements performed (over 50 isotope ratio measurements).

It is well known that mass bias must be corrected when isotope ratio measurements are carried out.[39,40] This effect could be ascribed to the better transmission of the heavier ions than the light ones through the ICP-MS instrument. For this work a linear mass bias was assumed as it was demonstrated previously for quadrupole ICP-MS[40]. The way the correction was performed involved the injection of a natural organotin standard at the beginning and between every triplicate sample. A slight increase in mass bias was detected during the whole analysis (typically from 0.9 to 1.5% per u) but it did not affect the results as mass bias was checked and corrected every three samples.

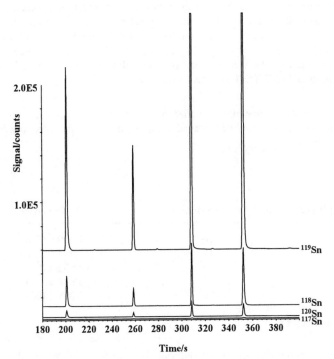

Figure 6 *Chromatogram of the synthesised enriched ^{119}Sn MBT, DBT and TBT mixture*

3.2.2 Synthesis and characterisation of a ^{119}Sn-enriched MBT, DBT and TBT spike. The initial objective of this work was to prepare a ^{119}Sn-enriched spike containing all MBT, DBT and TBT species so simultaneous isotope dilution analysis of the three organotin species could be carried out. Sisido et al[41] have reported that alkyltin chlorides were obtained in good yields by the direct reaction of alkyl chlorides with tin metal at relative low temperatures ($130^0 - 180^0$ C) when both an organic base and iodine were used as catalysts. Without either of these substances, alkyltin compounds were scarcely obtained, recovering almost all the tin metal used. After several experiments using natural tin the final synthesis conditions involved the use of 91.4 mg of [119]Sn-enriched metallic tin and a 1:3 molar excess of n-butylchloride, small amounts of triethylamine and iodine at 160^0 C for 14h in a sealed tube. Final reaction conditions are summarised under procedures. With regards to reaction yield, it was observed that from the 91.4 mg of metallic tin added to the reaction tube, 20.3 mg of residual non-reacted tin was recovered.

The analysis of the reaction product was performed after dilution with methanol-acetic acid (25:75) and customary ethylation plus detection using the coupling GC-ICP-MS. Figure 6 shows the chromatogram obtained at masses 117, 118, 119 and 120. In this chromatogram and the other shown later the profiles shown at each mass are shifted for clarity. The elution order is inorganic tin, MBT, DBT and, finally, TBT. As can be observed, similar amounts of DBT and TBT were obtained while MBT was produced with a lower yield. The first peak in the chromatogram corresponds to the inorganic (tetraethylated) tin. It is necessary to indicate that although the peak for inorganic tin is lower than those for the organotin species, it is present at higher concentration levels since its ethylation yield is much lower than for the other organotin species.

The mass bias corrected isotopic composition of the [119]Sn-enriched species is presented in Table 5. As can be observed, the 120/119 tin ratio obtained for the enriched species, 0.03603, differs greatly from the natural 120/119 tin ratio, 3.7928, allowing for the direct use of the synthesised mixture as spike solution in the simultaneously isotope dilution analysis of MBT, DBT and TBT in sediments.

Table 5 Isotopic composition of the enriched mixed spike solution computed from measured GC-ICP-MS isotope ratios for five independent injections. Uncertainty corresponds to 95 % confidence interval.

Isotopic composition	Sn isotopes	At. Abundances (%)
	116	0.029 ± 0.008
	117	0.114 ± 0.005
	118	14.33 ± 0.12
	119	82.40 ± 0.15
	120	0.069 ± 0.032

Atomic weight = 118.88 u.

The concentrations of every individual organotin species in the spike were determined by reverse isotope dilution analysis (using a natural standard of known concentration). This experiment was performed independently for each organotin species to study possible rearrangements reactions during derivation and analysis. In this vein, the mixture was first spiked with a natural TBT standard, derived and analysed by GC-ICP-MS; then with DBT and, finally, with MBT. No rearrangement reactions were observed in any of the reverse isotope dilution experiments. Figure 7 shows the chromatogram obtained for the determination of TBT in the spike solution. As can be observed, the MBT and DBT peaks exhibit tin abundances close to those obtained in the synthesised [119]Sn-enriched product while the tin isotope ratios have been drastically altered for the enriched TBT species in study.

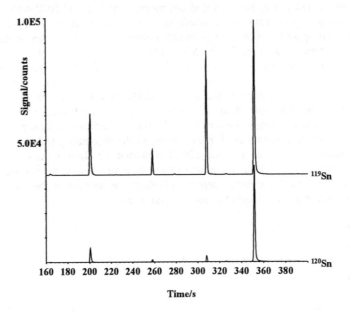

Figure 7 *Determination of enriched [119]Sn TBT by reverse isotope dilution analysis*

The final concentration and reaction yield results are summarised in Table 6. Each solution was injected three times to evaluate the precision on the tin isotope ratios and three independent reverse isotope dilution experiments were also carried out to obtain the overall uncertainty which is also included in the table. As can be observed, no significant differences were found between the precision on the isotope ratio and the overall concentration precision.

Table 6 Characterisation of the spike by reverse isotope dilution analysis. Results for MBT, DBT and TBT in ug/g (% RSD). Recovery expressed as % of the reacted metallic tin.

Replicate	**MBT** (n=3)	**DBT** (n=3)	**TBT** (n=3)
1	201 (1.1%)	1175 (0.52%)	1561 (0.79%)
2	205 (1.6%)	1160 (0.48%)	1581 (0.60%)
3	208 (0.54%)	1134 (0.62%)	1585 (0.69%)
Average	205 (1.8%)	1156 (1.8%)	1576 (0.81%)
Recovery	3.5%	19.6%	26.7%

3.2.3 Isotope dilution analysis of MBT, DBT and TBT in certified sediments. Once the methodology was developed, speciated isotope dilution analysis of the three organotin species was then carried out in two certified sediments: PACS-2 and BCR-646. Three independent spiking experiments and a blank analysis were made on each certified sediment, and each spiked sediment was injected three times. As an example, the measured 120/119 peak area ratios for the MBT, DBT and TBT in the 14 consecutive GC-ICP-MS injections are illustrated in Figure 8 for the PACS-2 certified sediment (the horizontal line corresponds to the theoretical 120/119 tin ratio[42]). Injections 1, 5, 9 and 13 correspond to the average ratios measured for a MBT, DBT and TBT mixed natural standard used for mass bias correction. Tin isotope ratio for the blank (injection 14) tuned out to be very close to the tin ratio in the spike mixture indicating, as expected, no organotin presence in the reagents used. Figure 9 shows the chromatogram obtained for the BCR-646 sediment at masses 119 and 120 (119 shifted for clarity). As it could be clearly observed also in Figure 8, an alteration of the natural tin abundances was detected for all the tin species and this quantity depended both on the amount of the tin species presented in the sample and the amount of mixed spike added.

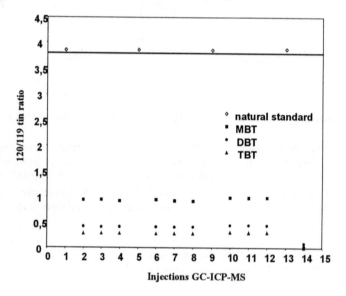

Figure 8 *Evaluation of mass bias for speciated isotope dilution analysis for PACS-2 using GC-ICP-MS*

Final individual results for PACS-2 and BCR-646 are shown in Tables 7 and 8 respectively. An excellent agreement between the certified and the found values were achieved for each individual organotin species for the two sediments. Moreover, the uncertainty associated (expressed as 95% confidence interval of the mean) to every result were always between 2.5 and 5 times lower than that for the certified values. So, as expected, the speciated isotope dilution method provided excellent accuracy and precision for the analysis of MBT, DBT and TBT in sediments.

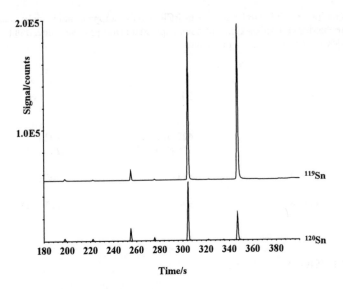

Figure 9 *Chromatogram obtained for the spiked BCR-646 sediment*

Table 7 Determination of MBT, DBT and TBT in PACS-2 by isotope dilution analysis using a mixed ^{119}Sn enriched butyltin spike. Concentration expressed in ug/g as Sn (uncertainty corresponds to 95% confidence interval of the mean).

Replicate	MBT (n=3)	DBT (n=3)	TBT (n=3)
1	0.51 (1.3%)	1.03 (1.4%)	0.88 (0.36%)
2	0.49 (1.7%)	0.99 (0.72%)	0.86 (0.37%)
3	0.49 (0.15%)	1.01 (0.45%)	0.86 (1.5%)
Average	0.50 ± 0.02	1.01 ± 0.03	0.87 ± 0.03
Certified	0.45 ± 0.05	1.09 ± 0.15	0.98 ± 0.13

Table 8 Determination of MBT, DBT and TBT in BCR-646 by isotope dilution analysis using a mixed ^{119}Sn enriched butyltin spike. Concentration expressed in ng/g as Sn (uncertainty corresponds to 95% confidence interval of the mean).

Replicate	MBT (n=3)	DBT (n=3)	TBT (n=3)
1	362 (1.2%)	415 (1.0%)	198 (0.8%)
2	387 (0.8%)	422 (0.2%)	202 (1.1%)
3	383 (2.0%)	418 (0.5%)	200 (1.0%)
Average	377 ± 24	418 ± 7	200 ± 4
Certified	410 ± 69	394 ± 36	195 ± 18

4 CONCLUSIONS

The use of isotope dilution analysis in trace metal speciation could be a powerful alternative to traditional quantification methods both using species-specific and species-unspecific spiking. In this paper we have observed that it could be applied on-line to the FPLC-double focusing ICP-MS for the quantification of Fe, Cu and Zn bound to proteins in human serum. Good recoveries were obtained for Fe and Zn speciation as compared with total IDA of sera. Differences between the speciation of healthy individuals and uremic patients were encountered for Fe speciation. This method could be used for diagnostic purposes once the causes of the differences are well known.

GC-ICP-MS in combination with species-specific isotope dilution analysis proved to be a suitable and reliable method for organotin speciation. Uncertainties associated to every sample triplicate turned out to be very close to the experimental isotope ratio precision. In addition, the accuracy obtained in this work using isotope dilution was excellent for every single tin specie in each certified sediment analysed.

In view of the above said, speciated isotope dilution could play an important role in the future for the development and validation both of reference materials and procedures in trace metal speciation in environmental and biological samples.

REFERENCES

1   J. Szpunar, Analyst, 2000, 125, 963
2   B. Gercken and R. M. Barnes, *Anal. Chem.*, 1991, **63**, 283.
3   A. Raab and P. Brätter, *J. Chromatogr B*, 1998, **707**, 17.
4   J. Wang, R. S. Houk, D. Dreessen and D. R. Wiederin, *JBIC*, 1999, **4**, 546.
5   T. Tomono, H. Ikeda and E. Tokunaga, *J. of Chromatography*, 1983, **266**, 39.
6   E. Blanco González, J. Pérez Parajón, J. I. García Alonso and A. Sanz-Medel, *J. Anal. At. Spectrom.*, 1989, **4**, 175.
7   M. Montes-Bayón, A. B. Soldado Cabezuelo, E. Blanco González, J. I. García Alonso and A. Sanz-Medel, *J. Anal. At. Spectrom.*, 1999, **14**, 947.
8   A. B. Soldado Cabezuelo, M. Montes Bayón, E. Blanco González, J. I. García Alonso and A. Sanz-Medel, *Analyst*, 1998, **123**, 865.

9   P. Gardiner, P. Brätter, B. Gercken and A. Tomiak, *J. Anal. At. Spectrom.*, 1987, **2**, 375.

10   K. Pomazal, C. Prohaska, J. Steffan, G. Reich and J. F. K. Huber, *Analyst*, 1999, **124**, 657.

11   G. K. Zoorob, J. W. Mckiernan and J. A. Caruso, *Microchim. Acta*, 1998, **128**, 145.

12   C. Sariego Muñiz, J. M. Marchante Gayón, J. I. García Alonso and A. Sanz-Medel, *J. Anal. At. Spectrom.*, 1998, **13**, 283.

13   J. M. Marchante Gayón, C. Sariego Muñiz, J. I. García Alonso and A. Sanz-Medel, *Anal. Chim. Acta*, 1999, **400**, 307.

14   F. Ariese, W. Cofino, J.L. Gómez-Ariza, G. Kramer and Ph. Quevauviller. *J. Environ. Monit.*, 1999, **1**, 191

15   M. Montes Bayón, M. Gutierrez Camblor, J.I. García Alonso and A. Sanz-Medel. *J. Anal. At. Spectrom.*, 1999, **14**, 1317

16   M. Abalos, J.M. Bayona, R. Compañó, M. Granados, C. Leal and M.D. Prat, *J. Chromatogr. A* **1997**, 788, 1.

17   R. Ritsema, T. de Smaele, L. Moens, A.S. de Jong and O.F.X. Donard, *Environ. Pollution* **1998**, 99, 271.

18   R.B. Rajendran, H. Tao, T. Nakazato and A. Miyazaki, *Analyst* **2000**, 125, 1757.

19   E. Graupera, C. Leal, M. Granados, M.D. Prat, and R. Compañó, *J. Chromatogr. A* **1999**, 846, 413.

20   O.F.X. Donard, B. Lalère, F. Martin and R. Lobinski, *Anal. Chem.* **1995**, 67, 4250.

21   M. Ceulemans, S. Slaets and F. Adams, *Talanta* **1998**, 46, 395.

22   R. Morabito, P. Massanisso and Ph. Quevauviller, *J. Trends Anal. Chem.* **2000**, 19, 113.

23   F. Adams and S. Slaets, *J. Trends Anal. Chem.* **2000**, 19, 80.

24   S. J. Hill, L. J. Pitts and A. S. Fisher, *TRAC*, 2000, **19**, (2/3), 120.

25   L. Rottmann and K. G. Heumann, *Anal. Chem.*, 1994, **66**, 3709.

26   L. Rottmann and K. G. Heumann, *Fresenius J. Anal. Chem.*, 1994, **350**, 221.

27   K. G. Heumann, L. Rottmann and J. Vogl, *J. Anal. At. Spectrom.*, 1994, **9**, 1351.

28   K. G. Heumann, S. M. Gallus, G. Rädlinger and J. Vogl, *Spectrochim. Acta Part B*, 1998, **53**, 273.

29   H.M. Kinston, D. Huo, Y. Lu and S. Chalk, *Spectrochim. Acta B*, 1998, **53**, 299

30   A.A. Brown, L. Ebdon and S.J. Hill, *Anal. Chim. Acta*, 1994, **286**, 391

31   S.M. Gallus and K.G. Heumann, *J. Anal. At. spectrom.*, 1996, **11**, 887

32   R.D. Evans and H. Hintelmann, *Fresenius J. Anal. Chem.*, 1997, **358**, 378

33   J. Ruiz Encinar, J.I. García Alonso and A. Sanz-Medel, *J. Anal. At. Spectrom.*, 2000, **15**, 1233

34   T. D. B. Lyon, S. Fletcher, G. S. Fell and M. Patriarca, *Microchem. J.*, 1996, **54**, 236.

35   K Inagaki, N. Mikuriya, S Morita, H. Haraguchi, Y. Nakahara and M. Hattori, *Analyst*, 2000, **125**, 197.

36   J.P. Valles Mota, J. Ruiz Encinar, M.R. Fernández De la Campa, J.I. García Alonso and A. Sanz-Medel, *J. Anal. At. Spectrom.* **1999**, 14, 1467.

37   C. Sariego Muñiz, J.M. Marchante Gayón, J.I. García Alonso and A.Sanz-Medel. *J. Anal. At. Spectrom.*, 1999, **14**, 1505

38   C.A. Burtis and E.R. Ashwood eds., *Tietz textbook of clinical chemistry*, W B

Saunders Company. Philadelphia:, 1994.

39   K.G. Heumann, S.M. Gallus, G. Rädlinger and J. Vogl, *J. Anal. At. Spectrom.*, 1998, **13**, 1001

40   J. Ruiz Encinar, J.I. García Alonso and A. Sanz Medel, *J. Anal. At. Spectrom.*, submitted for publication

41   K. Sisido, S. Kozima and T. Tuzi, *J. Organometal. Chem.* 1967, **9**, 109.

42   K.J.R. Rosman and P.D.P. Taylor, *J. Anal. At. Spectrom.* 1998, **13**, 45N.

ACKNOWLEDGEMENTS

The authors are grateful to the European Union for funding through the project UE-SMT4-CT98-2220. The Research Unit at the "Hospital Central de Asturias" is thanked for provision of serum samples from patients on hemodialysis .

PREDICTION OF INORGANIC AQUEOUS SPECIATION: A USEFUL TOOL FOR ICP-MS AND IC-ICP-MS ANALYSTS

D.A. Polya and P.R. Lythgoe

Department of Earth Sciences, The University of Manchester, Manchester, M13 9PL, UK.

1 INTRODUCTION

Few would question that the chemical reactivity and physical behaviour of many chemical components in aqueous solution may depend critically upon the speciation of those components. In particular, speciation may play an important role in determining the solubility, volatility and ease of adsorption of many chemical components. Explicit modelling of aqueous speciation is therefore essential for the meaningful understanding of the behaviour of, for example, many toxic and radioactive contaminants in the environment. Such speciation modelling may also be useful in understanding some potential fractionation effects during ICP-MS or coupled IC-ICP-MS analysis of aqueous solutions.

We present here a Turbo-Pascal code, PHOX, which facilitates the calculation of the equilibrium speciation of inorganic components for which suitable standard Gibbs free energy of formation data are available.. there exist numerous databases[1-4] for standard Gibbs free energies of formation at 25°C/1 atm whilst, amongst other databases[5] and packages, SUPCRT92[1] and its upgrades enable Gibbs free energy data at higher temperatures and pressures to be calculated. The PHOX output is graphical representation, for a specified component of the species predominance fields and relative solubility fields in Eh-pH space at fixed temperature, pressure and activity of specified ligands. These diagrams may consequently be used to predict, for specified conditions, (i) the predominant dissolved species of a given component and (ii) the potential for a component to precipitate or volatilise. A number of examples are presented to illustrate the utility of the code and speciation calculations to ICP-MS & IC-ICP-MS analysis.

2 CALCULATION OF SPECIES PREDOMINANCE DIAGRAMS USING PHOX

2.1 Theory

Details of the theory of the calculation of species predominance diagrams including calculation of relative mineral stability field and mineral or gas solubility boundaries may be found elsewhere [6-10] so only an outline of the theory particularly relevant to the computer code PHOX discussed in this paper is presented here.

2.1.1 Species Predominance Boundaries   The boundaries in redox parameter-pH space between the species predominance fields of two different species, Γ_1 (aq) and Γ_2 (aq) of the same chemical component can be defined as the set of redox-pH conditions under which the concentrations of the two species are equal. If thermodynamic equilibrium is assumed then the equation of such a boundary may, in general, be calculated from the equilibrium constant expression for the equilibrium below:

$$\Gamma_1 \text{ (aq)} \leftrightarrow \Gamma_2 \text{ (aq)} + h\text{H}^+ \text{ (aq)} + o\text{O}_2\text{(g)} + w\text{H}_2\text{O (aq)} + \sum_{n=n_{min}}^{n=n_{max}} l_n L_n (phase_n) \qquad (1)$$

$$K_{(1)} = \frac{\left(a_{\Gamma_2}^{aq}\right)\left(a_{H^+}^{aq}\right)^h \left(f_{O_2}\right)^o \left(a_{H_2O}^{aq}\right)^w}{\left(a_{\Gamma_1}^{aq}\right)} \prod_{n=n_{min}}^{n=n_{max}l}\left(a_{L_n}^{phase_n}\right)^{l_n} \qquad (2)$$

$$\log\left(f_{O_2}\right) = \frac{h}{o}pH + \frac{1}{o}\log\left(K_{(1)}\right) + \log\left(\frac{\gamma_{\Gamma_1}}{\gamma_{\Gamma_1}}\right) - \frac{w}{o}\log\left(a_{H_2O}^{aq}\right) - \frac{1}{o}\sum_{n=n_{min}}^{n=n_{max}} l_n \log\left(a_{L_n}^{phase_n}\right) \qquad (3)$$

where h, o and w are the stoichiometric coefficients of H^+(aq), O_2(g) and H_2O(aq) respectively; γi is the activity coefficient of the subscripted species; L_n (phase$_n$) refers to the nth species (and its host phase) selected to balance equation (1) and l_n is the stoichiometric coefficient for the nth such species; n_{max} is the number of such species as required, if $n_{max} = 0$ then $n_{min} = 0$ otherwise $n_{min} = 1$.

It is implicitly assumed that the stoichiometries of Γ_1 (aq) and Γ_2 (aq) are expressed in a form that enables the component for which the redox and pH dependent speciation is being determined to be conserved between the two species in equation (1), for example as illustrated in equation (4):

$$\text{VO}^{2+} \text{ (aq)} + 1/4\text{O}_2\text{(g)} + 2\text{H}_2\text{O (aq)} \leftrightarrow \text{VO}_{7/2}^{2-} \text{ (aq)} + 4\text{H}^+ \text{ (aq)} \qquad (4)$$

When treated to the highest degree of accuracy, in the most general case, activities of the solvent, activities of the species Γ_1 (aq) and Γ_2 (aq) and activities of the species L_n may be complex functions of both pH and redox. Particularly for dilute solutions, however, useful approximations of the location of aqueous species preodminance boundaries may be calculated by assuming unit activity of water and assuming that the activities of aqueous species may be approximated by their molal concentrations, thus enabling equation (3) to be simplified to:

$$\log(f_{O_2}) = \frac{h}{o}pH + \frac{1}{o}\log(K_{(1)}) - \frac{1}{o}\sum_{n=n_{min}}^{n=n_{max}} l_n \log(a_{L_n}^{phase_n})$$ (5)

or, in the absence of a requirement to balance the chemical equilibrium between Γ_1 (aq) and Γ_2 (aq) with species other than H^+(aq), O_2(g) and H_2O(aq), to:

$$\log(f_{O_2}) = \frac{h}{o}pH + \frac{1}{o}\log(K_{(1)})$$ (6)

A useful special case of equation (6) is where the activity of a balancing species, L_n, can be approximated to a constant value over a specified pH and redox parameter range, thus:

$$\log(f_{O_2}) = \frac{h}{o}pH + \frac{1}{o}\log(K_{(1)}) - \frac{1}{o}l_n \log(a_{L_n}^{phase_n})$$ (7)

provided that $\quad pH_{min} < pH < pH_{max}$

and $\qquad \log(f_{O_2})_{min} < \log(f_{O_2}) < \log(f_{O_2})_{max}$

where $pH_{min} - pH_{max}$ and $\log(f_{O_2})_{min} - \log(f_{O_2})_{max}$ represent the ranges of validity of the equation.

Such a treatment is particularly expedient when handling equilibria involving a secondary component which may itself exhibit pH-dependent or $\log(f_{O_2})$-dependent speciation: For example, the species predominance boundary between $Fe(OH)_3$(s) and $Fe(EDTA)^{2-}$(aq) may be determined by considering the following equilibria over appropriate restricted ranges of pH according to the pK_a values[11] of EDTA:

$$Fe(OH)_3(s) + H_4EDTA(aq) \leftrightarrow Fe(EDTA)^{2-}(aq) + 2H^+(aq) + \frac{5}{2}H_2O(aq) + \frac{1}{4}O_2(g)$$

for pH < 2.0 at 20°C / 1 atm (8a)

$$Fe(OH)_3(s) + H_3EDTA^-(aq) \leftrightarrow Fe(EDTA)^{2-}(aq) + H^+(aq) + \frac{5}{2}H_2O(aq) + \frac{1}{4}O_2(g)$$

for 2.0 < pH < 2.7 at 20°C / 1 atm (8b)

$$Fe(OH)_3(s) + H_2EDTA^{2-}(aq) \leftrightarrow Fe(EDTA)^{2-}(aq) + \frac{5}{2}H_2O(aq) + \frac{1}{4}O_2(g)$$

for 2.7 < pH < 6.2 at 20°C / 1 atm (8c)

$$Fe(OH)_3(s) + HEDTA^{3-}(aq) + H^+(aq) \leftrightarrow Fe(EDTA)^{2-}(aq) + \frac{5}{2}H_2O(aq) + \frac{1}{4}O_2(g)$$

for 6.2 < pH < 10.3 at 20°C / 1 atm (8d)

$$Fe(OH)_3(s) + EDTA^{4-}(aq) + 2H^+(aq) \leftrightarrow Fe(EDTA)^{2-}(aq) + \frac{5}{2}H_2O(aq) + \frac{1}{4}O_2(g)$$

for pH > 10.3 at 20°C / 1 atm   (8e)

2.1.2 Species predominance diagrams.

The construction of a species predominance diagram for a given component may be effected by firstly determining the loci of the species predominance boundaries between all possible pairs of the species of that component, and then by determining those boundaries or boundary segments that represent boundaries between, locally, two most predominant species in solution. Where N species are being considered, the number of combinations, C, of pairs of species is given by:

$$C = 0 + 1 + 2 + \ldots (N-2) + (N-1) = N(N-1)/2$$ (9)

This also represents the maximum number of species predominance boundaries that may be found. In cases where the relative predominance of two or more species is not a function of both pH and redox then the number of species predominance boundaries may be less than the number of possible combinations of pairs of species considered.

2.1.3 Mineral or gas solubilities and relative mineral stabilities.

The loci in redox-pH space of mineral or gas solubility boundaries may be determined by the use of analogous equations to (3) somewhat modified so that the condition for a mineral solubility boundary is:

$$\left(a_{species}^{phase}\right) = 1$$ (10)

and that for a relative mineral stability boundary is:

$$\left(a_{species_1}^{phase_1}\right) = \left(a_{species_2}^{phase_2}\right)$$ (11)

The phase rule requires that a variable, most usefully the total analytical concentration of the component being speciated, is specified for a mineral solubility boundary. Once the species predominance fields of a component have been determined, it is a computational convenience to approximate the concentration of the predominant species at a particular redox-pH condition to the total analytical concentration.

2.2 Thermodynamic data at ambient and elevated temperatures

The widespread tabulation of standard Gibbs free energy of formation data for a large range of inorganic aqueous, mineral and gaseous species at ambient temperatures and pressures mean that equilibrium constants required in equation (3) may be readily determined from:

$$K = \exp(-\Delta G_r^\circ / RT)$$ (12)

and

$$\Delta G_r^\circ = \sum_i v_i \Delta G_{f,i}^\circ \tag{13}$$

where the symbols have their usual meaning.

For higher than ambient temperatures, such as are typically found in ultrasonic nebulisers/membrane desolvators, thermodynamic data and, in particular standard Gibbs free energies of formation, may be calculated be a variety of techniques. Since for most ICP-MS and IC-ICP-MS applications the pressures do not vary substantially above atmospheric pressure, the pressure dependence of standard Gibbs free energies of formation is not considered here.

If constant enthalpy and entropy of formation are assumed, then a rough estimate of standard Gibbs free energies of formation can be made by:

$$\Delta G_{r,T}^\circ = \Delta G_{r,Tr}^\circ - \Delta S_{r,Tr}^\circ (T - Tr) \tag{14}$$

where Tr is the reference temperature (298.15 K).

More accurate estimates of standard Gibbs free energies of formation of mineral and gas species at elevated temperatures taking into account temperature dependent variations in heat capacity and of phase transitions can be obtained from Tanger and Helgeson's[12] expression for the apparent standard molal Gibbs free energy of formation:

$$G_{f,T}^\circ = G_{f,Tr}^\circ - S_{Pr,Tr}^\circ (T - Tr) + \sum_{i=1}^{i=\Phi} \int_{T_i}^{T_{i+1}} C_{Pr,i}^\circ dT - T \sum_{i=1}^{i=\Phi} \int_{T_i}^{T_{i+1}} C_{Pr,i}^\circ d\ln T - \sum_{i=1}^{i=\Phi} \frac{\Delta H_i}{T_i}(T - T_i) \tag{15}$$

where Pr is the reference pressure (1 atmosphere), Φ is the number of phase transitions from Pr,Tr to Pr,T; T_i and ΔH_i are respectively the temperature and the change in standard molal enthalpy associated with the subscripted phase transition

The corresponding expression to equation (15) for aqueous species is[1,12]:

$$G_{f,T}^\circ = G_{f,Tr}^\circ - S_{Pr,Tr}^\circ (T - Tr) - c_1 F_1(T) - c_2 F_2(T)$$
$$-\omega(Z_{Pr,T} + 1) + \omega_{Pr,Tr}\left[Z_{Pr,Tr} + 1 + Y_{Pr,Tr}(T - Tr)\right] \tag{16}$$

with

$$F_1(T) = \left[T\ln\left(\frac{T}{Tr}\right) - T - Tr\right] \tag{17}$$

and

$$F_2(T) = \left\{\left[\left(\frac{1}{T-\theta}\right) - \left(\frac{1}{Tr-\theta}\right)\right]\left[\frac{\theta-T}{\theta}\right] - \frac{T}{\theta^2}\ln\left[\frac{Tr(T-\theta)}{T(Tr-\theta)}\right]\right\} \tag{18}$$

where c_1, c_2, ω and $\omega_{Pr,Tr}$ are empirically derived constants characteristic of the aqueous species being considered; θ (228 K) is a constant characteristic of the solvent; and Z and Y are Born functions defined by Helgeson and co-workers[13] and are temperature and pressure dependent functions characteristic of the solvent.

2.3 PHOX software

The PHOX Version 2.0 software enables calculation and plotting of species preodimance diagrams in systems of up to 8 chemical components, up to 2 of which may be speciated as a function of redox and pH. The input requirements are (i) temperature, (ii) presssure, (iii) specification of the chemical system to be modelled and (iv) for each species considered, its chemical composition defined by integer stoichiometric coefficients and the standard Gibbs free energy of formation at the specified temperature and temperature. The code will process systems with up to 64 species involving up to 600 balanced equilibria for each speciated component.

The PHOX software assumes that (i) species for which data have been input constitute a complete set of species to be considered; (ii) complete equilibrium exists between all aqueous and non-aqueous species; (iii) activity coefficients on a molal concentration scale are unity for all aqueous species and the activity of water is unity; (iv) the concentration of any aqueous species of a speciated component is, in the predominance field of that species, equal to the analytical concentration of that component, provided that the solution is not saturated with respect to a solid, gaseous or liquid phase containing that component.

The last assumption constrains all constructed species predominance, relative mineral stability and mineral stability boundaries in Eh-pH space to be linear - this is clearly only an approximation, particularly near species predominance boundaries and for systems where the analytical concentration of the 2 speciated components are similar: under such conditions boundaries may actually exhibit significant curvature. Other limitations to the validity of the code include (i) phase relations and chemical activities may be modified by the presence in real systems of species and components not explicitly considered by the user; (ii) equilibrium may not be wholly or even substantially achieved in many real systems (this may be particularly relevant to ICP-MS introduction systems such as ultrasonic nebulisers and membrane desolvators); (iii) the assumption of unit activity coefficients leads to increasingly high errors in systems with increasingly high ionic strengths and may be substantial if modelling aqueous solutions with ionic strengths greater than 0.1 molal.

For all these reasons, the use of PHOX Version 2.0 might sensibly be limited to rapid preliminary assessments of the most likely predominant aqueous and non-aqueous species of a given component as a function of temperature, pressure, Eh, pH and the analytical concentrations of the specified components. As for any code calculating chemical equilibria, any meaningful application requires a satisfactory assessment of the validity of the model used. For more accurate and detailed modelling, particularly of systems with more than 8 significant components, the use of packages such as EQ3 or Geochemists Workbench may be indicated.[14-16]

PHOX Version 2.0 has been tested under DOS and WINDOWS 3.X operating systems. For reasonable speed of execution a Pentium processor with at least 16 MB RAM is recommended. Software options enable:

(1) a range of redox parameters to be selected (Eh, $\log\left(f_{O_2}\right)$, $\log\left(f_{H_2}\right)$, pε),

(2) suppression of species or phases, formation of which is kinetically inhibited, and

(3) selection of redox and pH ranges of the output plots. Published examples of the use of PHOX include those of Gize[17] and of Polya.[8-9]

An example of output of PHOX Version 2.0 is shown in figure 1 which illustrates the predominance fields in pε-pH space of various vanadium species and also shows the upper and lower stability limits of water.

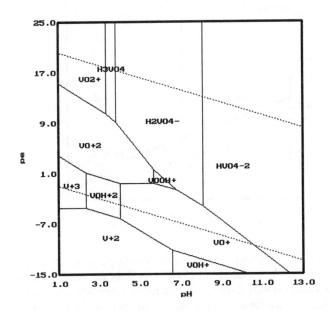

Figure 1 *pε/V vs pH V species predominance diagrams at 25°C and 1 atm. Total dissolved vanadium = 10⁻⁸ molal. Light dotted lines indicate upper and lower stability limits of water. Plots created by PHOX Version 2.0 using thermodynamic data from various sources.*[1,18-19]

3 EXAMPLES OF UTILITY TO ICP-MS & IC-ICP-MS ANALYSIS

3.1 Prediction of relative retention times in IC-ICP-MS analysis

One of the simplest and most straightforward applications of PHOX is facilitating the assignment of peaks obtained from ion chromatography, using ICP-MS or other detection techniques. For example, inorganic As(III) and As(V) have been separated on a CETAC ANX3206 column using a slightly alkaline (pH=8) eluent by Gault.[20] Inspection of a PHOX generated species predominance diagram for the system As-O-H shown on Fig. 2 indicates that at pH=8, As(III) occurs predominately as the neutral H_3AsO_3 (aq) species whilst As(V) occurs predominately as the doubly charged $HAsO_4^-$ (aq) species; this enables the low retention time peak to be identified as As(III) and the longer retention time peak to be identified as the more readily adsorbed As(V) species. Confirmation of the assignments of chromatographic peaks ideally requires several independent lines of evidence and species charge considerations aided, for example, by PHOX might usefully serve as one of these in many cases.

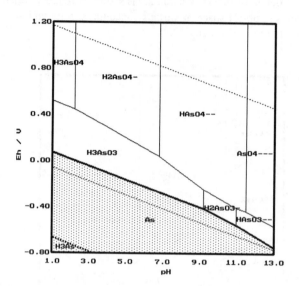

Figure 2 *Eh/V vs pH As species predominance diagrams at 25°C and 1 bar. Total dissolved arsenic = 10^{-6} molal. Solid thick lines indicate mineral saturation. Stipple indicates oversaturation with respect to a mineral or a partial pressure of arsine gas of 1 atmosphere. Light dotted lines indicate upper and lower stability limits of water. Plots created by PHOX Version 2 0 using thermodynamic data from various sources* [1,18-19]

3.2 Ensuring stabilisation of Fe-rich low pH natural waters during IC-ICP-MS

A significant problem in the IC-ICP-MS determination of metal speciation in low pH natural waters, such as acid rock drainage (ARD), is that relatively high concentrations of trace metals, and in particular iron, stabilised by low pH's may be unstable if eluents that are alkaline, neutral or even slightly acidic are used. PHOX may be used to rapidly assess whether or not dissolved metal components can be stabilised in the aqueous phase of the chromatographic column by the addition of various complexing agents. Fig 3 shows that precipitation of ferric hydroxide within the column might be expected if ARD waters with a total iron concentration of just 10^{-4} molal is passed through the column and eluted with slightly alkaline eluent, despite the model suppression of the formation of hematite, which is kinetically inhibited at near ambient temperatures. It can be seen from figure 4 (top) that even the addition of 1 molal chloride to such a sample would be insufficient to stabilise the iron in solution, however, from figure 4 (bottom) it can be seen that the addition of just 10^{-3} molal EDTA will prevent the precipitation of ferric hydroxide over a wide range of Eh conditions.

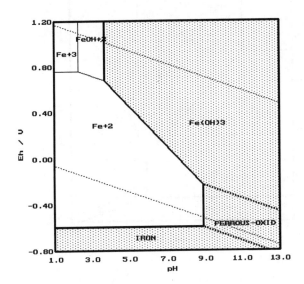

Figure 3 *Eh/V vs pH Fe species predominance diagrams at 20 °C and 1 bar. Total dissolved iron = 10^{-4} molal. Solid thick lines indicate mineral saturation (i.e. mineral solubility boundaries). Stipple indicates mineral over-saturation. Note that precipitation of hematite has been suppressed. Thick solid liLight dotted lines indicate upper and lower stability limits of water. Plots created by PHOX Version 2.0 using thermodynamic data from various sources* [1,18-19]

3.3 Species fractionation during USN/MDS

Various workers have commented on the possibility of fractionation of elements during nebulisation (including ultrasonic nebulisation/membrane desolvation) prior to the introduction of solutions to an ICP-MS.[21-24] It is generally possible to correct for such behaviour by the appropriate use of matrix-matched standards. More recently evidence for the fractionation of species of the same elemental component during USN/MDS has been found for inorganic arsenic species[25] and inorganic selenium species.[26] This behaviour is somewhat more difficult to correct for[25] but it does provide an opportunity to further our understanding of the fractionation processes that take place during USN/MDS.

Although Creed and his co-workers ascribe the fractionation of As(III) and As(V) during USN/MDS to the partial oxidation of As(III) particularly in the spray chamber,[25] there are several points in the USN/MDS process where significant species and elemental fractionation could potentially take place.

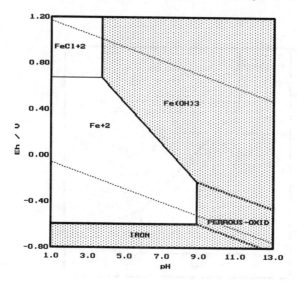

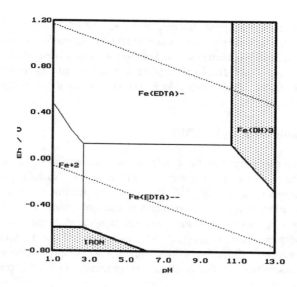

Figure 4 *Eh/V vs pH Fe species predominance diagrams at 20 °C and 1 bar in the presence of 1 molal chloride ion (top) or 10^{-3} molal EDTA (bottom). Total dissolved iron = 10^{-4} molal. See figure 3 for explanation of symbols and data sources* [1,18-19]

3.3.1 Nebulisation Nebulisation involves the creation of a spray of droplets of a range of diameters. The larger droplets tend to coagulate or fall out of suspension from the carrier gas and are ultimately are prefentially removed via a drain. The matrix-dependent effects on the droplet size distribution during nebulisation are well known, particularly as they exert a control on the overall efficiency of transport of solution into the plasma.[22,27] There is also, however, the possibility of elemental and species fractionation between the larger and small droplets. Borowiec and co-workers[21], for example, explained observed elemental fractionation effects during pneumatic nebulisation as being due to the preferential enrichment of highly charged species on the surface layers of droplets[28] and the preferential incorporation of surface layers into the finer droplets that ultimately were transported to the plasma. This provides a mechanism for fractionating elements whose predominant species are differently charged (e.g. Na^+ and Ca^{2+}) and also for fractionation differently charged species of the same elemental component (e.g. H_3AsO_3 and $H_2AsO_4^-$). Interestingly such a mechanism would tend to favour enhanced sensitivity for As(III) compared to As(V) and for Se(IV) compared to Se (VI); this is the opposite of the sensitivity contrasts actually observed [25-26] suggesting that the mechanism of surface layer enrichment of certain species, although theoretically sound, may, in practice, have relatively little impact on ICP-MS analysis.

3.3.2 Heating and subsequent condensation of solvent Heating and subsequent condensation during USN aims to remove the bulk of the solvent whilst essentially retaining the solutes as an aerosol. For volatile elements, this stage may result in the production of gaseous species which may subsequently be trapped and removed in the solvent condensate. Fassel and Bear, for example, noted that "the overall desolvation process may lead to analyte losses if an analyte is present in a form that has a significant vapor pressure at the applied desolvation tube temperatures" and cited boron and mercury as two elements that were particularly susceptible to this. The relatively low sensitivity observed for Se(IV) compared to Se(VI) might be the result of the relative ease of volatilisation of these species. Fig. 5 shows the range of stability of gaseous selenium as $Se_2(g)$ as a function of Eh and pH at temperatures typical of the maximum found in a USN as used by Lythgoe and co-workers.[26] The stability field of $Se_2(g)$ is immediately adjacent to that of the Se(IV) species H_2SeO_3 (aq) and $HSeO_3^-$ (aq) and at the relevant dissolved selenium concentrations it is apparent that the evolution of a selenium vapour from Se(IV) species seems likely to occur, thereby reducing the sensitivity of consequent ICP-MS analysis of Se(IV). It is noted that at the Eh-pH conditions given by the $Se_2(g)$ solubility boundary, the vapour pressure of Se(VI) species would theoretically be higher than that of Se(IV) species if these species occurred at the same concentrations, however, it is likely that the lower inferred volatilisation of Se(VI) is a result of (i) locally elevated Eh in the vicinity of the more oxidised Se(VI) species; and (ii) slower kinetics of the Se(VI) to Se(0) reduction compared to those of Se(IV) to Se(0).

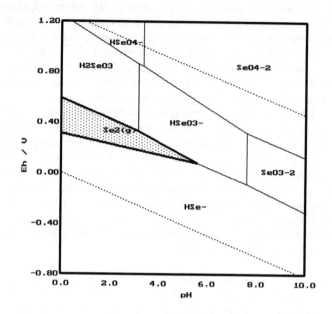

Figure 5 *Eh/V vs pH selenium species predominance diagrams at 140 °C and 1 bar. Total dissolved selenium = 10^{-6} molal. Light dotted lines indicate upper and lower stability limits of water. Stipple indicate saturation with respect to gaseous species. Se (s) not considered. Plots created by PHOX Version 2.0. Thermodynamic data at elevated temperatures calculated following the methods outlined in the text and in Tanger and Helgeson[12] using the data of Barin[5] for gaseous selenium species, of Johnson et al.[1] for $O_2(g)$ and other selenium species and of Helgeson and Kirkham[13] for the solvent*

3.3.3 Membrane desolvation   During this stage neutral volatile species may be efficiently stripped from the aerosol before it is carried into the ICP-MS plasma. The arguments put forward for the fractionation of species during the previous condensation stage are therefore also applicable here, with species with a greater tendency to volatilise likely to be partly or wholly removed from the aerosol.

3.3.4 Adsorption   Lastly, elemental and species fractionation may also take place within the USN because of preferential adsorption onto the walls of the nebuliser. Nakahara and Wasa,[24] for example, note that oxidation of iodide to iodine helps increase analytical sensitivity for I presumably because of an increased sample transport efficiency between the nebuliser and the plasma, and given the memory effects commonly reported for iodine it may be inferred that this increase in sensitivity is in part due to greater tendency for iodide to adsorb onto the nebuliser walls than iodine.

4 CONCLUSIONS

Understanding the aqueous speciation of natural and synthetic waters is not only an interesting end in itself, but under certain conditions and for particular instrumental configurations may be critical to obtaining accurate IC-ICP-MS and ICP-MS analysis of both species concentrations and total dissolved concentrations of elemental components.

The Turbo-Pascal programme PHOX facilitates the rapid calculation of species predominance and relative solubility diagrams and can be an effective and simple tool for understanding the species distribution of elemental components both during chromatographic separation and during ultrasonic nebulisation/membrane desolvation.

The rapid separation of aqueous, gaseous and solid phases during USN/MDS indicates the potential for fractionation of species in its operation. A consideration of speciation of elemental components using PHOX has been useful in attempting to understand the fractionation processes involved.

5 REFERENCES

1. J.W. Johnson, E.H. Oelkers and H.C. Helgeson, *Computers & Geosciences*, 1992, **18**, 899.
2. C.F. Baes, Jr. and R.E. Mesmer, *American J. Sci.*,1981, **281**, 935.
3. A.E. Martell and R.M. Smith, *Critical Stability Constants*, Plenum Press, 1974-6
4. CODATA, *CODATA Bull.*, **28**, 1978
5. I. Barin, *Thermochemical data of pure substances*, 3rd edition, VCH, 1995
6. M. Pourbaix, M. *Atlas of electrochemical equilibria in aqueous solutions*, Pergamon Press, 1966.
7. P. Fletcher. *Chemical Thermodynamics for Earth Scientists*. Longman, 1993.
8. D.A. Polya, In *Water-Rock Interaction* eds. G.B. Arehart and J.R. Hulston, Balkema, Rotterdam, 1998, 897.
9. D.A. Polya, In *Mineral Deposits: Prcoesses to Processing*, ed. C.J. Stanley et al., Balkema, 1999, 133.
10. D.A. Polya, PHOX Version 1.3.User Manual Issue 1.3, Rev 0, October 1998, Unpublished manuscript, Department of Earth Sciences, The University of Manchester.
11. J. Bassett, R.C. Denney, G.H. Jeffery and J. Mendham, *Vogel's Textbook of Quantitative Inorganic Analysis*, 4th edition, Longman, 1978
12. J.C. Tanger IV and H.C. Helgeson, *American J. Sci.*, 1988, **288**, 19.
13. H.C. Helgeson and D.H. Kirkman, *American J. Sci.*, 1974, **274**, 1089.
14. L.N. Plummer, In *Water-Rock Interaction 7* eds. Y.K. Kharaka and A.S. Maest, Balkema, **1**, 23, 1992
15. D.K. Nordstrom and J. Munoz, *Geochemical Thermodynamics*, Blackwell Scientific, 2nd Edition, 1994
16. C.M. Bethke, *Geochemical Reaction Modeling*, Oxford University Press,1996.
17. A.P. Gize, *Econ. Geol.*, 1999, **94**, 967,
18. R.M. Garrels and C.L. Christ, *Solutions, Minerals and Equilibria*, Freeman Cooper and Company, 1965
19. D.G. Brookins, *Eh-pH diagrams for geochemists*, Springer-Verlag, 1988
20. A.G. Gault, D.A. Polya and P.R. Lythgoe, contribution to this volume.

21. J.A. Borowiec, A.W. Boorn, J.H. Dillard, M.S. Cresser and R.F. Browner, 1980, *Anal. Chem.*, **52**, 1054.
22. M. Kodama and S. Miyagama, *Anal. Chem.*, 1980, **52**, 2358.
23. V.A. Fassel and B.R. Bear, *Spectrochim. Acta*, 1986, **41B**, 1089.
24. T. Nakahara and T. Wasa, *Applied Spectroscopy*, 1987,**41**, 1238.
25. J.T. Creed, T.D. Martin and C.A. Brockhoff, *J. Anal. At. Spectrom.*,1995, **10**, 443.
26. P.R. Lythgoe, D.A. Polya and C. Parker, In *Plasma Source Mass Spectrometry: New Developments and Applications*, eds. G. Holland and S.D. Tanner, Sp. Pub. Royal Society of Chemistry, 1999,. **241**,141.
27. M.R. Blocke and W. Luecke, *J. Geophys. Res.*, 1972, **77**, 5100
28. J.W. Olesik, I.I. Stewart, J.A. Hartshome and C.E. Hensman, In *Plasma Source Mass Spectrometry: New Developments and Applications*, eds. G. Holland and S.D. Tanner, Sp. Pub. Royal Society of Chemistry, 1999,. **241**, 3

6 ACKNOWLEDGEMENTS

The authors thank Alistair Bewsher and Andrew Gault for practical assistance and Andy Gize, John Olesik and Norbert Jakubowski for discussions. In particular NJ raised with one of us points regarding the relative solubility of Se(IV) and Se(VI) and this has directly led to us improving this manuscript. Any errors remaining, however, are of course our own.

INVESTIGATIONS INTO BIOVOLATILIZATION OF METAL(LOID)S IN THE ENVIRONMENT BY USING **GC-ICP-TOF-MS**

Jörg Feldmann, Karsten Haas, Laurent Naëls and Silvia Wehmeier

Department of Chemistry, University of Aberdeen, Old Aberdeen, AB24 3UE, Scotland, UK.

1 SUMMARY

A cryo-trapping GC-ICP-TOFMS method for the determination of volatile metal(loid) compounds (VOMs) in gases containing large amounts of CO_2 and CH_4 has been developed. Continuous internal standards (as nebulized indium or rhodium solution) added to the GC outlet are useful tools to identify plasma instabilities through elution of matrix gases. NaOH cartridges are efficient scavengers for CO_2, so that less blockage and more reliable measurements of VOMs are possible using -196°C as a cryo-trapping temperature. During this pre-concentration step, care has to be taken concerning the species' integrity if acidic compounds such as SbH_3 have to be determined. Anaerobic fermentors can be used to generate at least six different VOMs in the lab in order to study bio-volatilization processes. The use of the ICP-TOFMS as a detector for the cryo-trapping GC was shown to be beneficial because many different isotopes can be measured at the same time. Furthermore, the isotope ratios of individual VOMs can be measured with a precision of more than 0.5%, which is only limited by the counting statistics. However, it is not yet clear if this precision is good enough to identify isotopic fractionation of metal(loid)s through bio-volatilization processes.

2 INTRODUCTION

It is well known that certain heavier elements can form compounds which have a considerable vapour pressure at ambient temperature which allows them to volatilize and stay in the gas phase. These compounds are non-charged hydrides and/or methylated or alkylated compounds of main group elements of the groups 12 to 17 (e.g., dimethyl mercury (Me_2Hg), arsine (AsH_3)). The transition elements can only form rather stable neutral carbonyls (e.g., tetracarbonyl nickel ($Ni(CO)_4$). Most of them are solids at 20°C (e.g., hexacarbonyl molybdenum ($Mo(CO)_6$) but have considerable vapour pressure. The majority of these compounds can easily be synthesised. However, the conditions are very restricted: low temperature, a protective atmosphere and the exclusion of water are often necessary parameters to gain a sufficient yield in their synthesis. Despite these restrictive parameters it has been shown that a number of volatile compounds are formed in the environment. Our studies have revealed that compounds such as $Ni(CO)_4$, Me_2Se, Me_3As, Me_2AsH, $MeAsH_2$, AsH_3, $Mo(CO)_6$, Me_4Sn, Me_2Et_2Sn, Me_3Sb, Me_2Te, MeI, $W(CO)_6$,

Me_2Hg, Me_4Pb, Et_3MePb, Me_3Bi can be found in anthropogenic environments such as municipal waste deposits or anaerobic digestors of sewage sludge (Feldmann and Hirner 1995[1], Feldmann and Cullen 1997[2], Feldmann *et al.* 1994[3], Feldmann 1999[4]). Natural environments can also produce volatile metal(loid) compounds (VOMs) such as algae mats on hot spring pools. Besides different arsines, Me_3Sb and up to five different iodine species have been detected to be released by algae mats (Hirner *et al.* 1998[5]). In addition to the occurrence in the environment, laboratory experiments with pure strains of fungi and bacteria have shown to produce significant amounts of VOMs.

If all these bio-transformation reactions are enzymatically catalysed, it may be possible that an isotope fractionation effect of the metal(loid) could be observed in their volatile products. The aim of this study is to show how well the isotope ratio of individual volatile compounds can be measured in gas samples which contain VOMs in concentrations between ng to µg per m^3 gas.

This preliminary study has the following objectives:

- Cryo-trapping of process gases using NaOH cartridges to avoid CO_2 condensation
- Generation of volatile metalloid compounds using anaerobic cultures
- Isotopic measurements of volatile tin using an ICP-TOFMS as the element detector for capillary GC.

3 EXPERIMENTAL SECTION

3.1 Generation of volatile metal(loid)s from an anaerobic culture of microorganisms

A semi-automatic fermentor (Bioflo C30, New Brunswick Scientific Co., Inc) with a volume of about 1 litre equipped with a pH and temperature controller, an Eh-meter, a feeding unit and a set-up to measure the gas volume was filled with 500 mL sewage sludge sampled directly from an anaerobic digester on the sewage treatment plant (Inverurie, Aberdeenshire). 100 mL of 0.1 M sodium acetate was added to the sewage sludge. The cultures were kept at 35°C and monitored for over two weeks and the pH was maintained above pH 6.5 (steady increase pH 6.9 - 7.8). The redox potential decreased in the same time span from -170mV to -280mV. These conditions represent the ideal environment for mesophilic methanogenic bacteria. The headspace gas was sampled and the methane concentration of up to 50% was measured after a steady increase within 14 days. The headspace contained a similar amount of carbon dioxide. The headspace gas had undergone the procedure of pre-concentration applied for VOMs using cryotrapping at -80°C in order to avoid condensation of CO_2 and followed by GC-ICPMS determination.

3.2 Generation of volatile metal(loid) compounds by hydride generation methodology

Sodium arsenite, monomethylarsonic acid, dimethylarsinic acid, trimethylarsine oxide, trimethylantimony dichloride, dimethyltin dichloride, trimethyltin chloride, tetramethyl tin, *n*-butyltintrihydride have been diluted into 1 % HNO_3 and have been subjected to hydride generation by injecting 2 mL of 2 % $NaBH_4$ to the mixture (10 ng for all arsenic and tin species and 50 ng antimony as trimethylantimony dichloride. The purged gas (nitrogen or carbon dioxide) was used to transport the derived volatile arsenic, antimony and tin compounds into a Tedlar bag. The reaction is quantitative and the compounds show a good stability over the following 24 hours, so that a standard atmosphere in nitrogen or methane/carbon dioxide can be generated. Demethylation reactions were observed with Me_3SbCl_2, so that besides Me_3Sb, also Me_2SbH, $MeSbH_2$ and SbH_3 were generated. The procedure is described in more detail by Haas and Feldmann 2000[6].

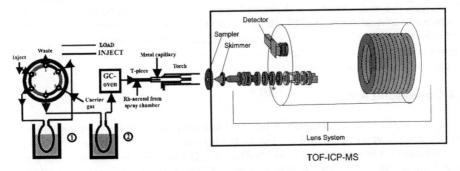

Figure 1 *Schematic of the set-up for the determination of VOMs in gas samples: cryo-trapping GC-ICP-TOF-MS. 1 fused silica-capillary cryo-trap and 2 on column-cryo-focussing*

3.3 GC-ICP-TOFMS method

Table 1 *Instrumental parameters of the GC-ICP-TOF-MS*

Gas chromatograph	
Instrument	GC 95 Ai Cambridge
Column	CP-Sil 5CB Ultimetall (25 m x 0.53 mm i.d., d=1 µm (Chrompack))
Cryotrapping	Liquid nitrogen (fused-silica capillary)
Crofocussing	Liquid nitrogen (on-column)
Transfer line	Ultimetal tubing, 0.53 mm i.d., methyl deactivated), heated (GC-outlet to torch 130°C, valve to cryofocussing step 70°C
Oven temperature	60°C, 1 min, 20°C/min, 180°C, 5 min.
He-flow	3.0 mL/min
ICP-unit	
Instrument	Renaissance, LECO Corporation Ltd.
Coolant, Auxillary, Nebulizer flow	14.9, 0.956, 0.907 L/min
RF power, frequency	1.29 kW, 40.68 MHz
Spray chamber	Hieftje spray chamber (LECO)
Nebulizer	Meinhard
Time-of-flight MS unit	
Sampling time (transient)	204 mS
Measured isotopes	^{12}C, ^{13}C, ^{75}As, ^{112}Sn, ^{115}In, ^{118}Sn, ^{119}Sn, ^{120}Sn, ^{121}Sb, ^{122}Sn, ^{123}Sb, ^{124}Sn, ^{125}Te, ^{126}Te, ^{129}Xe, ^{202}Hg, ^{209}Bi

The analytical method of capillary GC/ICPMS was realised by the on-line coupling of a chromatographic separation carried out in a commercially available GC (GC95, Ai

Cambridge Ltd. London) followed by transient multi-element detection using an ICP time-of-flight mass spectrometer (Renaissance, LECO Instruments, St. Joseph, MI). The ICP-TOFMS was run under standard operating conditions (Table 1).

Behind the spray chamber a tee-piece connects the heated transfer line with the gas stream from the spray chamber as shown in Figure 1. This gives us the opportunity to add a continuous internal standard (CIS, 10 ng/mL In) to the GC outlet in order to record the performance of the plasma and record matrix influences throughout the GC run when organic compounds are eluting. The parameters are shown in Table 1 and a more detailed description can be found elsewhere[6].

3.4 Cryotrapping method

Since CO_2 would block the capillary submerged into liquid nitrogen the gas samples from the fermentor have to be cleaned up. NaOH has been used in many applications and has been reported not to react with Hg and Me_2Hg, etc.[7]. Our cartridge is composed of a glass tube; 7 cm length and 1.5 cm i.d.. The two ends of the cartridge are 6 mm (o.d.) tubings which are directly coupled (Nylon, Swagelok) to the injection valve of the cryotrapping unit. The cartridge on both ends is secured with glass wool and filled completely with a NaOH pellets. Solid packing reduces the dead volume.

4 RESULTS AND DISCUSSION

A standard atmosphere containing volatile tin species, (SnH_4, $MeSnH_3$, Me_2SnH_2, Me_3SnH, $BuSnH_3$) in a mixture of CO_2, CH_4 and N_2 was cryo-trapped in liquid nitrogen without any NaOH cartridge. Only a very small volume of gas (10-20 mL) can be trapped at -196°C, because most of the CO_2 has blocked the capillary. In addition, the trapped gas CH_4 and CO_2 have an influence on the plasma stability. When CH_4 and CO_2 elute, indicated by the increase of the carbon traces (Figure 2), the continuous internal standard (nebulized indium solution of 10 ng/mL) shows a dramatic decrease to the baseline showing that the ionization zone in the plasma was shifted and no signal could be recorded. Volatile metal(loid) species such as $MeSnH_3$ co-elute with CO_2, which makes its determination impossible (indicated by the drop of the background signal on m/z 120 for Sn to nearly zero).

In order to avoid this effect two strategies can be followed: Firstly, the trapping temperature has to be increased so that only small amounts of CO_2 are condensed (e.g., -80°C). Secondly, the CO_2 can be trapped by NaOH. Methane is not trapped efficiently in the cryotrap and despite its effect on the plasma stability, it seems not to present a major problem, because the VOMs are by far less volatile than CH_4, so once the CH_4 is eluted the plasma can recover before the VOMs elute.

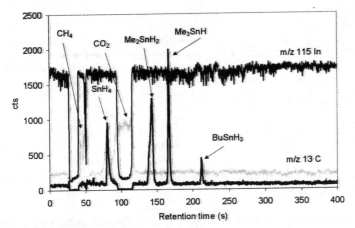

Figure 2 *GC-ICP-TOF-MS of a gas sample containing 100 pg of each tin species in a CO₂/CH₄/N₂ gas mixture cryotrapped at -196°C*

The headspace of the anaerobic microorganism culture was cryotrapped at -80°C in order to avoid the blockage of the column by CO_2. A chromatogram (Figure 3) shows that permethylated metal(loid) species, which have boiling points above 50°C can be trapped from the headspace of an anaerobic culture of a micro-organism consortium. However, it is known that the trapping efficiency of the more volatile species (boiling point below 2°C) is relatively low (<10 %).

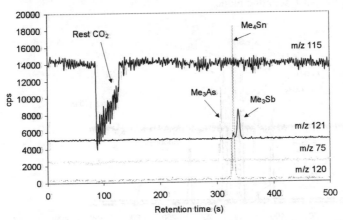

Figure 3 *GC-ICPMS chromatogram of a headspace gas sample trapped at -80°C from an anaerobic fermentor using a quadrupole ICP-MS (Spectromass 2000).* Indium was used as a continuous internal standard

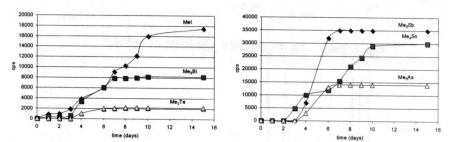

Figure 4 *Cumulative amounts given as peak areas of VOMs generated by an anaerobic culture of sewage sludge in the lab*

Figure 3 shows the production of VOMs by an anaerobic culture of a micro-organism consortium from sewage sludge amended with acetate. After a lag-phase of three days the generation of VOMs starts and levels off after 10 days (Figure 4). It seems that the generation of VOMs is followed by the production of methane. This set-up will be used in future work to generate VOMs by biomethylation in the lab and the VOMs will be subject to isotope ratio measurements.

As a NaOH cartridge was used for the CO_2 containing gas sample, 100 mL gas sample could easily be cryotrapped without any blockage. The CO_2 is very efficiently absorbed; the carbon as well as the continuous internal standard indium show relatively steady signals. Small amounts of $MeSnH_3$ (50 pg) or $MeAsH_2$ (500 pg) could easily be detected using the pulse counting mode.

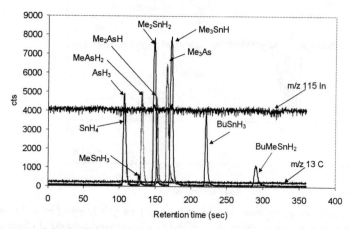

Figure 5 *GC-ICP-TOF-MS of a gas sample containing 0.5 ng As and Sn per species (except MeSnH₃: 0.05 ng) using a NaOH cartridge and -196°C as cryotrapping temperature. A rhodium or indium solution was used as a continuous internal standard and the carbon trace was monitored at m/z 13*

The disadvantage of using NaOH cartridges is that unstable compounds such as Me_3Sb or slightly acidic compounds such as SbH_3 can react with it. The recovery rate of some volatile arsenic, antimony and tin compounds has been measured using a NaOH cartridge and for 500 pg per species. Acceptable recovery rates were determined for the volatile tin compounds (Table 2). The arsenic compounds show slightly lower recovery rates, whereas the antimony compounds show an unacceptable level, in particular SbH_3. The fact that the low recovery rates are based on chemical reactions rather than adsorption phenomena is confirmed by performing memory tests. Here a blank gas sample was measured after different loads of VOMs on a used NaOH cartridge.

The memory effect has been determined for arsenic and tin species ranging between 0.1 and 8 %. The memory did not increase with the load of VOMs on the NaOH cartridge, which suggests that an absolute amount of VOMs sticks to the NaOH rather than a proportion of loaded VOMs, whereas no memory effect has been recorded for the volatile antimony compounds. This suggests that the low recovery rate of the volatile antimony compounds are due to reactions with the NaOH, which consequently fix the antimony on the solid phase. NaOH can eliminate CO_2 prior to cryotrapping of VOMs but there are some limitations about the use in particular if volatile antimony compounds are of interest.

Table 2 *Recoveries of VOMs in gas samples with and without NaOH cartridges. Number of replicates is five*

Element	Species	w/o cartridge RSD %	with cartridge RSD %	Recovery %
Arsenic	AsH_3	2.7	4.0	81.2
	$MeAsH_2$	6.7	6.6	84.7
	Me_2AsH	7.7	1.7	83.1
	Me_3As	4.4	5.4	89.0
Antimony	SbH_3	16.2	7.8	33.1
	$MeSbH_2$	25.2	10.9	69.7
	Me_2SbH	19.1	11.6	52.3
	Me_3Sb	11.9	8.6	74.2
Tin	SnH_4	7.2	5.3	85.4
	$MeSnH_3$	8.5	13.9	92.3
	Me_2SnH_2	7.8	4.2	97.9
	Me_3SnH	8.5	2.5	92.0
	$BuSnH_3$	9.1	2.7	91.0

In order to measure the isotope ratios of individual volatile tin and antimony species the precision and accuracy of standards has been to be measured. Both peak area and peak height acquisition have been tested in pulse counting as well as in analogue detection mode. The transient signals are baseline corrected by interpolation in contrast of previously published work in which Leach *et al.* [8] were using a baseline correction on m/z 108. Preliminary results on bulk analysis on tin compounds have revealed that integration time of 1 second can give a precision of 0.6 %, whereas 30 seconds give 0.1%; the precision is limited by the counting statistics. Therefore a precision of below 0.5% can be expected if a peak signal of about 2-3 seconds has been considered.

The precision calculated for Me_2SnH_2 and Me_3SnH in the analogue mode is the expected value. The signal for $BuSnH_3$ however is already too small to reach this precision in the analogue mode. In contrast the counting mode did not offer a particularly useful option either. The precision is not very good and the accuracy (16-17 %) shows undoubtedly that the linearity of the detector is already reached with a maximum peak height of about 7600 cts. Furthermore $BuSnH_3$ shows already the same effect with the maximum peak height of about 3700 cts. The accuracy of about 3 % is not satisfactory, but no mass bias correction has been used. Thus, a maximum of about 500 pg of each species is necessary to achieve precise isotope ratio measurements of less than 0.5 %. These results seem to be very encouraging. However, this precision is good enough to identify isotopic fractionation of metal(loid)s through biovolatilization can not yet be answered.

Table 3 *Isotope ratio A ($^{118}Sn/^{120}Sn$) determination on volatile tin compounds (200-500 pg Sn as Me_2SnH_2, Me_3SnH and $BuSnH_3$ in 50 mL air (n=5). No bias correction was used*

	Analogue Mode (mV)						
	Ratio 118/120 Areas	RSD %	Accuracy %	Peak height	Ratio 118/120 Height	RSD %	Accuracy %
Me_2SnH_2	0.714	0.2	3.9	18.1	0.727	0.02	2.2
Me_3SnH	0.716	0.4	3.6	21.4	0.727	0.02	2.2
$BuSnH_3$	0.721	1.5	2.8	2.3	0.746	5.2	0.3
	Counting Mode (cts)						
Me_2SnH_2	0.866	1.7	16.6	7614	0.956	2.5	27.8
Me_3SnH	0.875	0.7	17.8	7780	0.959	0.9	28.2
$BuSnH_3$	0.767	3.8	3.3	3689	0.785	1.8	4.9

REFERENCES

1   J. Feldmann, A.V. Hirner, *Intern. J. Environ. Anal. Chem.*, 1995, **60**, 339.
2   J. Feldmann and W.R. Cullen, *Environ. Sci. Technol.*, 1997, **31**, 2125.
3   J. Feldmann, R. Grümping, A.V. Hirner, *Fresenius J. Anal. Chem.*,1994, **350**, 228.
4   J. Feldmann, *J. Environm. Mon.*, 1999, **1**, 33.
5   A.V. Hirner, J. Feldmann, E. Krupp, R. Grümping, R. Goguel and WR Cullen, *Org. Geochem.*, 1998, **29**, 1765.
6   K. Haas and J. Feldmann, *Anal. Chem*, 2000, **72**, 4205.
7   C. Pecheyran, C.R. Quetel, F.M.M. Lecuyer and O.F.X. Donard, *Anal. Chem.*, 1998, **70**, 2639.
8   A. M. Leach, M. Heisterkamp, F.C. Adams and G. M. Hieftje *J. Anal. At. Spectrom.*, 2000, **15**, 151.

ACKNOWLEDGEMENTS

We thank EPSRC (GR/M 10755, GR/R04652/01), the Royal Society and LECO Corporation Ltd. for their financial support. Me_3SbCl_2 was kindly donated by Dr. WR Cullen.

TIME-OF-FLIGHT INDUCTIVELY COUPLED PLASMA MASS SPECTOMETRY FOR ULTRATRACE SPECIATION ANALYSIS OF ORGANOMETALLIC COMPOUNDS

Heidi Goenaga Infante[1], Monika Heisterkamp[1], Karen Van Campenhout[2], Xiaodan Tian[1], Ronny Blust[2] and Freddy C. Adams[1]

[1] University of Antwerp (UIA), Micro and Trace Analysis Center, Universiteitsplein 1, B - 2610 Wilrijk

[2] Department of Biology, University of Antwerp, (RUCA), Groenenborgerlaan 171, B-2020 Antwerp

ABSTRACT

Time-of-flight mass spectrometry (TOFMS) has recently been introduced as an alternative to scanning-based mass analyzers for use in elemental analysis. Coupled to Inductively Coupled Plasma (ICP) ion source, the TOFMS system generates a complete mass spectra more than 20,000 times every second. Because of this high spectral generation rate, multi-elemental simultaneous transient signal analysis can be performed without sacrifice in precision or detection limits regardless of how many isotopes are monitored or how short-lived the transient signal. Additionally, the simultaneous sampling of the plasma eliminates the time-phase variability experienced by sequential mass spectrometers. As a result, isotopic ratios, isobaric correction equations and internal standard corrections can be routinely performed with high precision. All these features makes ICP-TOFMS attractive for the measurement of transient signals such as those commonly encountered in speciation analysis. In this paper, results obtained by axial ICP-TOFMS are presented for ultra-trace speciation analysis of organometallic compounds using ICP-TOFMS coupled to gas chromatography (GC) as well as to high performance liquid chromatography (HPLC).

1 INTRODUCTION

The ultratrace speciation of organometallic compounds in biological and environmental samples has received increasing attention. These metal-containing substances occur at very low concentrations in highly polar solvents, such as biotic fluids or natural waters. It is thus necessary to determine the chemical form of the metal when examining the environmental impact of such compounds, and it is this that has given an impetus to trace metal studies[1] in recent years.

Inductively coupled plasma mass spectrometry (ICP-MS) is the preferred choice for trace metal analysis because:

a.   it offers part per trillion detection limits;

b.   a linear range of 6 to 7 orders of magnitude;

c.   multi-elemental and multi-isotopic measurement capability and limited spectral interferences;

d.   high sample throughput and almost complete elemental coverage[2-5].

Thus, elemental speciation by coupling different separation techniques such as GC or HPLC to ICP-MS is an attractive approach with excellent sensitivity and high selectivity[6].

Since the introduction of commercially available ICP-MS systems in 1983, most instruments incorporate quadrupole mass filters as these combine the necessary resolution with a reasonable cost[5]. However, some intrinsic limitations are still remaining in the quadrupole ICP-MS systems. It has been clearly demonstrated their inability to perform truly simultaneous multi-elemental and multi-isotopic analysis, particularly when fast transient or time-dependent signals are used to analyse a great number of isotopes in a single chromatographic peak. Furthermore, the total time of a measurement is directly proportional to the number of isotopes measured. Another less attractive feature of ICP-MS systems using quadrupole mass filters is the moderate signal stability in the single channel mode, typically 1-5 % RSD[4]. The instability has several origins such as changes in nebulization efficiency and variations in the plasma tail where the ions are sampled. This signal instability is additive whenever two signals are required for one piece of data, in other words, when using either isotope-ratioing or internal standardization techniques.

There are considerable advantages in the use of a faster mass spectrometer coupled to ICP, for example, time-of-flight mass spectrometry (TOFMS) as outlined by Hieftje et al.[2,4]. In such instrumentation, the time-of-flight mass spectrometer can be placed either orthogonal or axial to the plasma. During 1998, an axial inductively coupled plasma time of flight mass spectrometer was commercialized by LECO Corporation (St. Joseph, MI, USA) and the principle of operation of this axial ICP-TOFMS instrument has been previously described[7]. The ICP-TOFMS can produce a complete atomic mass spectrum in less than 50 µs. This high spectral generation rate is considered the main advantage of ICP-TOFMS, which leads to a much higher sample throughput compared with quadrupole ICP-MS. The factor 8 to 10 improvement in sample throughput provided by TOFMS in any multi-elemental analysis can consequently minimize the problem of orifice clogging caused by the introduction of high total dissolved solid samples (e.g. some environmental and biological materials) into the ICP and minute volumes of sample can be analysed. The high data acquisition speed also makes ICP-TOFMS ideally suited for the measurement of fast transient signals from flow injection, electrothermal vaporization (ETV), chromatographic or laser ablation systems. The nature of this detector should also enhance the precision for ratioing techniques since the ions determined are simultaneously extracted from the ionization source[2,4].

Considering the capability of ICP-TOFMS for multi-elemental simultaneous analysis in transient signals, the present work was aimed at investigating the potential of the coupling with GC or HPLC for ultra-trace speciation analysis of organo-metallic compounds in environmental and biological samples.

2 EXPERIMENTAL

2.1 Axial ICP-TOFMS Instrumentation

A Renaissance axial ICP-TOFMS system was used for the analysis. A detailed description of the instrument is given elsewhere[8,9].

2.2 GC - ICP-TOFMS

A gas chromatograph (Chrompack CP-9001) equipped with an on-column injection system was hyphenated to the ICP-TOFMS *via* an interface heated to 270°C as to prevent analyte re-condensation. The design of the interface is similar to that described by Leach et al.[10], but without heating of that part of the transfer line being inserted into the torch. The GC was programmed from 60°C at injection and then heated 30°C/min to 200°C. Hydrogen with 1 % xenon was used as carrier gas and ^{126}Xe was monitored to adjust the torch position. Moreover, ^{204}Pb, ^{206}Pb, ^{207}Pb and ^{208}Pb together with ^{12}C were also measured in transient signal mode with a 100 ms integration time.

Standards in hexane containing a mixture of derivative trimethyl- (TML), dimethyl- (DML), triethyl- (TEL) and diethyllead (DEL) species were injected (1 µl) onto a 25 m long 0.32 mm i.d. fused silica column coated with a HP-1 stationary phase (100 % poly-dimethylsiloxane).

Derivatization of the different ionic organolead species was performed *in-situ* using sodium tetrapropylborate with simultaneous liquid-liquid extraction of the derivatized species in hexane containing trimethylethyllead as internal standard (IS)[11].

2.3 HPLC-ICP-TOFMS

Size-Exclusion (SE) HPLC was performed with a Shimadzu (Kyoto, Japan) PEEK Solvent Delivery Module LC-10Ai equipped with a FCV-10AL quaternary valve, a Rheodyne Model 7125 PEEK sample injector (Rohnert Park, CA, USA) fitted with a 50 µl loop and a SUPELCO (Bornem, Belgium) TSK gel G 3000 PW$_{XL}$ gel-filtration column (7.8 mm i.d.×30 cm, 6 µm particle size). The HPLC system was directly coupled to the Meinhard nebulizer of the ICP-TOFMS via a PTFE tubing (0.5 mm i.d.×2.5 cm). The isotopes ^{137}Ba, ^{138}Ba, ^{209}Bi, ^{111}Cd, ^{112}Cd, ^{114}Cd, ^{59}Co, ^{63}Cu, ^{65}Cu, ^{115}In, ^{206}Pb, ^{207}Pb, ^{208}Pb, ^{88}Sr, ^{64}Zn, ^{66}Zn and ^{68}Zn were measured in transient signal mode using a 500 ms integration time.

2.3.1 HPLC - ICP-TOFMS analysis of gel-filtration standard. The gel-filtration column was calibrated using proteins of known relative molecular mass (Mr). Single standards of ferritin (Mr 540,000), bovine albumin (Mr 66,000), superoxide dismutase (SOD; Mr 31,200), rabbit liver metallothionein (MT; Mr ~10,000) and vitamin B$_{12}$ (Mr 1,355) were purchased from Sigma (St. Louis. MO. USA). The sample vial contents were dissolved in 2 mmol l⁻¹ Tris-HCl buffer solution (pH 7.4), filtered (0.45 μm Millipore filter) and applied directly to the column. The separation and elution of the standards was achieved by using 30 mmol l⁻¹ Tris-HCl buffer (pH 7.4; flow 0.8 ml min⁻¹) as mobile phase.

2.3.2 HPLC - ICP-TOFMS analysis of carp cytosols.   Carp were daily injected intraperitoneally with CdCl$_2$ at a dose of 2 mg/kg body weight during six days in order to induce the synthesis of MTs. Liver and kidney were homogenized in 3 volumes (v/w) of 10 mmol l⁻¹ Tris-HCl buffer (pH 7.4) containing 5 mmol l⁻¹ β-mercaptoethanol and 0.1 mmol l⁻¹ phenylmethanesulfonylfluoride (PMSF; protease inhibitor). The homogenates were ultracentrifuged at 100,000 g for 60 min at 4 °C and the supernatant (cytosol) was filtered (0.45 μm) just before applying to the column.

3 RESULTS AND DISCUSSION

3.1 Speciation analysis of organo-lead compounds using GC - ICP-TOFMS

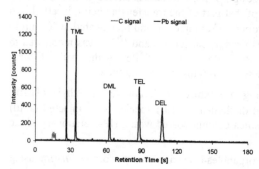

Figure 1 *Chromatogram of a mixture of propylated organolead standards; C signal: m/z = 12, Pb signal: m/z = 208*

A chromatogram of the different propylated organo-lead compounds obtained under optimum conditions is shown in Figure 1. All the species are baseline separated within a chromatographic run of only 3 min and absolute detection limits between 10 and 15 fg (as Pb) are obtained for the different organo-lead compounds. The solvent peak, which was evaluated by monitoring the signal of the ^{12}C isotope, is completely eluted before the most volatile organo-lead species (IS) is ionized in the plasma. This separation is mandatory since co-elution with the solvent results in decreased analyte signal intensity due to plasma quenching.

The optimized GC - ICP-TOFMS system was applied to the speciation analysis of organo-lead compounds in alpine snow. The snow was sampled at the Mont Blanc in France; details of the sampling are given elsewhere[12]. Only methyl-lead species were found in the alpine snow with concentrations in the fg g⁻¹ range. A characteristic example of a chromatogram monitored at ^{208}Pb is depicted in Figure 2. The peak with a retention time of 155 s is Pr$_4$Pb

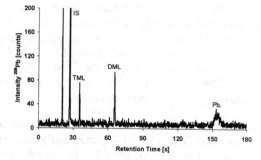

Figure 2 *Chromatogram of an alpine snow sample*

originating from the inorganic lead present in the alpine snow, which could not be masked completely.

The analyzed snow was part of a dated core, which was analyzed previously using gas chromatography hyphenated to microwave induced plasma atomic emission spectrometry (GC - MIP AES) after *in-situ* propylation.

Since organo-lead compounds are tracers of lead additives used as anti-knocking agents in leaded gasoline, and snow/ice cores are useful archives of environmental pollution, determination of these species in such archives can provide valuable information on their influence on the lead pollution. The speciation analysis of organo-lead compounds in the snow/ice core revealed an illustration of the impact of European automobile emissions from 1956 to 1994[12].

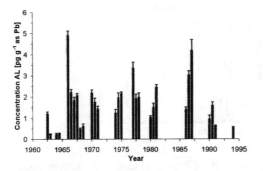

Time changes in total alkyl-lead concentrations of the snow/ice core samples are presented in Figure 3. Alkyl-lead concentrations are below the detection limits in samples dated before 1961. Only methyl-lead species were detected in the other samples but one sample correlated to the year 1974 also contained DEL.

Figure 3 *Time changes in total alkyl-lead concentration from 1962 to 1994 in Mont-Blanc snow and ice*

Methyl-lead was introduced as anti-knocking agent in Europe in 1960 to partly substitute ethyl-lead, which explains the fact that no methyl-lead was detected in samples dated before 1962. A significant increase in total alkyl-lead concentration was observed after 1965, but during the 1990s, concentrations dropped by about a factor of 3.

The augmentation observed after 1965 can probably be explained by the combined effect of the important increase of automobile traffic in all Western European countries and one local event: the opening of the Mont Blanc road tunnel. The tunnel was opened to road traffic in July 1965 and the traffic continuously increased during the following years. In 1967, about 540,000 automobiles and 60,000 trucks passed through the tunnel. A large fraction of these automobiles were gasoline-powered vehicles, which used leaded fuel. At that time, leaded gasoline contained 0.6 g l^{-1} of lead in France and Italy.

The drop in total alkyl-lead concentrations observed during the late 1980s and early 1990s is obviously linked with the reduction in the lead emissions from motor vehicles. The decline in the concentration of total alkyl-lead can be explained by the combined effect of lower lead concentrations in the gasoline and the fact that the use of unleaded gasoline has increased steadily. Since 1989, lead concentrations in leaded gasoline were restricted to a maximum of 0.15 g l^{-1} in France and Italy.

3.2 Speciation analysis of organo-lead compounds using multi-capillary GC - ICP-TOFMS

Preliminary experiments were done using a multi-capillary column (MC-1 ht; Alltech, Deerfield, IL, USA) in order to evaluate the feasibility of ICP-TOFMS as a detection system for very fast transient signals. This column consists of a bundle of narrow inner diameter (40 µm) wall-coated glass capillaries with a length of 1 m and a film thickness of 0.2 µm.

In Figure 4 two chromatograms are depicted operating at an a integration time of 12.75 ms with one using a capillary and the other using a multi-capillary column for separation of the organo-lead species.

Using a multi-capillary column the different organo-lead species could be isothermally baseline separated within less than 20 s, instead of ca. 2 min with capillary GC. An additional gain in time in comparison with capillary GC results from the absence of a need to cool the column between subsequent runs. Hence, employing multi-capillary GC the analysis time necessary for separating the different organo-lead species can be decreased by more than one order of magnitude if the cooling time of the oven is taken

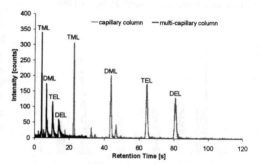

Figure 4 *Chromatograms of organo-lead compounds obtained by GC using capillary and multi-capillary columns*

into account. An integration time of 12.75 ms (equal to a data acquisition rate of 78 Hz) provides more than 35 measurement points for the TML peak with a full width at half maximum of 0.18 s. This data acquisition rate provides a very adequate peak definition.

3.3 Characterization of metal-binding proteins using SE HPLC - ICP-TOFMS

The organo-metallic speciation of trace elements has been studied in different media by some authors[13]. A particular aspect of this topic concerns the study of the low molecular weight organo-metallic compounds called metallothioneins (MTs). These cysteine-rich metal-binding proteins have been characterized as responsible for metal detoxification in living organisms since their biosynthesis can be induced by the presence of high content of heavy metals[14,15]. Therefore, they have been studied particularly in aquatic organisms that can be exposed to elevated levels of trace metals through both bio-geochemical processes and pollution.

The coupling of HPLC with ICP-MS has become a popular hyphenated technique for metallic speciation[16] and particularly Size-Exclusion HPLC - ICP-MS has been the most widely used technique for the quantification of MTs[17].

Based on the advantages afforded by the use of the ICP-TOFMS as multi-element, multi-isotopic and fast detector for transient signal analysis, we have exploited here the coupling SE HPLC - ICP-TOFMS for characterization of metalloproteins in biological materials.

A typical SE HPLC - ICP-TOFMS profile of a gel-filtration standard mixture under optimum conditions is shown in Figure 5. As can be seen from the chromatogram, a good separation of the proteins was achieved within a chromatographic run of 15 minutes. The retention time (t_R) of the protein and their Mr values are given in Table 1.

Table 1: Comparative SE HPLC-ICP-TOFMS and uv-vis (λ=254 nm) retention times of gel-filtration standards and elements associated with the individual proteins

Protein standard	Relative Mr	uv-vis t_R / min	Isotope	Elements associated with protein and mean retention time, n=3 SE HPLC-ICP-TOFMS t_R / min
Ferritin	540,000	6.5	^{63}Cu	6.51 ± 0.02
			^{64}Zn	6.54 ± 0.01
			^{88}Sr	6.54 ± 0.03
			^{111}Cd	6.53 ± 0.02
			^{137}Ba	6.51 ± 0.01
			^{208}Pb	6.53 ± 0.03
			^{209}Bi	6.54 ± 0.02
Bovine albumin	66,000	7.2	^{63}Cu	7.23 ± 0.03
			^{64}Zn	7.28 ± 0.01
SOD	31,200	7.9	^{63}Cu	7.93 ± 0.04
			^{64}Zn	7.93 ± 0.03
Rabbit liver MT 1	~10,000	8.7	^{63}Cu	8.99 ± 0.04
			^{64}Zn	8.73 ± 0.03
			^{111}Cd	8.70 ± 0.01
Vitamin B_{12}	1,355	14.5	^{59}Co	14.54 ± 0.01

Relative standard deviations for retention times were always lower than 0.5 % for the gel-filtration standards by the hybrid technique proposed here. The isotopes likely to be associated with each of the proteins, based on the time at which the element eluted from the column, are also given in Table 1. As expected, Co appears to be only bound to vitamin B_{12} (t_R = 14.5 min) while the isotopes of Pb, Cu, Zn, Cd, Ba and Sr eluted at similar time to the ferritin peak (t_R = 6.5 min) (Cu isotopes profile can not be clearly appreciated from the figure because of the relatively low Cu content associated to ferritin at the protein concentration used). Moreover, Bi bound to ferritin was also encountered. As shown in this Figure, no "spectral skew" is observed when monitoring 7 isotopes simultaneously in a chromatographic peak (ferritin peak) that is of about 5 seconds width (50 % height).

Elution profiles of the isotopes of Zn, Cu and Cd, Figure 5, indicate that these analytes are found in the fraction corresponding to the MT peak (t_R = 8.7 min) as well as Zn and Cu appear to be associated to both albumin (t_R= 7.2 min) and SOD (t_R = 7.9 min). Additionally, there was no evidence that the elution of each of the proteins was adversely affected by the presence of the other proteins at the concentrations used.

The potential of the hybrid technique proposed here was preliminary evaluated for the MT characterization in cytosols of carp liver and kidney. The presence of increased levels of metallothioneins in such aquatic organisms has been proposed as a biomarker for heavy metal exposure and effects in environmental monitoring since the synthesis of MT should be induced by metals as a possible detoxifying process[18]. Figure

6 shows the comparative chromatograms observed for liver cytosols of control (Fig. 6a) and Cd injected (Fig. 5b) carp obtained by SE HPLC - ICP-TOFMS. Three protein fractions (retention times of 6.4, 7.3 and 8.7 minutes) were detected for the control liver. These retention times are very similar to those of the ferritin, albumin and MT fractions. ^{64}Zn and ^{63}Cu profiles indicate that both Cu and Zn are associated to the three protein fractions. These metals are essential elements and it can explain their relatively higher concentration compared to the total Cd amount detected for the control cytosol. Cd elution profile clearly indicates that a very small amount of Cd appears to be associated to the MT fraction.

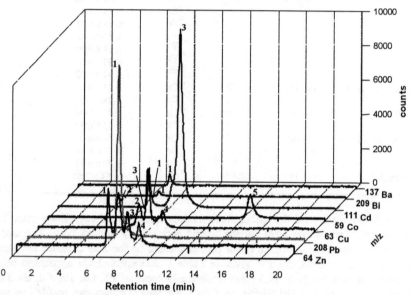

Figure 5 *Typical SE HPLC-ICP-TOFMS profile of a gel-filtration standard mixture; 1: ferritin, 2: bovine albumin, 3: SOD, 4:MT and 5: vitamin B_{12}*

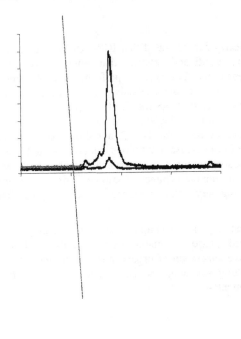

Retention time (min)

Figure 6 *SE HPLC - ICP-TOFMS profile of a control (a) and Cd injected (b) carp liver cytosol*

The profiles of [63]Cu for both control and injected carp cytosol (Figs. 6a and 6b) are almost identical: most of the Cu is found in the fraction corresponding to the MT peak (t_R = 8.7 min). However, the distribution of [64]Zn in the cytosol fractions (Figs. 6a and 6b) is different: no Zn associated to the relatively high molecular mass fraction (t_R = 6.4 min) is observed for the injected fish while the amount of Zn bound to MT seems to increase with the increasing of MT content (MT induction due to the Cd injection) in the liver tissue. Moreover, the peak area of the Cd specie detected in the control cytosol increased for the injected carp. These results lead to confirmation of previous findings about the Cd, Cu and Zn-MT induction in the liver of some aquatic organisms by Cd injection[19].

Results obtained for a kidney cytosol of the same Cd injected carp body used during the experiments discussed

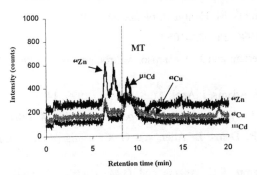

Figure 7 *Chromatogram of a Cd injected carp kidney cytosol*

above are illustrated in Figure 7. As can be observed, relatively small amounts of [63]Cu and [111]Cd elute at retention times corresponding to the high molecular mass (t_R = 6.4 min) and MT (t_R = 8.7 min) fractions. However, no [64]Zn associated to the MT peak is observed. It appears that the relatively small amount of MT induced in the kidney of the carp due to the Cd injections is saturated with Cd and Cu, according to the general order previously found for the binding constant of inorganic thiolates ($Hg^{II} > Ag^I \sim Cu^I > Cd^{II} > Zn^{II}$)[20].

4 CONCLUSIONS

The results presented here have clearly demonstrated that ICP-TOFMS can detect fast transient signals with peak widths significantly shorter than one second with high temporal fidelity and ultratrace sensitivity. The need to perform multi-elemental and multi-isotopic measurements on very fast transient signals such as those encountered when using multi-capillary GC coupled to ICP-MS for ultratrace speciation analysis of organo-metallic compounds drive continued development in the field of TOFMS. Additionally, using ICP-TOFMS as detector for HPLC not only the simultaneous distribution of multiple trace metals in the cytosolic fractions can be studied, but also up to 65 isotopes can be simultaneously measured in a chromatographic peak without affecting the time of analysis by using minute volumes of sample. This fact becomes of significant importance when analysing very limited volumes of real biological matrices such as fish tissues.

Further investigations based on this technique (particularly isotope dilution analysis on-line with the reversed phase, or anion-exchange HPLC-ICP-TOFMS system) which allow the quantitative speciation of organo-metallic compounds in fish cytosols, are being carried out taking advantage of the enhanced precision of ICP-TOFMS for isotope ratios measurements.

5 REFERENCES

1   O. F. X. Donard and J. A. Caruso, Spectrochim. Acta B, 1998, **53**, 157.

2   G. M., Hieftje, D. P., Myers, G., Li, P. P., Mahoney, S. J., Burgoyne, S. J., Ray, and J. P., Guzowski, *J. Anal. At. Spectrom.*, 1997, **12**, 287.

3   A. Montasar (ed.), *Inductively Coupled Plasma Mass Spectrometry*, Wiley-VCH, New York, USA, 1998.

4   P. P, Mahoney, S. J., Ray and G. M, Hieftje, *Appl. Spectrosc.* 1997, **51**, 16A-28A.

5   R. S., Houk, *Anal. Chem.*, 1986, **58**, 97A-105A.

6   G. K. Zoorob, J. W. McKiernan and J. A. Caruso, Mikrochim Acta, 1998, **128**, 145.

7   H. Emteborg, X. Tian and F. Adams, J. Anal. At. Spectrom., 1999, **14**, 1567.

8   X.Tian, H. Emteborg and F.C. Adams, *J. Anal. At. Spectrom.*, 1999, **14**, 1807.

9   F. Vanhaecke, L. Moens, R. Dams, L. Allen and G. Georgitis, *Anal. Chem.*, 1999, **71**, 3297.

10 A.M. Leach, M. Heisterkamp, F.C. Adams and G.M. Hieftje, *J. Anal. At. Spectrom.*, 2000, **15**, 151.

11 M. Heisterkamp and F.C. Adams, *J. Anal. At. Spectrom.*, 1999, **14**, 1307.

12 M. Heisterkamp, K. Van De Velde, C. Ferrari, C. F. Boutron and F.C. Adams, *Environ. Sci. Technol.*, 1999, **33**, 4416.

13 D. J. Mackey, Mar. Chem., 1983, **13**, 69.

14   M. Nordberg, Talanta, 1997, **46**, 243.

15   G. Roesijadi, Aquatic Toxicol., 1992, **22**, 81.

16   B. Welz, J. Anal. At. Spectrom., 1998, **13**, 413.

17   J. Szpunar, Analyst, 2000, **125**, 963.

18   A. Mazzucotelli, A. Viarengo, L. Canesi, E. Ponzano and P. Rivaro, Analyst, 1991, **116**, 605.

19   A. Castaño, G. Carbonell, M. Carballo, C. Fernández, S. Boleas and J. Tarazona, Ecotox. Environm. Safe., 1998, **41**, 29.

20   M. J. Stillman, Coord. Chem. Rev., 1995, **144**, 461.

AN APPETITE FOR ARSENIC: THE SEAWEED-EATING SHEEP FROM ORKNEY

Jörg Feldmann, Thorsten Balgert, Helle Hansen and Paramee Pengprecha

Department of Chemistry, University of Aberdeen, Old Aberdeen, AB24 3UE, Scotland, UK.

1 SUMMARY

The toxicity of arsenic depends strongly on its species. Cation and anion exchange chromatography has been coupled to ICP-MS in order to separate and detect 15 different arsenic species in seaweed extracts and in urine and blood samples of a seaweed-eating sheep from Northern Scotland. This rare breed of North Ronaldsay sheep live the entire year on the beach and eat up to 3 kg seaweed daily, which is washed ashore on this remote Orkney island. Seaweed, especially brown algae (e.g., *Laminaria* spp.) contains about 20-100 mg arsenic per kg dry weight, mainly as dimethylarsinoylriboside species commonly called arsenosugars. The daily intake is approximately 50 mg arsenic. In order to investigate the metabolism of the sheep, urine, blood and wool samples were taken from 20 individual sheep, and from another 11 sheep samples of liver, kidney and muscle were extracted and analysed for their arsenic content.

The arsenic concentration in the tissue and the wool is at least two orders of magnitude higher than for non-exposed sheep, but does not reach the maximum allowed arsenic level in foodstuffs according to a UK guideline (1 mg As per kg fresh weight). The arsenic speciation in the urine reveals that the arsenosugars are taken up by the sheep and excreted mainly as dimethylarsinic acid (DMAA) into the urine. The proposed metabolism includes an enzymatic side chain elimination and a ß-elimination reaction of the non-substituted riboside to give DMAA. In addition to DMAA minor metabolides such as tetramethylarsonium, and monomethylarsonic acid (MMAA) have been identified. The occurrence of MMAA in the urine and the large amount of arsenic bound to the wool suggests that in addition to the arsenosugars intake, the sheep take up non-methylated arsenic species which may occur in other seaweed. This leads to the question of whether this particular sheep is resistant to large quantities of inorganic arsenic, since the tradition of having sheep on the beach is several hundred years old and they have shown no adverse effects.

2 INTRODUCTION

North Ronaldsay (most northern Island of the Orkney Archipelago, 59° 22'N 2° 26'W) is home to the rare seaweed-eating sheep. This unique breed of sheep lives entirely on a diet of seaweed with the ewes, at lambing time, being taken inside the dyke to feed on grass for three to four months. The sheep appear to feed mainly on *Laminaria* species. During low water spring tide the sheep have access to sub-littoral growing Laminariales. When these are not available the sheep mainly feed on the seaweed washed ashore of which *L. digitata* and *L. hyperborea* are most abundant. It has been noticed that the sheep have also been feeding on 14 other different species of seaweed. These include 10 brown seaweeds (*Alaria esculenta, Ascophyllum nodosum, Chorda filum, Fucus serratus, Fucus spiralis, Fucus vesiculosus, Himanthalia elongata, Laminaria saccharina, Himanthalia elongata, Saccorhiza polyschides*), one red seaweed (*Palmaria palmata*) and three green seaweed (*Codium fragile, Enteromorpha intestinalis, Ulva lactuca*). *L. digitata* contains 72 mg arsenic per kg dry weight. Based on this brown algae their daily intake in the summer is with 40 mg arsenic extremely high. Since the toxicity of arsenic is highly depend on its species (the LD_{50} value can vary almost about a factor 1000; arsenite (34.5 mg/kg body weight) and arsenobetaine (>10,000 mg/kg))[1], it is very important to establish how much of the arsenic take in by the sheep is bioavailable and how is it excreted.

The aims of this study are to identify the major arsenic species in the diet, how they are metabolised and to identify major routes of arsenic excretion and storage.

3 EXPERIMENTAL PROCEDURES

3.1 Sampling

The urine, wool and blood was taken directly after slaughtering the sheep and sampled into PE vials and kept cooled 4°C until analysis. This was a regular process and no sheep were sacrificed for the sake of the study. The meat was regularly consumed by the islanders. 10g of tissue were extracted from each organ, packed in a plastic vial and immediately frozen (-20°C).

3.2 Total acid digestion of tissues

To 0.1 g of wool and 0.6 g of whole blood respectively 3 mL HNO_3 and 2 mL H_2O_2 were added in PTFE bombs (Microwave-assistant heating 15 min. at 470 W). The clear solution was put into a PE vial and diluted to 25 mL dionized water with 2 mL of 100 µg/L rhodium acting as an internal standard. The three tissues were ground at subzero temperatures and three replicates of 0.25 g were taken from each sample and placed a in PTFE pressure bomb. The liver and kidney samples were digested into 5 mL HNO_3 and 1 mL of H_2O_2 with microwave assistants. For the muscle a more vigorous method had to be applied: 9 mL HNO_3, 2mL H_2O_2 and 2 mL of H_2SO_4 were necessary to digest the fat-containing neck muscle samples. A more detailed description of the digestion methods is described by elsewhere[2].

3.3 Fat extraction and subsequent digestion

3 g (fresh weight) of the tissue was homo-genized using an ULTRA-Turrax T8 (IKA, Labortechnik) placed in a thimble (cellulose 10 x 50 mm). A Soxleth extraction was carried out with 40 ml of n-hexane for 6 h. After the extraction the n-hexane was removed and the mass of the fat residue was determined. The fat residue was digested with 8 mL HNO_3, 5 mL H_2SO_4 heated under reflux for 8 h. After the addition of a further 5 mL HNO_3 and 2 mL H_2SO_4 on a hot plate, the fat was fully digested when white fumes occurred. The samples were diluted to 25 mL in a PE vial.

3.4 Extraction of Seaweed

The main food source of the sheep are the leafy parts of the marine brown macroalgae *Laminaria digitata*. This algae was collected on a beach in North East Scotland and immediately freeze-dried. The dried seaweed was then subjected to total acid digestion and gentle methanol/water extraction: 1g was placed in a centrifuge vial with 10 mL methanol/water (1:1) sonicated for 10 min and then centrifuged. This procedure was repeated three times and the supernatant solutions were combined and the solvent evaporated at 30°C. The residue was taken up by 10 mL deionized water and stored at 4°C prior to analysis.

3.5 Arsenic speciation methods

Cation and anion exchange chromatography were used for the species identification of the seaweed extracts as well as the serum and the urine samples. The parameters are shown in Table 1. A HPLC (Spectra Physica 400P) with a Rheodyne 4 way valve and a sample loop of 20 μL was used. Spectromass 2000 (Spectro Analytical Instruments) was tuned on arsenic (m/z 75) and monitored the additional m/z 77 for $ArCl^+$ interference (dwell time: 100 mS). Standard conditions with Meinhard nebulizer 0.9 L/min nebulizer flow and a water-jacked cyclonic spray chamber were used. Spikes of the baseline observed using the analog mode of the detection system gave reason to switch to a pulse-counting mode which eliminated the baseline spikes.

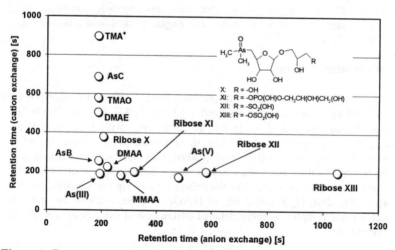

Figure 1 *Retention time map of anion exchange versus cation exchange chromatographic method for 13 different arsenic species using HPLC-ICP-MS*

Table 1 *HPLC parameter used for arsenic speciation*

Method	Column	Mobile phase
Anion exchange	Hamilton PRP X-100, 250 mm x 4.1 mm, 5 μm	30 mM H_3PO_4, pH 6.0 adjusted with NH_3, 1.3 mL/min
Cation exchange	Supelcosil, SCX, 250 mm x 4.6 μm, 5 μm	20 mM pyridine, pH 2.9 adjusted with formic acid, 1.0 mL/min, 35 °C

3.6. Reagents and standards

All reagents for digestion and extractions were analytical reagent grade. Stock solutions for all arsenic species were made in dionized water (1000 ng/mL). Sigma AAS standard was used for As(III); Na_2AsO_4 and dimethylarsinic acid (DMAA) were purchased from SIGMA. Monomethylarsonic acid (MMAA), arsenobetaine (AsB), arsenocholine (AsC), and tetramethylarsonium iodine (TMA^+) were kindly donated by Dr. XC Le and trimethylarsine oxide (TMAO) was provided by Dr. WR Cullen. The four arsenosugars (see Figure 1) and dimethylarsinoyl ethanol (DMAE) were a gift from Dr. W Goessler.

4 RESULTS AND DISCUSSION

The macroalgae *Laminaria digitata* is the major food source of the NR sheep. The results of the total arsenic content of wool, blood, liver, kidney muscle and urine are shown in Table 2. When the amounts are compared to non-exposed sheep (all tissues below 10 ng/g and wool 100 ng/g) the concentration of arsenic in the body fluids and the tissue is elevated by a factor of approximately 100 in all samples. Although the arsenic level is elevated by two orders of magnitude (Feldmann 2000)[2], the tissue meets quite comfortably the UK regulation for foodstuff (1 mg per kg fresh weight) and poses no risk for the consumers. The amount of arsenic in the urine, wool and blood suggests that a large proportion of the arsenic in the seaweed is actually taken up by the sheep and distributed in their body. The only arsenic species identified in the blood serum was DMAA. So far only limited information is available on how much arsenic in seaweed is actually absorbed by the sheep. Preliminary results of absorption rates in humans after a seaweed-meal suggests that approximately 1/3 of the arsenic eaten as arsenosugars in seaweed is excreted into the urine (Smith 2000)[3]. The methanol/water extraction of the major food source (*Laminaria digitata*) gave an extraction efficiency of about 85 %, from which the most species are the three different arsenoribosides (X, XI, XII) shown in Figure 2. The non-extractable arsenic is not specified yet.

Figure 2 *Arsenic Speciation in Laminaria digitata and the urine of NR sheep*

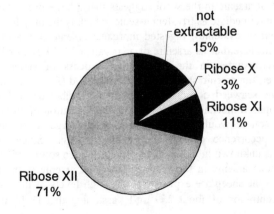

Table 2 *Amounts of total arsenic in tissues and body fluids of the North Ronaldsay (NR) sheep. Number of subjects in paraphrases*

Sample	Min.-max	Mean ± std. Deviation
Wool	2380 - 20660 µg/kg (20)	10470 ± 5690 µg/kg
Blood	19 - 88 µg/kg (20)	44 ± 19 µg/kg
Urine	457 - 9335 µg/L (20)	3179 ± 2667 µg/L
Liver	145 - 477 µg/kg (11)[a]	292 ± 99 µg/kg
Kidney	281 - 916 µg/kg (11)[a]	565 ± 193 µg/kg
Muscle	414 - 1117 µg/kg (11)[a]	680 ± 224 µg/kg

[a] dry weight

Table 3 *Arsenic fractionation in the tissue from 11 NR sheep*

Tissue	Total arsenic. (mg/kg fresh weight)	Fat content (%)	Arsenic in fat (mg/kg fat)	Arsenic in rest tissue (mg/kg)
Muscle	0.242	19.8 ± 11.7	0.754 ± 0.482	0.094
Liver	0.087	2.10 ± 1.66	2.74 ± 0.85	0.029
Kidney	0.112	2.14 ± 2.24	1.32 ± 0.76	0.084

UK-guideline for food: 1mg arsenic per kg fresh weight.

Table 4 *Arsenic Speciation in urine of NR sheep*

Species	Mean conc. (µg/L)	Min. - Max. (µg/L)
Arsenite	n.d.	n.d.
Ribose X	44	< 1 - 44
DMAA	3088	457 - 9116
MMAA	72	< 1 - 259
Arsenate	n.d.	n.d.
Unidentified	12	< 1 - 144
TMA$^+$	5.1	< 1 - 56
Total	3179	457 - 9335

The amount of arsenic in the wool suggests that a large proportion of arsenic has occurred maybe as intermediate as trivalent arsenic, which is the only form which binds to keratin (high content of cysteine). Ingested inorganic arsenic is the obvious source of these trivalent arsenicals, either as arsenite or as trivalent MMA(III) or DMA(III), which are formed as intermediates in the methylation process of arsenic in mammalian organisms[4]. However, the arsenic speciation in the seaweed (Figure 2) suggests that most of the arsenic in the seaweed occurs as dimethylarsinyl-ribosides (ribose X, XI, XII). Considering the proposed mechanism in Figure 3, the formation of an DMA(III) is quite unlikely, the only source would be the ingestion of inorganic arsenic. This is also confirmed by the occurrence of MMAA(V) in the urine. Since demethylation of methylated arsenic is unknown in mammalian organisms, the source of MMAA(V) would also point to inorganic arsenic in the food. If inorganic arsenic is the precursor of the excreted MMAA(V) the sheep are exposed to very high levels of inorganic arsenic with the maximum concentration of about 259 µg/L, assuming that DMAA(V) is the major metabolite of inorganic arsenic. The elevated levels of arsenic in the tissues point in the same direction. Accumulation of arsenic in tissues are known to increase if the methylation rate decreases so that more arsenic (III) compounds are formed in the organism which can bind to the sulfurylgroups of proteins in the tissue.

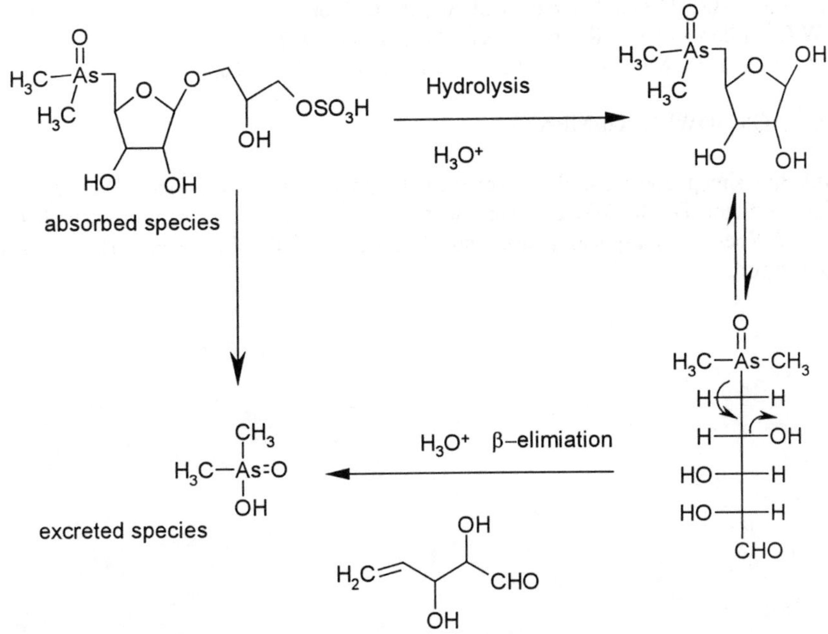

Figure 3 *Proposed metabolism of dimethylarsinoylribosides (arsenosugars) in mammalian organism*

However, the fractionation of the tissue showed that the most of the accumulated arsenic can be found in the fat fraction and consequently the tissue with higher amounts of fat shows the highest total arsenic concentration. This suggests that lipophilic arsenic species are either taken up or formed during the metabolic process and consequently stored in the fat fraction of the tissues. It is known that some seaweed species (*Undaria pinnatifida* an edible seaweed known as Wakame) contain significant amounts of lipid soluble arsenicals (Shibata *et al.* 1992)[5]. The question if the NR sheep have developed a new way to detoxify arsenic by formation of fat soluble arsenic compounds or if the diet contains arsenolipids (e.g., in the non-extractable fraction) which easily can be absorbed and stored in the fat, cannot be answered. Further arsenic speciation in the food and the body fluids are necessary to establish the full picture of arsenic speciation in the variety of food and its consequences on the sheep's adaptation to the seaweed diet.

5 REFERENCES

1   T. Kaise and S. Fukui, *Appl. Organomet. Chem.*, 1992, **6**, 155.
2   J. Feldmann, K. John and P. Pengprecha, *Fresenius J. Anal. Chem.*, 2000, **368**, 116.
3   J. Smith, BSc Thesis, University of Aberdeen, 2000.
4   W.R. Cullen and K.J. Reimer, *Chem. Rev.*, 1989, **89**, 713.
5   Y. Shibata, M. Morita and K. Fuwa, *Adv. Biophys.*, 1992, **28**, 31.

6 ACKNOWLEDGEMENTS

We thank the sheep court for the permission to do research with their sheep and in particular we thank Dr. K. Woodbridge and Mr. W. Stewart for providing us with the samples and Spectro Analytical Instruments UK Ltd. and the Carnegie trust for their financial support.

HYPHENATED IC-ICP-MS FOR THE DETERMINATION OF ARSENIC SPECIATION IN ACID MINE DRAINAGE

A.G. Gault, D.A. Polya and P.R.Lythgoe

Department of Earth Sciences, The University of Manchester, Manchester M13 9PL, UK.

1 INTRODUCTION

Acid mine drainage (AMD) and acid rock drainage (ARD) are problems that afflict both the developed and developing world. As such, there is considerable interest in the biogeochemistry and environmental impact of these waters.[1-8] AMD and ARD arise when sulphide rich minerals are exposed to the atmosphere. The oxidation and subsequent solubilisation that occurs creates highly acidic drainage waters that contain significant heavy metal / metalloid concentrations of which arsenic is an important constituent. Although the chemical components of AMD vary between sites (Table 1), the principal characteristics such as low pH and high dissolved iron and sulphate concentrations are common features.

These waters present formidable analytical challenges due to their low pH, high total dissolved solids (TDS) and elevated content of trace components whose solubilities are strongly pH dependent. Whilst there have been several studies of arsenic speciation in natural waters using IC-ICP-MS,[9] HPLC-ICP-MS,[10] IC-HG-ICP-MS,[11] HPLC-HG-AAS[12] and knotted reactor-ICP-MS,[13] few have addressed the particular problems of determining arsenic speciation in such problematic matrices as AMD or ARD.

We present here the results of work on the development of a robust hyphenated IC-ICP-MS technique for the analysis of arsenic speciation in AMD or ARD. The key matrix compositional parameters in these waters are pH and sulphate concentration, hence the influence of HNO_3 and H_2SO_4 on analytical sensitivity is reported. Although the chloride content of AMD and ARD is generally relatively low (Table 1), the effect of NaCl and HCl matrices on analytical sensitivity has also been investigated due to the importance of $^{40}Ar^{35}Cl^+$ interference in the determination of monoisotopic $^{75}As^+$ by ICP-MS. We present examples of the utility of the technique for the analysis of natural waters with a wide range of pHs from the Carnon River system in Cornwall.

Table 1 *Representative chemical analyses of Acid Mine Drainage. A dash indicates that no data was available for this site*

Analyte	Zeehan, Tasmania[14]	West Squaw Creek, California[15]	Parys Mountain, Wales[16]
PH	2.6	2.4	2.0
Conductivity / $\mu S\ cm^{-1}$	3800	4500	-
SO_4^{2-} / mg l^{-1}	2420	5100	2470
Cl^- / mg l^{-1}	10.1	4.4	-
Fe / mg l^{-1}	199	1600	433
Zn / mg l^{-1}	50.1	156	60
Mn / mg l^{-1}	54.7	1.9	9
Cu / $\mu g\ l^{-1}$	7.6	190	34
As / $\mu g\ l^{-1}$	4.64	0.45	1.36
Cd / $\mu g\ l^{-1}$	0.50	-	0.15
Pb / $\mu g\ l^{-1}$	3.55	-	0.28

2 EXPERIMENTAL

2.1 Instrumentation

Separation of arsenic species was obtained using a CETAC ANX3206 anion exchange column fitted to a Dionex 2000i unit, using a 20 mM $(NH_4)_2(CO_3)$ eluent adjusted to pH 8. The column was connected to the ICP-MS via 0.25mm I.D. PEEK tubing. A Fison's PlasmaQuad II$^+$ ICP-MS fitted with a Fassel-type quartz torch was used in conjunction with a 1mm nickel sample cone and a 0.7mm skimmer cone. Arsenic was detected at m/z = 75 with mass interference from $^{40}Ar^{35}Cl^+$ corrected through monitoring of m/z = 77 and 82 using TJA Solutions' PlasmaLab software. Further instrumental operating conditions are summarised in Table 2. Peak areas were determined using an in-house Turbo Pascal program, TRPEAK.[17]

2.2 Reagents

Standard aqueous stock solutions of arsenite [As(III)], arsenate [As(V)], monomethylarsonic acid [MMAA] and dimethylarsinic acid [DMAA] were prepared in 18.2 $M\Omega$ deionised water (Elga, U.K.) from the appropriate solid sodium salts (all BDH except MMAA which was supplied by Argus Chemicals, Italy). The arsenic concentration of these standards was verified by ICP-AES (Fison's Horizon). No detectable As was found in the $(NH_4)_2(CO_3)$ (BDH, Analar) solution used. The maximum arsenic impurity stated by the manufacturer for stock solutions of HNO_3 (BDH, Aristar) and H_2SO_4 (BDH, Aristar) is 5 $\mu g\ l^{-1}$. This corresponds to a maximum As background concentration of 0.09 $\mu g\ l^{-1}$ and 0.04 $\mu g\ l^{-1}$ for the highest concentration solutions used of HNO_3 (0.28 M) and H_2SO_4 (0.14 M) respectively. It should, however, be noted that As(III) impurities of 4.5 $\mu g\ l^{-1}$ and 14.7 $\mu g\ l^{-1}$ were detected in solutions of 0.28 M HNO_3 and 0.14 M H_2SO_4 respectively, far exceeding the manufacturer's specifications. No other arsenic species impurities were detected. The level of arsenic impurity in the NaCl (BDH, Analar) and HCl (BDH, Aristar) solutions used was not determined, however, the manufacturer's data suggests maximum contaminant levels of 0.66 $\mu g\ l^{-1}$ and 0.02 $\mu g\ l^{-1}$ for 1000 mg l^{-1} (as Cl) solutions of NaCl and HCl respectively.

Table 2 *IC-ICP-MS operating parameters*

Ion Chromatography-	
Column	Cetac ANX3206 column
Eluent	20mM $(NH_4)_2CO_3$
Eluent flow rate / ml min^{-1}	0.37
Elution mode	Isocratic
ICP-MS-	
Instrument	VG PlasmaQuad PQ II^+
Instrument power / W	1350
Reflected power / W	< 0.1
Coolant gas flow rate / l min^{-1} 13.5	
Auxiliary gas flow rate / l min^{-1}	1.0
Nebulizer gas flow rate / l min^{-1}	0.95
Nebulizer	Concentric Glass
Spray Chamber	Water Cooled Impact Bead (3°C)
Total acquisition time / s	500
TRA / m/z	51, 53, 75, 77, 78, 82, 83
Detection mode	Pulse counting

2.3 Sampling Procedure

Approximately 100 ml of the natural waters sampled were immediately filtered through a 0.45 μm paper (Whatman cellulose nitrate membrane) and acidified by the addition of 1 ml of 0.1 M HNO_3 (Aristar, BDH). The pH (WATERCHECK®, HANNA Instruments) conductivity (DiST WP4 / 3, HANNA Instruments), redox potential (ORP Redox Meter, HANNA Instruments) and temperature (HI 93551, HANNA Instruments) of the water were measured on site. The acidified samples were analysed for arsenic species by IC-ICP-MS within 48 hours of collection.

3 RESULTS AND DISCUSSION

3.1 Chromatographic Separation and $^{40}Ar^{35}Cl^+$ Correction

A typical chromatogram for a mixed arsenic species standard is displayed in Figure 1. Excellent separation is achieved for arsenite [As(III)], arsenate [As(V)] and monomethylarsonic acid [MMAA] whilst the dimethylarsinic acid [DMAA] peak was not fully resolved from the As(III) peak.

Chloride (measured at m/z = 51 corresponding to $^{35}Cl^{16}O^+$) elution occurs as a sharp peak well separated from the inorganic arsenic forms (Figure 2), thereby minimising the $^{40}Ar^{35}Cl^+$ mass interference corrections for these species. It should, however, be noted that the elution of chloride from the column coincides with that of MMAA. Since MMAA typically occurs in natural waters at low µg l^{-1} levels, this complicates its measurement.

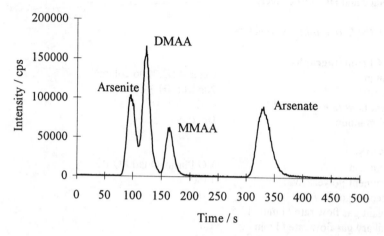

Figure 1 *Chromatogram of arsenite (90 µg l^{1}), DMAA (107 µg l^{1}), MMAA (62 µg l^{1}) and arsenate (107 µg l^{1}) in deionised water obtained by IC-ICP monitoring at m/z = 75*

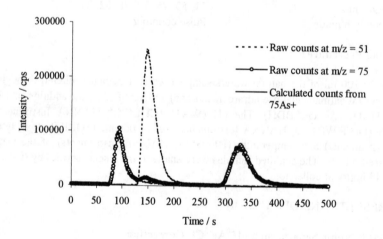

Figure 2 *Chromatogram of 90 µg l^{1} As(III) and 107 µg l^{1} As(V) in 1000 mg l^{1} chloride (NaCl) matrix obtained by IC-ICP-MS monitoring at m/z = 75*

The raw counts were corrected for $^{40}Ar^{35}Cl^{+}$ using the standard correction equation (1)[18]

$$I(^{75}As) = I(75) - 3.127*[I(77) - (0.874*I(82))]$$ (1)

where I(x) denotes the count rate in cps measured at m/z = x and I(^{75}As) denotes the count rate in cps calculated to be due to ^{75}As$^+$ alone. As demonstrated in Figure 2, the correction equation appears efficient at removing ^{40}Ar^{35}Cl$^+$ interferences in high chloride matrices.

3.2 Matrix dependence of analytical sensitivity

The sensitivity of the IC-ICP-MS set-up for As(III) and As(V) in synthetic solutions of varying pH, chloride and sulphate concentration was systematically assessed. Each solution matrix was spiked with 90 μg l^{-1} As(III) and 107 μg l^{-1} As(V).

3.2.1 NaCl

A series of NaCl solutions ranging from 1 to 1000 mg l^{-1} as Cl$^-$ were spiked and analysed by IC-ICP-MS (Figure 3(a)). The raw counts were corrected for ^{40}Ar^{35}Cl$^+$ using equation (1). Even after correction, the addition of chloride increases the calculated ^{75}As$^+$ signal, however, it is still within the limits of analytical precision. Other than the effect previously noted, the addition of chloride as NaCl on the measured As(III) and As(V) concentration is negligible. This represents an improvement over the significant matrix dependencies encountered during previous IC-ICP-MS work performed in this laboratory using a different column coupled to a USN / MDS system.[19]

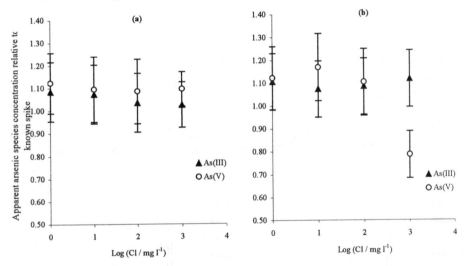

Figure 3 *Influence of (a) NaCl and (b) HCl on IC-ICP-MS analysis of As(III) (90 μg l^{-1}) and As(V) (107 μg l^{-1})*

3.2.2 HCl

The same range of chloride concentration solutions were prepared using HCl, spiked and analysed by IC-ICP-MS (Figure 3(b)). The raw counts were corrected for ^{40}Ar^{35}Cl$^+$ using equation (1). A similar increase in the calculated ^{75}As$^+$ signal to that noted in the NaCl solutions was observed, however, a dramatic reduction in the As(V) response occurs at a chloride concentration of 1000 mg l^{-1}. Since no such decrease was evident at the corresponding NaCl chloride concentration, this effect must be pH driven.

3.2.3   HNO₃

A series of HNO₃ solutions ranging from 17.4 to 17360 mg l⁻¹ NO₃⁻ (equivalent to a nominal pH range from approximately 3.55 to 0.55) were spiked and analysed by IC-ICP-MS. The ⁷⁵As⁺ chromatograms of these solutions are shown in Figure 4. The integrated peak areas for the peaks observed in Figure 4 are listed in Table 3.

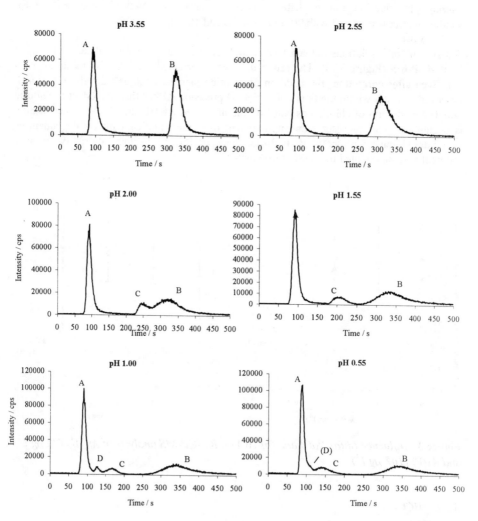

Figure 4 *Influence of sample pH on peak shape and retention time of As(III) (90 μg l⁻¹) and As(V) (107 μg l⁻¹) (HNO₃ matrix) determined by IC-ICP-MS monitoring at m/z = 75*

Based on their retention times, peaks A and B are assigned to As(III) and As(V) respectively. The peak area of the As(III) signal remains unaffected throughout the pH range with the exception of pH 1 where it is diminished. The area of the As(V) peak is markedly reduced at pH ≤ 2.00 where it undergoes peak splitting. This is accompanied by the formation and upfield migration of an additional peak, labelled C in Figure 4. At pHs ≤ 1 a second additional peak, labelled D, is also observed in the $^{75}As^+$ chromatogram. This pH dependent peak splitting and migration behaviour observed may be rationalised by considering the temporal and spatial variation of pH within the column during separation and the stability of As(III) and As(V) species in Eh-pH space.

Table 3 *Measured peak area of peaks in spiked HNO_3 solutions obtained by IC-ICP-MS illustrated in Figure 4. *pH calculated assuming full dissociation and molal concentration scale activity coefficients of unity. A dash indicates that no peak was observed or the peak is not resolvable from a larger peak.*

	Peak area / cps			
Nominal pH*	A	B	C	D
3.55	1.355×10^6	1.582×10^6	-	-
2.55	1.307×10^6	1.522×10^6	-	-
2.00	1.305×10^6	7.906×10^5	2.966×10^5	-
1.55	1.306×10^6	9.317×10^5	2.016×10^5	-
1.00	1.153×10^6	9.301×10^5	1.053×10^5	6.468×10^4
0.55	1.306×10^6	8.396×10^5	2.379×10^5	-

The pH of the column effluent from the HNO_3 spiked solutions was measured using an in line micro-pH meter (Cole-Palmer) fitted between the column and the ICP-MS (Figure 5).

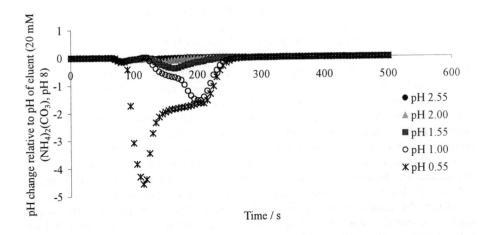

Figure 5 *Change in pH of column effluent from HNO_3 spiked solutions relative to pH of 20 mM $(NH_4)_2(CO_3)$ eluent (pH 8)*

It should be noted that the pH change at the start of the column is likely to be more pronounced than that measured at the end. Therefore, the full difference between the sample and eluent pH is not recorded since the pH of the slug of sample solution will rise due to its interaction with the eluent as it travels along the column.

The pH 1.55, 2.00 and 2.55 solutions all exhibit similar trends. A small decrease in pH is noted after approximately 70 s, corresponding to the almost unretarded migration of the front of the acidic sample slug. This is followed by a slow rise in pH due to buffering by the $(NH_4)_2CO_3$ eluent. There follows a second fall in pH, more pronounced for the more acidic sample solutions, which is ultimately buffered by the bicarbonate in the as shown in equation (2).

$$H^+ + HCO_3^- \rightleftharpoons H_2CO_3 \qquad\qquad\qquad\qquad\qquad (2)$$

The structure of the pH 1.00 solution plot differs to the previous samples as the pH dips markedly again at ~175 s after appearing to plateau. The bicarbonate and nitrate anions will compete for sorption sites on the column, thus if the nitrate anions have a greater affinity for the packing material then there will be more HCO_3^- available to bind the free H^+ cations. The pH begins to drop at ~130 s as observed earlier for the less acidic sample solutions, however, the further decrease may be caused by desorption of nitrate anions from the column. This would increase the number of sorption sites available to the bicarbonate anions. As more bicarbonate is adsorbed to the packing material, the equilibrium of equation (2) will shift to the left thus accelerating the release of H^+ ions and causing the further reduction in pH.

The pH 0.55 sample displays a sharp decline in pH as a significant proportion of the sample H^+ ions pass straight through the column without complexation by the eluent. The ensuing rise and plateau in pH are due to buffering according to equation (2) as the concentration of H^+ ions approaches that of free HCO_3^- ions.

Figure 6 shows the Eh-pH stability field for inorganic arsenic species. As(III) remains as the neutral H_3AsO_3 form throughout the pH range studied, however, As(V) is found as either of two species. Between pH 2 and 7, As(V) exists predominantly as the monoanionic $H_2AsO_4^-$ species whilst at pHs < 2, neutral H_3AsO_4 is the dominant form. Since the neutral As(V) species will have a shorter column retention time to that of the charged form, this may explain the development of peak C for samples with pHs ≤ 2.

At pH 2, As(V) is predicted to be present as equal concentrations of the charged and neutral arsenate forms. The pH rise obtained by mixing of the acidic sample solution with the alkaline eluent will cause As(V) to revert to the monoanionic form. However, since the sample solution is flushed into the column as a slug, propelled by the eluent, the deprotonation does not occur immediately. As more eluent is pumped through the column more bicarbonate is available to buffer the pH of the solution. Since more time is required to pass sufficient bicarbonate to buffer the lowest pH solutions, the neutral As(V) species will travel further down the anion exchange column before it is deprotonated, yielding peak C. Hence, the retention time of peak C will decrease with decreasing pH of the sample solution. A second "extra" peak, D, appears in the pH 1.00 and pH 0.55 (partially hidden in the shoulder of the As(III) peak) chromatograms (Figure 4). This is also thought to be due to As(V), however, the reason for its separation from peak C is not fully understood. Work is ongoing to establish the deprotonation-protonation, adsorption-desorption and mixing kinetics of the low pH sample solutions within the column.

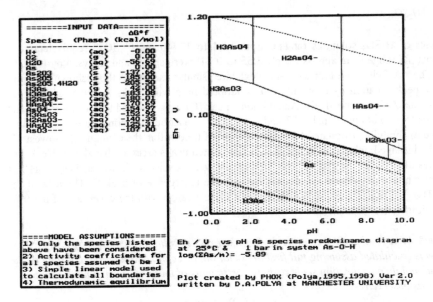

Figure 6 *Eh-pH stability plot for inorganic arsenic species in the system As-O-H*[20]

Using the assignments deduced for peaks A, B, C and D (Figure 4), the ratio of apparent arsenic species concentration relative to that spiked was calculated (Figure 7). The ratio for As(III) remains close to 1 throughout the calculated pH range investigated. The ratio for As(V) is around 1 for sample pHs ≥ 2.5 but for sample pHs ≤ 2, the ratio drops to approximately 0.7 despite including the area of peaks C and D. This apparent loss is consistent between sample pHs 2.00 and 0.55 but the mechanism is undetermined.

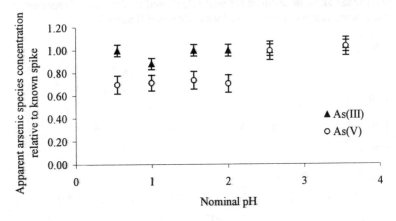

Figure 7 *Influence of HNO$_3$ on IC-ICP-MS analysis of As(III) (90 µg l^1) and As(V) (107 µg l^1). Note that As(V) concentration for samples of pH ≤ 2 results from the combined area of peaks B, C, and D (see Figure 4)*

3.2.4 H_2SO_4

A series of H_2SO_4 solutions ranging from 13.4 to 13440 mg l^{-1} SO_4^{2-} (equivalent to a nominal pH range from approximately 3.55 to 0.74) were spiked and analysed by IC-ICP-MS. The $^{75}As^+$ chromatograms obtained exhibited similar features to those shown in Figure 4. The peak notation used in Figure 4 is adopted here with peaks A and B ascribed as As(III) and As(V) respectively. An additional peak (C) was observed between peaks A and B in sample solutions of pH 1.57 and 1.13. This is ascribed to monoanionic As(V) that existed as the fully protonated form for a period of time within the column. The lowest pH sample displayed only peaks A and B. The integrated peaks areas are listed in Table 4. For samples of pH > 2, the As(V) peak area is relatively consistent, however, below pH 2 it becomes noticeably depressed, coinciding with the emergence of peak C. The area of the As(III) signal remains unaffected throughout the pH range with the exception of pH 0.74 where it increases dramatically.

Table 4 *Measured peak area of peaks in spiked H_2SO_4 solutions obtained by IC-ICP-MS. *pH was calculated assuming full first dissociation, a second dissociation constant of $10^{-1.92}$ and molal concentration scale activity coefficients of unity. A dash indicates that no peak was observed*

Nominal pH*	Peak area / cps		
	A	B	C
3.55	1.944×10^6	2.231×10^6	-
2.55	1.969×10^6	2.152×10^6	-
2.10	1.939×10^6	2.118×10^6	-
1.57	1.998×10^6	5.693×10^5	8.728×10^5
1.13	1.972×10^6	6.788×10^5	1.367×10^6
0.74	5.059×10^6	8.768×10^4	-

Using the assignments deduced for peaks A, B, and C, the ratio of apparent arsenic species concentration relative to that spiked was calculated (Figure 8).

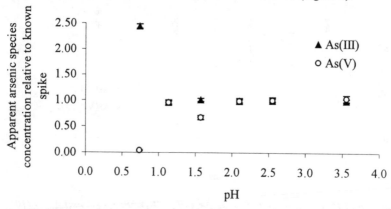

Figure 8 *Influence of H_2SO_4 on IC-ICP-MS analysis of As(III) (90 µg l^{-1}) and As(V) (107 µg l^{-1}). Note that As(V) concentration for samples of pH < 2 results from the combined area of peaks B and C (see text)*

The As(V) ratio is close to 1 for sample pHs > 2 and 1.13, however, it declines markedly for the samples of pH 1.57 and 0.74. The As(III) ratio is steady at approximately 1 in all the samples analysed with the exception of the lowest pH sample where it apparently increases to 2.44. This coincides with a further reduction in apparent As(V) concentration, and the absence of peak C (Table 4) suggests that the As(III) peak is actually due to both the inorganic arsenic species. Scaling of this peak to the proportions of As(III) and As(V) known to be present in solution yields ratios of 1.12 for both species. At present, the mechanism responsible for these apparent pH dependent shifts in sensitivity for As(V) (and to a smaller extent As(III)) is unclear.

3.3 Analysis of AMD from Wheal Jane Mine, Cornwall

The Carnon River, situated in S.W. Cornwall, (Figure 9) was selected as a suitable site for method testing of the IC-ICP-MS set up.

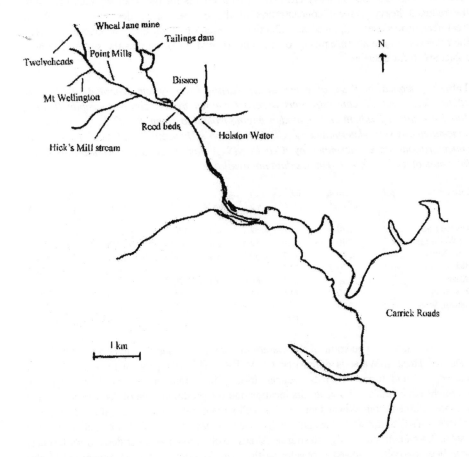

Figure 9 *Sketch map of the Carnon River sampling area, Cornwall, S.W. England*

The waters towards the head of the Carnon River are subject to significant acid mine drainage from a number of disused mines around the region. The waters are typically acidic (pH 3-6) and have elevated heavy element concentrations (Σ heavy elements > 1 mg l^{-1}). In 1992, an adit leading from the recently closed Wheal Jane tine mine burst, flooding the Carnon River and the estuary downstream with heavily contaminated mine water. Hunt and Howard conducted a seasonal survey on arsenic speciation in the Carnon River system in 1992,[12] following the Wheal Jane incident, recording a maximum dissolved arsenite concentration of 240 μg l^{-1} close to the discharge point. Following the execution of a contingency plan, the burst adit was plugged and the heavy element levels in the Carnon River gradually returned to their pre-incident values.

One sampling visit was made to the Carnon River between 10 –12 April 2000. Table 5 lists the major chemical components of the sites sampled. The waters towards the head of the Carnon River are generally more acidic, reflecting the mine drainage input whilst the water from the Bissoe site, sourced from the Wheal Jane tailings dam, is near neutral. The tailings dam contains mine water that is limed and treated with flocculants to precipitate out the bulk of heavy elements. This accounts for the relatively high pH whilst the reduced heavy element concentration of the Bissoe waters demonstrates that the treatment procedures employed are effective. The elevated Zn and Cu concentrations in the river waters are a consequence of the mine related pollution that the system has been subjected to for decades.[12]

Table 5 *Chemical analyses of waters in the Carnon River system, sampled 10-12 April, 2000. SO_4^{2-} and Cl^- concentrations were determined by ion chromatography (Dionex 4000I utilizing a gradient elution mode ramping from 0 to 100mM NaOH). Fe, Zn and Cu concentrations were determined by ICP-AES (Fison's Horizon). Total As and As species concentrations were determined by IC-ICP- MS. A dash indicates concentrations below the limits of detection; n.a. denotes data not available*

Location	pH	Conductivity / $\mu S\ cm^{-1}$	SO_4^{2-} / $mg\ l^{-1}$	Cl^- / $mg\ l^{-1}$	Fe / $mg\ l^{-1}$	Zn / $\mu g\ l^{-1}$	Cu / $\mu g\ l^{-1}$	As (total) / $\mu g\ l^{-1}$	$As(III)$ / $\mu g\ l^{-1}$	$As(V)$ / $\mu g\ l^{-1}$
Twelveheads	6.2	210	29.4	32.7	0.1	470	54	22.1	4.2	17.9
Mt Wellington	2.7	340	63.0	32.4	2.1	2990	759	3.2	3.2	-
Point Mills	4.4	240	57.2	30.7	3.1	2390	539	2.3	2.3	-
Hick's Mill	6.1	200	27.7	27.7	0.1	740	69	40.0	16.7	23.3
Bissoe	7.2	n.a.	227	69.0	0.1	170	-	-	-	-
Reed beds	4.8	330	70.0	33.6	1.3	2590	419	34.2	29.9	4.3
Helston Water	6.0	200	0.0	27.8	0.8	100	-	53.8	50.1	3.7

The arsenic speciation of the waters of Mount Wellington adit, Point Mills, Reed beds and Helston Water is dominated by As(III). This is typical of mine water where arsenic is found primarily in its reduced form. The origin of the dissolved arsenic at Twelveheads is unclear, however, its location and the predominance of As(V) suggest that it is not sourced from Wheal Jane. Hick's Mill stream contains substantial concentrations of both As(III) and As(V). This arsenic may be supplied from As(III) rich mine water and a second, unidentified As(V) rich source. Alternatively, the mine water derived As(III) may have been partially oxidised. Considering the slow kinetics of As(III) oxidation,[21,22] this would probably have to be biologically mediated or catalysed by Mn and / or Fe hydrous oxides.[23,24]

4 CONCLUSIONS

A hyphenated IC-ICP-MS technique has been used to determine arsenic speciation in acid mine drainage (AMD). Separation of the arsenic species was effected by a CETAC ANX 3206 anion exchange column and the arsenic detected at m/z = 75 using a PlasmaQuad PQ II⁺ ICP-MS with mass interference from $^{40}Ar^{35}Cl^+$ corrected through monitoring of m/z = 77 and 82.

The determination of inorganic arsenic speciation in synthetic solutions containing up to 1000 mg l⁻¹ chloride was unaffected by $^{40}Ar^{35}Cl^+$ interference since chloride is chromatographically resolved from As(III) and As(V).

Peak splitting of the As(V) peak, observed for spiked HNO_3 and H_2SO_4 matrices with pH ≤ 2, was found to occur due to the change in arsenate species predominance from monoanionic to neutral form within the column. This may present problems for the analysis of the low pH waters associated with AMD, however, it may be overcome through pre-treatment of sub pH 2 samples with the minimal addition of eluent in order to raise the sample pH and also reduce the introduction of contaminants. Less acidic natural water samples should be acidified to no lower than pH 3 to avoid this complication in As(V) determination. The results indicate that acidification of samples for arsenic speciation work to 1-2% nitric acid, a common practice in trace element analysis, may cause analytical problems.

Arsenic in the Carnon River is supplied primarily from mine water inputs towards the head of the river and is composed largely of As(III). The two sites where As(V) is the major species are thought to be supplied by an additional, unidentified source other than Wheal Jane. The treatment processes in place at the Wheal Jane tailings dam appear effective at removing much of the toxic load of the contaminated mine water.

5 REFERENCES

1. K.J. Edwards, P.L. Bond, T.M. Gihring and J.F. Banfield, *Science*, 2000, **287**, 1796.
2. M.O. Schrenk, K.J. Edwards, R.M. Goodman, R.J. Hamers and J.F. Banfield, *Science*, 1998, **279**, 1519.
3. T.M. Williams and B. Smith, *Environ. Geol.*, 2000, **39**, 272.
4. M.F. Hochella, J.N. Moore, U. Golla and A. Putnis, *Geochim. Cosmochim. Acta*, 1999, **63**, 3395.
5. K.A. Hudson-Edwards, C. Schell and M.G. Macklin, *App. Geochem.*, 1999, **14**, 1015.
6. J.G. Webster, P.J. Swedlund and K.S. Webster, *Environ. Sci. Technol.*, 1998, **32**, 1361.
7. M. He, Z. Wang and H. Tang, *Sci. Total Environ.*, 1997, **206**, 67.
8. N.F. Gray, *Environ. Geol.*, 1997, **30**, 62.
9. P. Teräsahde, M. Pantsar-Kallio and P.K.G. Manninen, *J. Chromatogr. A*, 1996, **750**, 83.
10. R. Pongratz, *Sci. Total Environ.*, 1998, **224**, 133.

11.   M.L.Magnuson, J.T. Creed and C.A. Brockhoff, *J. Anal. Atom. Spectrom.*, 1996,
       11, 893.L.E. Hunt and A.G. Howard, *Mar. Poll. Bull.*, 1994, **28**, 33.
12   X-P. Yan, R. Kerrich and M.J. Hendry, *Anal. Chem.*, 1998, **70**, 4736.
13.   T.E. Parr and D.R. Cooke, in *Water Rock Interaction 9*, eds. G.B. Arehart and J.R.
       Hulston, Balkema, Rotterdam, 1998, p. 71.
14   L.H. Filipeck, D.K. Nordstrom and W.H. Ficklin, *Environ. Sci. Technol.* 1987, **21**,
       388.
15   K.C. Walton and D.B. Johnson, *Environ. Poll.*, 1992, **76**, 169.
16.   D.A. Polya, TRPEAK, unpublished manuscript, Department of Earth Sciences,
       University of Manchester, 1998.
17.   J.L.M. de Boer, in *Plasma Source Mass Spectrometry*, eds. G. Holland and
       S.D.Tanner, Royal Society of Chemistry, Cambridge, 1999, p. 27.
18.   P.R. Lythgoe, D.A. Polya and C. Parker, in *Plasma Source Mass Spectrometry*,
       eds. G. Holland and S.D. Tanner, Royal Society of Chemistry, Cambridge, 1999, p.
       141.
19.   D.A. Polya, in *Water Rock Interaction 9*, eds. G.B. Arehart and J.R. Hulston,
       Balkema, Rotterdam, 1998, p. 897.
20.   J.A. Cherry, A.V. Shaikh, D.E. Tallman and R.V. Nicholson, *J. Hydrol.*, 1979, **43**,
       373.
21.   L.E. Eary, *Abstr. Programs Geol. Soc. Am.*, 1987, **19**, 650.
22.   D.W. Oscarson, P.M. Huang, C. Defosse and A. Herbillon, *Nature*, 1981, **291**, 50.
23.   R. de Vitre, N. Belzile and A. Tessier, *Limnol. Oceanogr.*, 1991, **36**, 1480.

6 ACKNOWLEDGEMENTS

The authors thank Andy Gize and Dave Cooke for discussions and Alistair Bewsher for
practical assistance. AGG is supported by a NERC/CASE PhD studentship with CETAC
Technologies.

SELENIUM SPECIATION IN HUMAN URINE

B. Gammelgaard, L. Bendahl and O. Jøns

Department of Analytical and Pharmaceutical Chemistry, The Royal Danish School of
Pharmacy, Universitetsparken 2, DK-2100 Copenhagen, Denmark

1 INTRODUCTION

Selenium is an essential element and functions through selenoproteins. Several selenoproteins
have been identified. The majority of these are involved in antioxidative processes in the body
but the function of all selenoproteins are not fully understood[1]. In the selenoproteins, selenium
is incorporated as the amino acid selenocysteine. Apart from the selenoproteins, selenium is
incorporated unspecifically as selenomethionine in other proteins e.g. albumin.

The metabolism of selenium in the human body is far from fully understood. Inorganic
as well as organic compounds are easily absorbed[2]. In the body, selenate and selenite are
reduced to selenide by cellular glutathione and selenoproteins are metabolised in the liver and
erythrocytes to selenide. Selenide is either incorporated into selenoproteins as selenocysteine
or methylated to metabolites as monomethylselenol, dimethyl selenide, dimethyl diselenide or
trimethylselenonium ion. The volatile methyl selenides are expired while trimethylselenonium
is excreted in urine[2]. Methylation is considered to be a detoxification process in the body[3].
Thus, most naturally occurring selenium compounds probably transfer a shared metabolic
selenide pool before incorporation into selenoproteins or elimination as methylated
compounds[2].

The selenium content of urine reflects the daily dietary intake and has together with
plasma selenium been used as an indication of short term selenium status, while the selenium
concentration in erythrocytes reflects long term selenium status[3]. When selenium is ingested
as selenomethionine, the retention is larger compared to retention of inorganic selenium
compounds. This effect, however, can be ascribed to the unspecific incorporation of
selenomethionine into proteins and results in storage of the element rather than increased
synthesis of selenoproteins[2].

In recent years, selenium has attracted attention as certain selenium compounds may
have a protective function against certain forms of cancer[4]. The doses of selenium given in the
clinical trial were larger than the dose required to maintain the selenoproteins functional, hence
the effect could be due to compounds different from the selenoproteins. It has been suggested
that the anticarcinogenic effect is due to methylated selenium compounds that enters the
metabolic chain after the selenide pool[5]. The selenium supplementation given in the cancer
study was selenized yeast, which apart from the major compound, selenomethionine, contains
methylated selenium compounds[6].

Hence, the interest in selenium speciation in biological material has increased during recent years as more knowledge is needed on the metabolism of different selenium compounds in the human body.

The purpose of this paper is to give an overview of our experiments on selenium speciation, including development of different chromatographic systems, the development of an interface for coupling of capillary electrophoresis to ICP-MS and application of these systems to urine samples.

2 EXPERIMENTAL

The ICP-MS system was a Perkin-Elmer SCIEX Elan 6000 equipped with an AS 90 autosampler, a cross-flow nebulizer and a Ryton Scott-type double-pass spraychamber (Perkin-Elmer, Norwalk, CT, USA). As an alternative a MicroMist AR30-1-F-02 glass microconcentric nebulizer in combination with a cyclonic spraychamber (Glass Expansion, Romainmotier, Switzerland) was used for some chromatographic applications. An ultrasonic nebulizer, a U-6000AT$^+$ (without membrane desolvator) (CETAC Technologies Inc., Nebraska, USA) was used for some experiments.

The HPLC pump was a Jasco 880-PU (Jasco, Gross-Umstadt, Germany). The pump was connected directly to the nebulizers with PEEK tube (Upchurch Scientific, Oak Harbor, WA, USA). The instrument was optimized with respect to nebulizer gas flow, ion lens voltage, and rf power when aspirating eluents spiked with selenium.

A Waters Quanta 4000 CE instrument (Millipore Corp., Milford, MA USA) was used for capillary electrophoresis. The interface for connection of the CE instrument to the ICP-MS was based on direct injection nebulization and was operated in self-aspirating mode with a sheath liquid uptake of 10 μl min^{-1} as described elsewhere[7].

All reagents were of analytical grade. Purified water obtained from a Milli-Q deionization unit (Millipore, Bedford, MA, USA) was used throughout. 10 mg l^{-1} selenium stock solutions of selenite, selenate, selenomethionine (SeMet), selenocystine (SeCys)$_2$ and trimethylselenonium (TMSe) were prepared in water and diluted further prior to use.

3 RESULTS AND DISCUSSION

3.1 Selenium Measurements in General

The determination of selenium in biological material by ICP-MS is complicated for two reasons: the high ionization potential of selenium results in a degree of ionization of only 30%, leading to poor sensitivity; and there are several interferences possible on all selenium isotopes. The many polyatomic argon-containing interferences are dependent on plasma conditions as well as sample matrix and can not be overcome by blank correction[8]. The interferences on the selenium isotopes are shown in Table 1.

Different approaches have been used to overcome these problems. The interference from salts has been removed by removal of the interferent[9] or by use of the hydride generation technique[10]. Mathematical corrections for interferences have also been used[11] and in recent years the use of collision or dynamic reaction cells has made it possible to obtain interference free measurements even on ^{80}Se[12].

The sensitivity of selenium can be improved by adding carbon containing solutes[13,14]. Different mechanisms of the effect have been proposed. One suggestion being a charge transfer from ionized carbon to selenium bringing selenium into the ionized first excited state as the ionization energy of this state and carbon are similar[15].

Table 1 Interferences on selenium isotopes

^{74}Se	(0.89 %)	^{37}Cl$_2^+$, ^{40}Ar^{34}S$^+$
^{76}Se	(9.36 %)	^{36}Ar^{40}Ar$^+$, ^{38}Ar^{38}Ar$^+$, ^{40}Ar^{36}S$^+$, ^{31}P$_2$^{14}N$^+$
^{77}Se	(7.63 %)	^{40}Ar^{37}Cl$^+$, ^{40}Ar^{36}Ar^{1}H$^+$
^{78}Se	(23.78 %)	^{40}Ar^{38}Ar$^+$, ^{31}P$_2$^{16}O$^+$
^{80}Se	(49.61 %)	^{40}Ar$_2^+$
^{82}Se	(8.73 %)	^{12}C^{35}Cl$_2^+$, ^{34}S^{16}O$_3^+$, ^{82}Kr$^+$, ^{40}Ar$_2$H$_2^+$, ^{1}H^{81}Br$^+$

In practice, only the ^{76}Se, ^{77}Se, ^{78}Se and ^{82}Se isotopes can be used as the large argon interference on ^{80}Se makes this isotope unmeasurable on an instrument without reaction cell and the abundance of ^{74}Se is too small to achieve adequate sensitivity.

In quantitative measurements, low and constant blank values are preferred. If the blank level is not low it should be constant and independent of the sample matrix. In urine samples for instance, the sodium and potassium content can vary between 50 and 150 mmol l^{-1} and the chloride concentration can vary between 100 and 250 mmol l^{-1}. Thus, blank independence of matrix is crucial in urine samples. In a study on the influence of different salts and salt concentrations on the different isotopes, we observed that blanks of the different isotopes varied considerably when different amounts of synthetic urine were added to an aqueous solution. When adding increasing amounts of synthetic urine, the ^{76}Se blank decreased 50% (from 75000 cps to 35000 cps) probably due to a decreased efficiency in argon dimer formation in high salt concentrations. The ^{77}Se blank increased (from 500 to 6000 cps) due to the increasing chloride interference. The ^{78}Se signal varied (between 14000 and 20000 cps) due to a combination of reduced argon dimer formation and a potassium related increase in blanks at large potassium concentrations (maybe ^{39}K$_2$). Furthermore, the blanks were different in acetic acid and nitric acid. Only the ^{82}Se blank was low and constant (between -150 and 200 cps). Thus, it was concluded that only ^{82}Se was suitable for direct measurements in samples with varying salt contents[16].

When different carbon containing solutes were added to improve sensitivity, no difference was found in the effect of methanol, ethanol, propanol, butanol, glycerol, acetonitril and acetic acid when the nebulizer gas flow rate and rf power was optimized in each solution. Sensitivity enhancement between 5 and 6 times was obtained[16].

3.2 Nebulization

Ultrasonic nebulizers (USN) are used to improve sensitivity and attenuate the formation of polyatomic ions containing oxygen. The effect of the USN is production of larger amounts of aerosol with smaller drop sizes compare to pneumatic nebulization[17,18].

In an attempt to improve sensitivity, an ultrasonic nebulizer was compared to the cross-flow nebulizer. The influence of instrumental parameters as well as composition of the solvent on the selenium species, selenite, selenate, SeMet and TMSe were studied. There was no difference in sensitivity of the four species when the cross-flow nebulizer was used and the species responded similarly to all changes. In a 3% methanol solution, the increase in sensitivity when using the USN corresponded to a factor of 2.5 for selenite, 8.4 for selenate, 6.0 for SeMet and 8.7 for TMSe, respectively. The general lower sensitivity of selenite could be explained by loss of SeO_2 from H_2SeO_3 formed in acidic solution. The species behaved differently on changing the heat temperature on the USN and the pH of the solution. Blank signals of ^{82}Se were low and stable when using the the the cross flow nebulizer (CFN) while blank values were higher using the USN and memory effects were extremely large. After flushing a 50 µg l^{-1} selenium solution in 3% methanol, it took several hours to reach a blank level below 3000 cps. Relative standard deviations obtained with the CFN were below 1% and were generally doubled using the USN. Thus, it was concluded that for direct analysis of selenium, the USN was inferior to the CFN.

The higher blanks are of minor importance in chromatography if they are constant and can be compensated for by baseline correction. Thus, the USN can be used for transient signals. When applying the USN to a chromatographic system separating the species on an anion exchange column, detection limits between 0.15 and 0.24 µg l^{-1} were obtained. There was no peak broadening compared to the CFN. The detection limits using the CFN varied between 0.43 and 1.6 µg l^{-1}. Hence, although the sensitivity improved, the detection limits did not improve correspondingly. The results were obtained by monitoring the ^{82}Se isotope. Although the sensitivity of ^{78}Se is three times higher, signal to noise ratios were similar for the two isotopes owing to the noise on ^{78}Se isotope[19]. It is our impression that the USN is not very stable and the improvement of detection limits was to small to justify the time-consuming stabilization of the USN.

The use of a microconcentric nebulizer in combination with a cyclonic spray chamber improved sensitivity 2-3 times compared to the CFN. Furthermore, this system has optimum flow rate around 0.2 ml min^{-1} which is compatible with the flow of chromatographic columns with inner diameters of 2 mm. As the tendency goes towards decreasing inner diameters of columns, such nebulization systems are more suitable than the CFN with optimum sample flow rates around 1 ml min^{-1} and the USN with recommended sample uptake rates larger than 2 ml min^{-1}.

3.3 Speciation in urine by ion chromatography

According to the theories on selenium metabolism, selenium should be excreted as methylated compounds in urine[2]. However, several selenium species are present in human urine, but the identity of the species has not been established. The presence of selenite[20], TMSe[20-22] and a monomethylselenol-like species[23] have been suggested.

Chromatography of urine samples is not straightforward owing to the high and varying salt content of the samples and only a limited number of studies have been performed. The separation systems include reversed phase chromatography[22,23], vesicle mediated chromatography[24], ion-pairing chromatography[25] and ion chromatography. In the chromatograpic systems developed by Lafuente et al.[23] and Quijano et al.[22], three selenium containing species were separated in urine.

In our experiments, a cation exchange column also posessing some anion exchange properties was used for separation. Thus, it was possible to separate the four species selenite, selenate, SeMet and TMSe in aqueous solution on this column. When the standards were added to urine, selenate and SeMet co-eluted, while TMSe was separated from the other species. A chromatogram of a urine sample in an eluent based on oxalic acid is shown in Figure 1.

It appears that TMSe was separated from the other species in urine. A large interference on ^{78}Se appeared close to the TMSe signal in in urine, thus ^{82}Se was monitored. When this system was used to analyse TMSe in 9 urine samples, only 2 samples contained

TMSe in concentrations above the detecion limit of 0.8 μg l⁻¹ [26].

In an anion exchange system, selenite and selenate were separated from SeMet in a 25 mM sodium hydroxide eluent. A chromatogram of a urine sample is shown in Figure 2. In this system, SeMet eluted in the front together with the majority of the selenium content. When analysing 23 urine samples, the selenite mean was 1.8 μg l⁻¹ while the total selenium mean was 47.2 μg l⁻¹ . No detectable amounts of selenate were found in any of these samples. The detection limits for selenite and selenate were 0.4 and 0.8 μg l⁻¹, respectively[27].

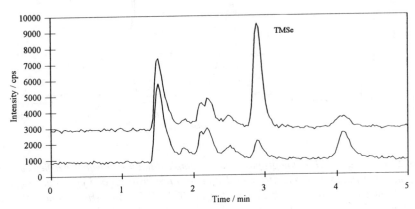

Figure 1 *Ion exchange chromatography of a urine sample diluted 1+1 with water before (lower curve) and after addition of 10 μg l⁻¹ TMSe (offset by 2000 cps). Injection volume: 50 μl; column: Ionpac CG5 + CS5 (Dionex) 50 + 250 mm x 4 mm i.d.; eluent: 10 mM oxalic acid - 20 mM potassium sulphate - 2 % methanol, pH 3; flow rate: 1.0 ml min⁻¹; nebulization: cross-flow; nebulizer gas flow rate 0.95 l min⁻¹; rf power 1300 W; isotope monitored:* ^{82}*Se*

From these results it can be concluded that TMSe and selenite are present in human urine samples but only as minor constituents.

Several solid phase extraction procedures including ion-pair adsorption to C_8 and C_{18} cartridges have been examined for selective extractions of urine samples without much success. The competition of the selenium species with the many other small ionic compounds present in urine results in poor recoveries for the selenium species.

In attempt to remove the sodium and potassium content of urine, samples were extracted with crown ethers. The benzo-15-crown-5-ether complexes sodium and the sodium-crown ether complex was extracted into chloroform as a picrate salt, while potassium was precipitated as potassium picrate. In order to spare the chromatographic column from excess picrate, the extracted sample was anion exchanged with formate. This extraction improved separation considerably and an example of an extracted sample is shown in Figure 3.
Two of the species co-eluted with SeMet and TMSe, respectively, while at least three other unidentified species were present. Urine samples from different individuals showed different elution profiles.

The species eluting just prior to SeMet was present in all extracted samples, while the unresolved signal eluting around 5 min could be a degradation product as it increased with heating of the sample. However, it is not known if some species are lost by this procedure or some species are decomposed. The procedure is laborious and not suitable for routine analysis, but it may allow preconcentration of samples which is not possible with untreated urine.

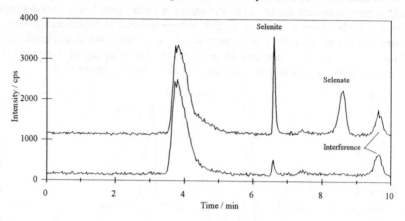

Figure 2 *Ion exchange chromatography of urine sample diluted 1+1 with water before (lower curve) and after addition of selenite and selenate in concentrations of 5 μg l⁻¹ each (offset by 1000 cps). Injection volume: 20 μl; column: Ionpac AG11-HC + AS11-HC (Dionex) 50 + 250 mm x 2 mm i.d; eluent: 25 mM sodium hydroxide - 2% methanol; flow rate: 0.2 ml min⁻¹; nebulization: Micromist glass concentric - cyclonic spraychamber (Glass Expansion); nebulizer gas flow rate 1.025 l min⁻¹, rf power 1300 W, isotope monitored: ^{82}Se*

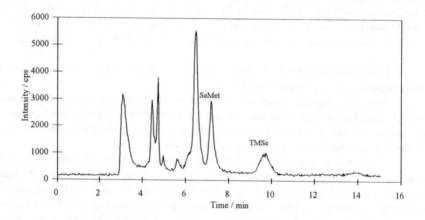

Figure 3 *Ion exchange chromatography of urine sample after crown ether extraction and anion exchange. Injection volume: 20 μl; column: Ionpac CG5 + CS5 (Dionex) 50 + 250 mm x 4 mm i.d.; eluent: 30 mM ammonium formate - 2 % methanol, pH 3; flow rate: 0.5 ml min⁻¹; nebulization: Micromist glas concentric - cyclonic spraychamber (Glass Expansion); nebulizer gas flow rate: 0.95 l min⁻¹; rf power 1300 W; isotope monitored: ^{82}Se*

3.4 Capillary Electrophoresis

CE-ICP-MS could be an alternative for selenium speciation. Interfacing CE to ICP-MS is not straightforward and was first reported by Olesik et al.[28]. The technique has recently been reviewed[29]. The interface should provide a stable electrical connection to the outlet end of the capillary, prevent introduction of laminar flow in the capillary that could compromise the efficiency of CE and offer a high sample transfer efficiency. The limited amount of sample injected into the capillary, often less than 100 nl, demands a highly efficient sample introduction system. Sample transfer efficiencies of 100% can be obtained by direct injection nebulizers, thus this sample introduction system seems promising for the hyphenation.

An interface based on direct injection was developed with special emphasis on obtaining optimal sensitivity at low sample uptake rates. The electrical circuit was established by aspirating sheath liquid through a grounded platinum tube. The sheath liquid was self-aspirated from a levelled reservoir resulting in a flow of 10 μl min^{-1} [7].

Four anionic species, selenite, selenate, SeMet and (SeCys)$_2$ were separated in a system run with reversed polarity. The relative standard deviations with respect to migration time, peak heights and peak areas were below 6% for all species. Detection limits using hydrostatic sample injection for 125 nl sample volumes were calculated to be 0.2 μg l^{-1} for selenite and selenate, 0.5 μg l^{-1} for SeMet and 1.0 ug l^{-1} for (SeCys)$_2$ corresponding to absolute detection limits between 25 and 125 fg. These absolute detection limits are at least one order of magnitude lower than previous reported values[30,31].

When electrokinetic sampling was used, detection limits of 25 ng l^{-1} were obtained for selenite and selenate and 100 ng l^{-1} for SeMet and (SeCys)$_2$, respectively. An electropherogram of standards in a concentration of 100 ng l^{-1} each is presented in Figure 4. Detection limits in this order of magnitude for selenium species using CE-ICP-MS have not been seen before.

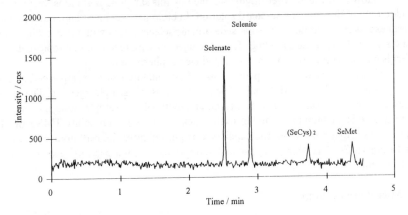

Figure 4 *CE-ICP-MS separation of four selenium standards in concentrations of 100 ng l^{-1} each. Sampling: electrokinetic sample injection at -10 kV for 40 s; capillary: fused silica 75 cm x 50 μm i.d.; electrophoresis medium: 50 mM ammonium nitrate - 0.5 mM hexadecyltrimethylammonium hydroxide, pH 9.25; voltage: -30 kV; sheath liquid, 50 mM nitric acid - 5 % acetic acid; nebulizer gas flow rate: 0.2 l min^{-1}; rf power: 1100 W; isotope monitored: ^{82}Se*

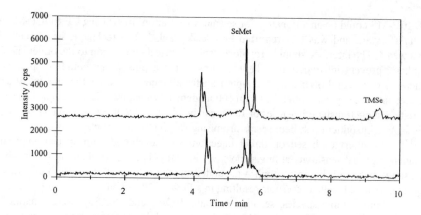

Figure 5 *CE-ICP-MS separation of a urine sample before (lower curve) and after addition of SeMet and TMSe standards in concentrations of 10 μg l^{-1} each (offset by 2500 cps). Sampling: hydrodynamical injection - 1bar for 3 s (corresponding to 80 nl); electrophoresis medium: 50 mM TAPSO (2-hydroxy-N-tris(hydroxymethyl)methyl)-3-aminopropanesulfonic acid sodium salt) - 0.01 % hexadimetrine nitrate, pH 8.2; voltage: -20 kV; sheath liquid: 50 mM nitric acid - 5 % acetic acid; nebulizer gas flow rate 0.2 l min^{-1}; rf power: 1200 W; isotope monitored: ^{82}Se*

However, the high sensitivity obtained by this sampling method is owing to the low conductivity of the aqueous standard solution. In biological samples, high mobility inorganic ions in excess concentration will supersede organic selenium ions with lower mobility resulting in a discrimination during sampling. Thus, the injected sample will not be representative. This effect is clearly demonstrate in the presented electropherogram.

Figure 5 shows electropherograms of an undiluted urine sample before and after spiking with 10 μg l^{-1} of SeMet and TMSe using hydrostatic sample injection. The system was based on reversed polarity. The higher conductivity of the sample zone compared to the electrophoresis medium prevented efficient stacking of the high mobility TMSe ion. Thus, it appeared as broad peak when the capillary was purged after the run. Four selenium species apart from TMSe were separated, although the species were not baseline separated. One of the peaks co-migrated with SeMet.

3.5 Microbore Columns

The direct injection nebulizer developed for the the hyphenation of CE and ICP-MS works very well at low sample flow rates and is for the time being used in combination with microbore columns with a flow rate of 50 μl min^{-1}. This low sample introduction flow rate allows at least 50% of methanol in the eluent without formation of carbon deposits on the sampling and skimmer cones. Thus, chromatographic systems with large amounts organic solute in the eluent can be hyphenated with the ICP-MS without a desolvation unit.

A chromatogram of a urine sample in an ion-pairing chromatographic system is shown in Figure 6. The shapes of the signals for the ^{77}Se, ^{78}Se and ^{82}Se isotopes were identical after 100 s, where the inorganic salts had been eluted. Thus, at least 10 selenium species were separated in this system.

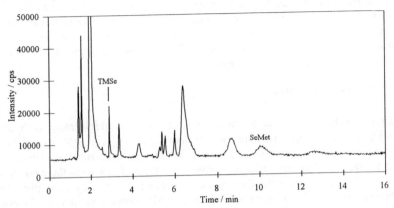

Figure 6 *Microbore chromatography of a urine sample adjusted to pH 2. Injection volume: 3 μl; column: Luna 3μm C8 (Phenomenex) 100 mm x 1.0 mm i.d.; eluent: 0.1% heptafluorobutanoic acid - 20% methanol; flow rate: 50 μl min⁻¹; nebulizer gas flow rate: 0.2 l min⁻¹; rf-power 1500 W; isotope monitored: ^{78}Se*

A benefit of using microbore columns is the high separation efficiency and large sensitivity can be obtained with small sample volumes. Besides, the reduced solvent consumption allows the use of expensive solvents and buffer additives. By using narrow bore tubing for the connection of the microbore column to the direct injection nebulizer, the entire dead volume from column to plasma is less than a microliter. Additionally, direct sample introduction prevents losses of volatile species which can be a problem using desolvation systems.

3.6 Future Aspects

One of the major challenges in selenium speciation is to identify the separated species. The obvious choice of mass spectrometry is however not straightforward as LC-MS systems demands volatile eluents as ammonium formate or acetate and large amounts of organic solvents are preferred. Furthermore, the sensitivity of MS is much lower than for ICP-MS. Our experiments showed that concentrations of 10 μg l⁻¹ (sample volume 50 μl) of SeMet and TMSe were detectable when monitored in the single ion monitoring mode after direct injection of the sample. The sensitivity of (SeCys)₂ was an order of magnitude lower. However, when applying the cation exchange column and the eluent required to elute the species, 30 mM ammonium formate, the SeMet and TMSe signals in aqueous standards could still be seen in the single ion monitoring mode, while the sensitivity was lost in the scanning mode. In negative ionization mode, selenite and selenate could be detected in concentrations of 1 mg l⁻¹ in the single ion monitoring mode. Thus, measurement of selenite and selenate in relevant concentrations for biological samples by MS are for the time being impossible. If selenium species in urine are to be identified by LC-MS, instruments less restrictive towards eluents and more sensitive in general are needed. Furthermore, precleaning and preconcentration of samples will be needed.

In the meantime, the need for commercially available standards is growing. Some of the species present in urine could be small amino acid like compounds and/or methylated forms of these. Several of these compounds have been synthesized, but are not commercially available. The use of microbore columns seems quite promising for improvement of separation as well as compatibility with MS measurements.

4 CONCLUSIONS

In different ion chromatographic systems, TMSe and selenite and selenate could be separated from other species in urine. Selenate was not detected in any urine samples and TMSe and selenite were only minor constituents of urine. When samples were extracted with crown ether, at least 5 different selenium containing species were separated. Two of these co-eluted with TMSe and SeMet, respectively; thus SeMet is present in some urine samples. This was confirmed by ion-pair chromatography and CE-ICP-MS. Using a microbore column and ion-pair chromatography at least 10 different selenium containing species were separated, although some of these were in very low concentrations. Thus, a lot of work still remains before selenium metabolism in the human body is understood.

5 REFERENCES

1.   M. P. Rayman, *Lancet,* 2000, **356**, 233.
2.   L. H. Foster and S. Sumar, *Crit. Rev. Food Sci. Nutr.,* 1998, **37**, 211.
3.   D. G. Barceloux, *Clin. Toxicol.,* 1999, **37**, 145.
4.   L. C. Clark, G. F. Combs, B. W. Turnbull, E. H. Slate, D. K. Chalker, J. Chow, L. S. Davis, R. A. Glower, G. F. Graham, E. G. Gross, A. Krongrad, J. L. Lesher, H. K. Park, B. B. Sanders, C. L. Smith and J. R. Taylor, *JAMA,* 1996, **276**, 1957.
5.   H. E. Ganther and J. R. Lawrence, *Tetrahedron,* 1997, **53**, 12299.
6.   C. Ip, M. Birringer, E. Block, M. Kotrebai, J. F. Tyson, P. C. Uden and D. J. Lisk, *J. Agric. Food Chem.,* 2000, **48**, 2062.
7.   L. Bendahl, B. Gammelgaard, O. Jøns, O. Farver and S. H. Hansen, *Accepted for publication in J. Anal. At. Spetrom.,* 2000.
8.   I. Llorente, M. Gomez and C. Camara, *Spectrochim. Acta Part B,* 1997, **52**, 1825.
9.   T. D. B. Lyon, G. S. Fell, R. C. Hutton and A. N. Eaton, *J. Anal. At. Spectrom.,* 1988, **3**, 601.
10.   M. A. Quijano, A. M. Gutierrez, M. C. Perez-Conde and C. Camara, *J. Anal. At. Spectrom.,* 1995, **10**, 871.
11.   J. Goossens, L. Moens and R. Dams, *Talanta,* 1994, **41**, 187.
12.   J. J. Sloth and E. H. Larsen, *J. Anal. At. Spectrom.,* 2000, **15**, 669.
13.   E. H. Evans and L. Ebdon, *J. Anal. At. Spectrom.,* 1990, **5**, 425.
14.   H. P. Longerich, *J. Anal. At. Spectrom.,* 1989, **4**, 665.
15.   F. R. Abou-Shakra, M. P. Rayman, N. I. Ward, V. Hotton and G. Bastian, *J. Anal. At. Spectrom.,* 1997, **12**, 429.
16.   B. Gammelgaard and O. Jøns, *J. Anal. At. Spectrom.,* 1999, **14**, 867.
17.   M. A. Tarr, G. Zhu and R. Browner, *Appl. Spectrosc.,* 1991, **45**, 1424.

18. R. H. Clifford, P. Sohal, H. Liu and A. Montaser, *Spectrochim. Acta Part B*, 1992, **47B**, 1107.
19. B. Gammelgaard and O. Jøns, *J. Anal. At. Spectrom.*, 2000, **15**, 499.
20. A. J. Blotcky, G. T. Hansen, N. Borkar, A. Ebrahim and E. P. Rack, *Anal. Chem.*, 1987, **59**, 2063.
21. X. F. Sun, B. T. G. Ting and M. Janghorbani, *Anal. Biochem.*, 1987, **167**, 304.
22. M. A. Quijano, A. M. Gutierrez, M. C. Perez-Conde and C. Camara, *Talanta*, 1999, **50**, 165.
23. J. M. G. Lafuente, J. M. Marchante-Gayon, M. L. F. Sanchez and A. Sanz-Medel, *Talanta*, 1999, **50**, 207.
24. J. M. G. Lafuente, M. Dlaska, M. L. F. Sanchez and A. Sanz-Medel, *J. Anal. At. Spectrom.*, 1998, **13**, 423.
25. K. Yang and S. Jiang, *Anal. Chim. Acta*, 1995, **307**, 109.
26. B. Gammelgaard, K. Jessen, F. Kristensen and O. Jøns, *Anal. Chim. Acta*, 2000, **404**, 47.
27. B. Gammelgaard and O. Jøns, *J. Anal. At. Spectrom.*, 2000, **15**, 945.
28. J. W. Olesik, J. A. Kinzer and S.V. Olesik, *Anal. Chem.*, 1995, **67**, 1.
29. K. L. Sutton and J. A. Caruso, *LC-GC*, 1999, **17**, 36.
30. B. Michalke and P. Schramel, *Electrophoresis*, 1998, **19**, 270.
31. A. Prange and D. Schaumlöffel, *J. Anal. At. Spectrom.*, 1999, **14**, 1329.

SENSITIVE DETECTION OF PLATINUM-BOUND DNA USING ICP-MS AND COMPARISON TO GRAPHITE FURNACE ATOMIC ABSORPTION SPECTROSCOPY (GFAAS)

A. Azim-Araghi[1], C. J. Ottley[2], D. G. Pearson[2], and M. J. Tilby[1].

[1]Cancer Research Unit, Medical School, University of Newcastle, Newcastle-upon-Tyne, NE2 4HH, UK.

[2]Department of Geological Sciences, University of Durham, Science Laboratories, South Road, Durham DH1 3LE, UK

1 INTRODUCTION

At present, two platinum compounds *cis*-Diamminedichloroplatinum (II) (cisplatin) and *cis*-Diammine-1,1-cyclobutanedicarboxylato platinum (II) (carboplatin), are in regular use in the treatment of cancer[1] while several others are currently undergoing trials. Cisplatin and carboplatin are effective in the treatment of ovarian, testicular and small cell lung cancers[2]. These drugs appear to kill cells as a result of interactions with genomic DNA[3]. Cisplatin is a square planar platinum II complex with two chloride and two ammonia ligands in the *cis* conformation (Figure 1). The ammine ligands remain attached to the platinum atom in the majority of the reactions. In contrast, the loss of the chloride atoms is central to the anti-tumour and toxic effects of cisplatin.

The anti-cancer potential of cisplatin was recognised in the mid-1960s when Rosenberg, who was studying the effect of electronic currents on the growth of bacterial cultures, realised that an alternating voltage led to a reduction in the rate of growth of the bacteria and that the remaining cells displayed a filamentous growth pattern[4]. These effects were due to dissolution of platinum from the electrodes and formation of platinum complexes in the growth medium. Further studies revealed that whereas charged complexes were potent bacteriocides, neutral species such as cisplatin had a more selective effect on cell division[5-7]. Cisplatin was subsequently tested as an anticancer drug and was found to have substantial anticancer activity[8,9]. A number of studies have shown that for several platinum complexes that are effective anti-cancer agents and in which the chloride atoms are in the *cis* configuration, the corresponding *trans* isomers are ineffective[10]. Furthermore, analogues in which one of the two reactive groups is blocked are also ineffective.

Under neutral conditions, platinum binds to the N7 atom of guanine, the N7 and N1 atoms of adenine and the N3 atom of cytosine[11,12]. In DNA, N7 of guanine is exposed on the surface of the major groove making it very accessible to metal binding. Cisplatin initially binds by one bond first, when it reacts with the DNA, and then it reacts with the second position to form a cross-link[13-15].

Despite the overall usefulness of these drugs, certain types of tumours, and many individual tumours fail to respond to them. An important tool in studies aimed at understanding the cause of treatment failure and how tumours can be resistant, or can develop resistance to platinum agents, is measurement of the extent to which these drugs react with DNA and also the rate at which the resulting products are removed by cellular processes.

In this paper we describe the effectiveness of ICP-MS in measuring how much platinum becomes bound to DNA within two lung cancer cell lines as the result of treatment with cisplatin. The results are compared to results from the other widely used method, GFAAS.

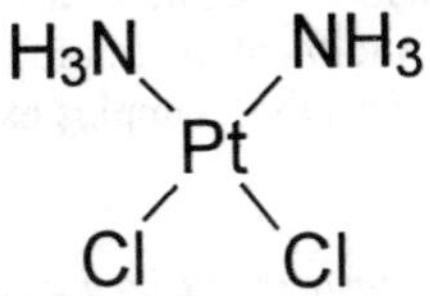

$$H_3N \diagdown \diagup NH_3$$
$$Pt$$
$$Cl \diagup \diagdown Cl$$

Figure 1 *Cisplatin has a simple structure with two ammonia ligands and two chloride ligands bound to platinum in the cis conformation*

2 EXPERIMENTAL

2.1 DNA Extraction

After exposing small cell lung cancer lines, H69, and adenocarcinoma cells, MOR[16], to different concentrations of cisplatin, DNA was extracted, using the technique developed by Tilby *et al.*, 1991[17]. Using this method, good recoveries of DNA are achieved from a small number of cells. For these experiments, up to 2×10^7 cells were lysed in 2 mL of lysis buffer (80 mM potassium phosphates, 10 mg/mL sarkosyl, 10 mM EDTA, pH 6.8), sonicated to minimise the viscosity of the DNA, and then incubated (37 °C, 15 min) with RNAase. The resulting solution plus an additional 3 mL of lysis buffer was mixed with 5 mL of phenol reagent at room temperature for 20 minutes to remove lipids and proteins. After centrifugation (600 g, 15 minutes, 4 °C) the aqueous phase was removed and the organic layer was washed with 0.5 mL of the lysis buffer. 25 mL of 6 M urea/0.08 M potassium phosphate (pH 6.8) and 0.5 g of hydroxyapatite (DNA-grade Biogel-HTP from bio-rad) were added to the DNA solution and mixed for 15 minutes at room temperature. The solution was then transferred into a spun column device, constructed from a 10 mL plastic syringe barrel in which a circle of glass fibre filter paper glass had been placed, supported by a stainless steel gauze disc. After centrifugation (All centrifugations were at 50 g for 5 min.), the eluates were discarded. The column was washed by centrifugation successively, with 2×10 mL of 5 M urea/0.08 M potassium phosphate (pH 6.8) and then with 3×10 mL 0.08 M potassium phosphate (pH 6.8). The DNA was finally recovered from the hydroxyapatite in 2 mL of 0.5 M potassium phosphate (pH 6.8). The resulting solution was then desalted and concentrated using Centricon centrifugal ultrafiltration devices, with a 10,000 MW cut-off (Amicon). The 2 mL of DNA solution was concentrated to about 60 µL. 1 mL of distilled water was then added to this solution and centrifugation was continued until the volume was again reduced to about 60 µL. 400 µL of distilled water was then added and the device was centrifuged in the inverted orientation to collect the concentrated and desalted DNA.

Before determining the levels of platinum in each DNA solution by the ICP-MS, a 50 µL aliquot of the DNA solution was diluted × 20 in 3.5 % nitric acid and left at 70 °C over night for the DNA to become hydrolysed. 1 mL of this solution was aspirated into the ICP-MS. For analysis by GFAAS, a 120 µL aliquot of the DNA solution was made to 1 M with HCl and incubated over-night at 70 °C, prior to injection of 40 µL aliquots into a Perkin Elmer Analyst 600 AAS equipped with a THGA graphite furnace and AS-800 autosampler.

To improve handling procedures, separate working areas and incubators were designated to lower the background and blank platinum levels. Table 1 shows some of the typical readings for background and for DNA samples extracted from cells not exposed to cisplatin.

Table 1 *Typical platinum levels for instrument background (high purity nitric acid) and control samples for 4 total procedural blanks, using ICP-MS, when care was taken to improve handling procedures*

Back ground (3.5% nitric acid) Pt (ppt)	DNA extracted from cell lines not treated with cisplatin and hydrolysed in 3.5% nitric acid. (DNA concentration = ca 100 µg/mL) Pt (ppt)
0.07	1.0
0.3	1.2
0.1	1.0
0.2	1.2

2.2 ICP-MS analytical protocols

Pt was analysed at the Department of Geological Sciences, Durham University, on a Perkin Elmer Sciex Elan 6000 ICP-MS using a standard cross-flow nebuliser and Scott-type double-pass spray chamber. RF power was generally 1150 W. Nebuliser gas flow rates varied between 0.8 to 1.0 L/minute and were optimised to keep the production of CeO^+ less than or equal to 3 % of the total Ce^+ signal.

Pt-bearing solutions from the above DNA extraction procedure are low in dissolved solids and can be analysed using calibrations based on standard solutions. Standard solutions of 0.5, 1, 2 and 5 ppb were made from Johnson Matthey 1000 ppm stock solutions. Most samples from this study gave signals within this calibration range. These solutions were run at the beginning of the analytical session, during, and at the end of the session, to check for instrumental drift. No significant instrumental drift was observed. Four isotopes of Pt were monitored, ^{194}Pt (32.97 % abundance), ^{195}Pt (33.83 % abundance), ^{196}Pt (25.24 % abundance) and ^{198}Pt (7.18 % abundance) to evaluate possible isobaric interferences. Corrections for isobaric interferences from Hg on ^{196}Pt and ^{198}Pt were made on-line by monitoring ^{202}Hg. Concentrations calculated using each Pt isotope agree within analytical error, indicating that other potential interferences are negligible. Hf-oxide ion species are the commonest potential isobaric interferences on the Pt mass spectrum[18]. The very low Hf levels in the DNA solutions mean that these interferences are insignificant and no oxide corrections are made.

Concentrations are calculated from average values produced by the 4 separate Pt isotopes. Procedural blanks of approximately 1 ppt, (Table 1) give limits of quantitation (10* standard deviation of the blank) of close to 1.1 ppt in the final solution, hence, no blank corrections were made to the data.

Initial tests using a CETAC Aridus desolvating nebuliser to obtain increased sensitivity revealed sporadic memory effects for both standard and sample solutions. Further work is planned using patient-derived cells where overall amounts of Pt are likely to be substantially lower than in the experiments reported here, possibly in the femto-molar range or below. For this purpose we have evaluated multi-collector, magnetic-sector ICP-MS as a means of obtaining improved sensitivity. Initial tests on a Finnigan Neptune instrument indicate at least a factor of 50 increased signal intensity. This increased sensitivity, combined with the ability to make simultaneous measurements of different isotopes using multiple electron multipliers, should allow greatly improved analytical sensitivity, providing that sample processing blanks can be improved below the 10^{-12} g level.

3 RESULTS

At the inception of this study, GFAAS was used to measure platinum-bound DNA. However, during analysis, there were a series of problems including the lack of sensitivity of the machine and occasional high background readings. Using ICP-MS overcomes these problems. Commercially available quadrupole ICP-MS instruments generally have instrumental detection limits of around 0.001 µg/L or better, compared to only 2.0 µg/L for GFAAS. ICP-MS analysis is also much faster than GFAAS and more reproducible measurements were obtained.

The high sensitivity and low memory of ICP-MS make it the method of choice for this type of study. However, given the wide spread use of GFAAS in clinical institutions, we conducted a comparative study for some samples, whereby they were analysed using both techniques.

For samples processed using the techniques described above, GFAAS appeared to give slightly higher concentrations for most samples, but this could have been due to memory effects, or the use of different standard solutions for calibration of the GFAAS compared to the ICP-MS. The routine analysis of platinum in geological samples, in the

Durham Laboratory, using similar methods and verified by isotope dilution[18] and using international standards indicate that the ICP-MS data is probably the most reliable. There is relatively close comparability between GFAAS and ICP-MS at higher concentrations of cisplatin (Figure 2A) but at lower concentrations, the results are less comparable (Figure 2B).

The cisplatin concentrations used in this study varied between 10 and 100 μM. The total platinum bound to DNA was mostly between 20 to 250 nmol/g DNA in the resistant cell lines of MOR, and 65 to 450 nmol/g DNA in the sensitive cell lines of H69, depending on the drug dose used, after two hours incubation of the cell lines with cisplatin. For a typical sample containing about 1000 ppt of platinum, the standard deviation would be between 10-20 ppt.

The samples analysed here were treated with relatively high levels of cisplatin as part of studies into the mechanisms of drug action. Consequently, the levels of Pt bound to the DNA in these samples are higher than would be expected in, for example, samples of DNA extracted from tissues of cancer patients receiving platinum-based chemotherapy. In DNA from blood cells from such patients, the levels of bound Pt are typically below 5 nmol/g DNA[19]. The limit of detection of drug-induced Pt binding is determined by the background Pt levels in DNA from control cells, since this was always higher than in pure diluent (Table 1). Levels of 1 ppt of Pt in a sample containing 100 μg DNA/ml were observed. These equate to a level of Pt-bound to DNA of about 50 pmoles/g DNA. Therefore, even with the existing instrumentation and techniques the use of ICP-MS should permit an accurate determination of Pt bound to DNA in clinical samples.

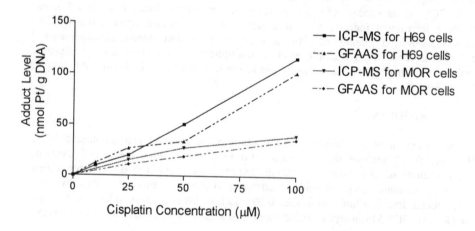

Figure 2A *Cisplatin-DNA adduct levels at different concentrations of cisplatin, using GFAAS and ICP-MS methods*

4 CONCLUSION

The high sensitivity of ICP-MS allows detection of platinum-DNA adducts treated at lower doses of cisplatin than the levels routinely analysable by GFAAS. This is of great significance, as cell lines treated at lower doses of cisplatin are more representative of samples from blood and tumours taken from patients receiving platinum-based therapies. In these samples the levels of platinum-bound DNA are too low to be analysed by GFAAS.

During measurements of platinum bound DNA, using GFAAS, a major limitation was that highly concentrated cisplatin-DNA samples were needed in order to detect Pt levels. Therefore, almost all the sample had to be used, which made it difficult to repeat each analysis, or to carry out further studies on these samples. Using ICP-MS solves this problem because of its greater sensitivity. ICP-MS does not require aspiration of highly concentrated Pt-DNA samples, thereby reducing sample memory and allowing more replicates to be analysed if needed, and so more reliable data could be obtained. The ICP-MS method allows considerably better sample throughput than GFAAS, largely because of the low levels of background memory.

Development of multi-collector, magnetic sector ICP-MS instruments, together with the application of isotope dilution analysis to biological samples, will allow greatly improved detection limits, thereby widening the scope of future clinical studies. For example, this will permit the analysis of smaller samples of tissue than is currently possible.

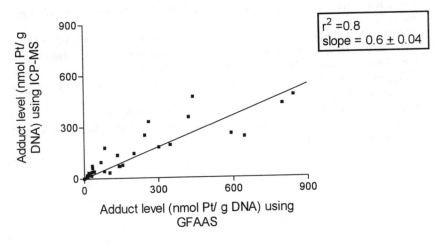

Figure 2B *Comparison of the adduct levels at the same drug dose, when measured by the ICP-MS and the GFAA*

5   REFERENCES

1.   Ozols, R.F. *Curr. Probl. Cancer,* 1992, **16,** 65.
2.   Loehrer, P.J. and Einhorn, L.H. *Ann. Intern. Med.,* 1984, **100,** 704.
3.   Roberts, J.J., Knox, R.J., Friedlos, F., and Lydall, D.A. In *Biochemical Mechanisms of Platinum Antitumour Drugs,* ed. D.C.H. McBrien and T.F. Slater, IRL Press: Oxford, 1986, 29.
4.   Rosenberg, B. *Interdisciplinary Science Review,* 1978, **3,** 134.
5.   Renshaw, E. and Thompson, A.J. *Escherichia coli. Journal of Bacteriology,* 1967, **94,** 1915.
6.   Rosenberg, B., Renshaw, E., VanCamp L., Hartwici, J., and Drobnik, J. *Escherichia coli. Journal of Bacteriology,* 1967 a, **93,** 716.
7.   Rosenberg, B., VanCamp L., Grimley, E.B., and Thompson, A.J. *Journal of Biological chemistry,* 1967 b, **242,** 1347.
8.   Rosenberg, B., VanCamp L., Trosko, J.E., and Mansour, V.H., *Nature, London,* 1969, **222,** 385.
9.   Rosenberg, B. In *Cisplatin: current status and new developments,* ed. A.W. Prestayko, S.T. Crooke, and A.K. Carter, Academic Press, New York, 1980, 9.
10. Connors, T.A., Cleare, M.J., and Harrap, K.R. *Cancer Treatment Reports,* 1979, **63,** 1499.
11. Eastman, A. *Biochemistry,* 1983, **22,** 3927.
12. Eastman, A. *Pharmacol. and Ther.,* 1987, **34,** 155.
13. Pinto, A.L., and Lippard, S.J. *Biochimica Biophysica Acta,* 1985, **780,** 167.
14. Knox, R.J., Friedlos, F., Lydall, D.A., and Roberts, J.J. *Mutation Research,* 1986, **46,** 1972.
15. Roberts, J.J., and Friedlos, F. *Pharmacology and Therapeutics,* 1987, **34,** 215.
16. Barrand, M., Heppel-Parton, A.C., Wright, K.A., Rabbits, P.H., and Twentyman, P.R. *Journal of National Cancer Institute,* 1994, **86,** 110.
17. Tilby, M.J., Johnson, C., Knox, R.J., Cordell, J., Roberts, J.J., and Dean, C.J. *Cancer Research,* 1991, **51,** 123.
18. Pearson, D.G. and Woodland, S.J., *Chem. Geol.,* 2000, **165,** 87.
19. Peng, B., Tilby, M.J., English, M.W., Price, L., Pearson, A.D.J., Boddy, A.V., and Newell, D.R. *Br. J. Cancer,* 1997, **76,** 14.

Author Index

Subject Index